Disaster Resiliency

Natural disasters in recent years have brought the study of disaster resiliency to the forefront. The importance of community preparedness and sustainability has been underscored by such calamities as Hurricane Katrina in 2005 and the Japanese tsunami in 2011. Natural disasters will inevitably continue to occur, but by understanding the concept of resiliency as well as the factors that lead to it, communities can minimize their vulnerabilities and increase their resilience.

In this volume, editors Naim Kapucu, Christopher V. Hawkins, and Fernando I. Rivera gather an impressive array of scholars to provide a much needed re-think to the topic disaster resiliency. Previous research on the subject has mainly focused on case studies, but this book offers a more systematic and empirical assessment of resiliency, while at the same time delving into new areas of exploration, including vulnerabilities of mobile home parks, the importance of asset mapping, and the differences between rural and urban locations. Employing a variety of statistical techniques and applying these to disasters in the United States and worldwide, this book examines resiliency through comparative methods which examine public management and policy, community planning and development, and, on the individual level, the ways in which culture, socio-economic status, and social networks contribute to resiliency. The analyses drawn will lead to the development of strategies for community preparation, response, and recovery to natural disasters.

Combining the concept of resiliency, the factors that most account for the resiliency of communities, and the various policies and government operations that can be developed to increase the sustainability of communities in face of disasters, the editors and contributors have assembled an essential resource to scholars in emergency planning, management, and policy, as well as upper-level students studying disaster management and policy.

Naim Kapucu is a Professor at the School of Public Administration at the University of Central Florida (UCF). He is the founding director of the Center for Public and Nonprofit Management (CPNM) at UCF (2008-2011). He has developed Emergency Management and Homeland Security graduate certificate and undergraduate minor programs at UCF. Dr. Kapucu has published widely in areas of public policy and administration, crisis leadership, nonprofit management, and disaster management.

Christopher Hawkins is an Assistant Professor in the School of Public Administration at the University of Central Florida. His research has focused on metropolitan governance, local economic development, smart growth, sustainability and application of network analysis to understanding planning and policy decisions

Fernando I. Rivera is an Associate Professor of Sociology at the University of Central Florida. His research interests and activities fall under staple areas of sociology, primarily the sociology of mental health, race and ethnicity, medical sociology and sociology of disasters.

Routledge Research in Public Administration and Public Policy

1 **Leadership and Policy Innovations**
From Clinton to Bush
Joseph R. Cerami

2 **Network Governance in Response to Acts of Terrorism**
Naim Kapucu

3 **A Complexity Theory for Public Policy**
Goktug Morcol

4 **Disaster Resiliency**
Interdisciplinary Perspectives
Edited by Naim Kapucu, Christopher V. Hawkins, and Fernando I. Rivera

Disaster Resiliency

Interdisciplinary Perspectives

Edited by Naim Kapucu, Christopher V. Hawkins, and Fernando I. Rivera

Routledge
Taylor & Francis Group
NEW YORK LONDON

First published 2013
by Routledge
711 Third Avenue, New York, NY 10017

Simultaneously published in the UK
by Routledge
2 Park Square, Milton Park, Abingdon, Oxon OX14 4RN

Routledge is an imprint of the Taylor & Francis Group, an informa business

Library of Congress Cataloging-in-Publication Data

Kapucu, Naim.
Disaster resiliency : interdisciplinary perspectives / Naim Kapucu, Christopher V. Hawkins, Fernando I. Rivera. — 1st ed.
p. cm. — (Routledge research in public administration and public policy ; 4)
Includes bibliographical references.
1. Disasters—Social aspects. 2. Community organization. 3. Preparedness. 4. Resilience (Personality trait) 5. Emergency management. I. Hawkins, Christopher V. (Christopher Vincent) II. Rivera, Fernando I. III. Title.
HV553.K37 2012
363.34—dc23
2012022525

ISBN: 978-0-415-62689-7 (hbk)
ISBN: 978-0-203-10245-9 (ebk)

Typeset in Sabon
by IBT Global.

Contents

Figures

Tables

Foreword

Within the fields of disaster research and policy, no concept has caught on as fast or diffused as widely as the concept of disaster resilience. Scarcely discussed prior to the beginning of this century, community and societal resilience were highlighted in the 2005 Hyogo Framework for Action and in subsequent efforts of the International Strategy for Disaster Reduction (ISDR). In the U.S., research has increasingly focused on the conditions for and characteristics of resilience, on strategies for improving resilience, and on outcomes resulting from resilience-enhancing activities. The importance of community resilience has also been emphasized in a variety of policy initiatives and agency activities. Under the Obama administration, resilience was emphasized as a national goal in the 2010 National Security Strategy, and elements of the National Security Council that had been responsible for preparedness, protection, and response policy issues have been merged into the National Security Council's Resilience Directorate. Resilience was highlighted in the 2010 Quadrennial Homeland Security Review as one of three key dimensions of a comprehensive approach to homeland security. Updating the Bush administration's Homeland Security Presidential Directive 8, Presidential Policy Directive 8 (National Preparedness) emphasizes the need for collaboration among governmental, private sector, and civil society institutions and organizations in achieving societal resilience. A parallel effort, the Voluntary Private Sector Preparedness Accreditation and Certification Program, known as PS-Prep, focuses on activities to improve business disaster resilience. This comprehensive approach to resilience is also evident in the Federal Emergency Management Agency's emphasis on a "whole community" approach to disaster risk reduction. In 2011, the U.S. also took a leading role in the APEC (Asian Pacific Economic Cooperation) High Level Policy Dialogue on Disaster Resiliency.

A host of other activities signal a growing emphasis on resilience as a fundamental strategy for disaster loss reduction. The National Academy of Sciences and National Research Council have recently carried out two studies on disaster resilience; one study discussed in several chapters in this volume focuses on private-public partnerships as vehicles for improving community resilience, while the other focuses on general issues related to resilience conceptualization and measurement. Through its Coastal Services Center and Sea Grant programs, National Oceanic and Atmospheric Administration (NOAA) has sponsored research and supported the development of coastal and regional resilience networks. The U.S. Department of Agriculture (USDA) has partnered with the National Science Foundation (NSF) to sponsor studies on rural disaster resilience; findings from several of those studies are discussed in this volume. Tools such as the Coastal Resilience Index (CRI) and the Community Assessment of Resilience Tool (CART) have been developed to aid communities in self-assessments that can identify resilience-related strengths and weaknesses and suggest areas where improvements are needed. The Community and Regional Resilience Institute has sponsored an important series of white papers on such topics as social and economic resilience and has initiated a series of programs designed to improve resilience in communities around the country.

Efforts at resilience definition, conceptualization, and measurement, as well as application of theories of resilience in real-world settings are ongoing, but a general consensus on the concept has emerged along several lines. First is the notion that disaster resilience encompasses both properties and processes that enable social units—individuals, households, organizations, communities, and societies—to resist and absorb disruptions originating from natural, technological, or human-caused disaster agents and properties and processes that enable social units to "bounce back," cope with, readjust to, and overcome such shocks.

Second is the idea that resilience is a key element in all phases of what is traditionally framed as the "hazards cycle": mitigation, preparedness, response, and recovery. Prior to disaster, *anticipatory resilience* consists of activities that enable communities to assess risks, form communities of interest, exercise foresight, and enact mitigation and preparedness measures to manage risks. When disasters strike, *responsive resilience* makes it possible for social and organizational entities to mobilize resources through emergent interpersonal and interorganizational networks, to carry out plans, and to improvise and exercise creativity in instances where plans fall short. After disasters, *adaptive resilience* enables social units to reassess their circumstances, learn from their disaster experiences, and adjust their strategies in light of the "new normal" ushered in by disaster. The chapters in this volume focus in a balanced way on these temporally bounded dimensions of resilience.

A third theme that is present in current resilience scholarship and also in these contributions is the idea that resilience capacities are in some respects pre-structured by the nature of the social order and in other respects capable of being altered through human agency and institutional design. Vulnerability science seeks to develop a better understanding of the social forces that influence the potential for disaster-related mortality, morbidity, damage, and loss, on the one hand, and the ability to absorb and cope with disaster-induced shocks, on the other. Axes of stratification such as social class, race and ethnicity, and gender contribute to differential vulnerability, as do other factors such as age and physical ability and disability. Many of the chapters that follow also explore rural/urban differences in vulnerability and resilience, emphasizing that the vulnerabilities and coping capacities of rural populations and communities to disasters has often been overlooked. Authors remind us that rural communities in the U.S. are no longer the small, self-reliant farm towns that we often romanticize. Rather, less densely populated areas are often home to polluting hog farms, fracking and mountaintop removal, workers who migrate to seek low-paying jobs in the sprawling agribusiness sector, poor and elderly residents, and former city-dwellers seeking to escape government regulation. Rural communities are also under-resourced places in which the capacity to anticipate, cope, and adapt has been seriously compromised.

At the same time, it is clear that individuals, groups, organizations, and communities can consciously overcome societally induced deficits and enhance their resilience capacities. As the selections in this volume show, when enabling conditions are present, communities can enact appropriate governance mechanisms and design institutions that enable them to reduce disaster impacts. During disasters the diverse organizations that comprise emergent response networks can coordinate their activities in the absence of overarching command-and-control authority. By applying principles of adaptive management, communities can enhance their disaster resilience and also increase their capacity to cope with the slower-onset impacts of climate change, such as sea level rise. High levels of disaster vulnerability do not automatically translate into higher disaster losses precisely because individuals, households, businesses, communities, and societies can avail themselves of knowledge, strategies, tactics, and tools that increase anticipatory, responsive, and adaptive resilience.

Chapters in this volume also allude to conundrums that have arisen in resilience theory and applications. Both research and policy emphasize the idea that resilience implies a time dimension; the process of bouncing back after a disaster should not be so prolonged that disruptions and their attendant losses continue to mount. However, rapid restoration of pre-disaster levels of functioning can take place at the expense of longer-term adaptation and sustainability. What could appear to be a resilient response in the short term—rapid rebuilding of residential structures, for example—might increase vulnerability to future disasters. The community of Greensburg, Kansas had the political will and foresight to reinvent itself as a model sustainable community following a devastating tornado, but most communities settle for far less. Responsive and adaptive resilience exist in tension with one another, and communities must find ways of balancing that tension. A second source of tension is the need to develop laws, policies, and programs that are sufficiently robust and well-institutionalized that they can endure in the face of public apathy and the ups and downs of the political issue attention cycle, while also remaining flexible and capable of absorbing new information and exploiting the policy windows that disasters frequently open.

Current resilience policies give rise to a third conundrum that centers on the respective roles of the public and private sectors in disaster risk reduction. Various contributions to this volume make the point that government action alone is insufficient for bringing about community and societal resilience. Rather, resilience involves intensive and far-reaching collaborations among public, for-profit, and nonprofit entities. This idea forms the basis for current "all community" and "all nation" policies on resilience. However, as several authors in this volume indicate, while non-governmental actions and engagement are crucial for resilience across the hazards cycle, disaster risk reduction is a public good that cannot be achieved without strong governmental leadership and support. This raises the question of how public sector entities can carry out their critical resilience-enhancing roles during difficult economic times in which many interest groups advocate for a diminished role for government in all aspects of national life. A research participant discussed in Chapter 17 is quoted as stating that emergency managers are "practitioners with starved budgets who, despite becoming more and more efficient year after year, simply cannot keep up with rising professional standards and the additional risk caused by crumbling infrastructure, climate change, urban sprawl, and growing poverty." If that is the case across the board—and it may well be—then there is genuine cause for concern. During a time of rising physical, social, and environmental vulnerability, efforts to increase disaster resilience should be redoubled, not weakened, and the lofty aspirations enshrined in current policy discourses must be realized on the ground.

Kathleen Tierney
Department of Sociology and Institute of Behavioral Science
Natural Hazards Center
University of Colorado Boulder
May 2012, Boulder, Colorado

Foreword

It is a privilege for me to set the stage for this book on *Disaster Resiliency: Interdisciplinary Perspectives*. I only became interested in disaster resilience after the catastrophe of Hurricane Katrina, although the vulnerabilities of rural communities have been high on my professional agenda for a long time. Traditionally, the definition of "rural" has been based on administrative, land use, or economic concepts. But depending on the definition used, rural America is home to 20–49% of the nation's population and comprises 75% of the nation's land.

Although there is considerable variation in socio-economic characteristics and well-being, most rural communities are burdened with significant liabilities which make them uniquely vulnerable to natural disasters—droughts, earthquakes, floods, hurricanes, tornadoes, tsunamis, and volcanic eruptions as well as accidents at facilities that handle dangerous materials such as explosive chemicals. Furthermore, there tends to be an imbalance between rural and urban communities with respect to their capacity to manage disasters. Factors such as geography, infrastructure, history, economy, leadership, civic engagement, and institutional capacity affect the ability of rural communities to mobilize their assets to address disaster preparedness, relief, recovery, and redevelopment.

The severity and perhaps the number of critical events that affect rural areas may continue to increase, but their consequences need not be disastrous if these communities and residents reduce their vulnerabilities and increase their resilience. There is much research on vulnerability and resilience in urban communities, but much less about how rural communities and their populations are responding to natural and man-made hazards. More importantly, even the limited research that exists on rural resilience is narrowly focused, and does not sufficiently address the multifaceted problems of these communities. Furthermore, even to people who witnessed the effects of Hurricane Katrina on national television, through pictures of New Orleans and the surrounding urban communities, the need for more multi-disciplinary research focused on rural communities may not be obvious. To demonstrate the significance of Hurricane Katrina to agriculture and rural communities, a few examples are: (i) the total loss to the forestry industry from Hurricane Katrina is calculated to be approximately $5 billion; (ii) the Farm Bureau has estimated that the direct cost to agriculture because of the hurricane could be about $1 billion and the indirect costs could be an additional $1 billion because of waterway transportation costs and rising fuel prices; and (iii) in terms of social consequences, Hurricane Katrina redistributed over one million people from the central Gulf coast elsewhere across the United States.

Since the 2005 Atlantic hurricane season, USDA has expanded their efforts to enhance our knowledge of building disaster-resilient communities. One such effort is the new extramural grants program on Disaster Resilience for Rural Communities that some of my colleagues and I at the National Institute of Food and Agriculture (NIFA) jointly developed with the National Science Foundation (NSF). The goal of this program was to promote multi-disciplinary research that included social sciences and engineering on disaster resilience for rural communities. The program launched in April 2010 has funded thirteen projects that address the multifaceted problems of rural communities. Although this program will no longer be offered, NIFA and NSF will continue to fund projects focused on disaster resilience through other program areas.

This book contains the papers discussed at a workshop entitled "Disaster Resilience" held at University of Central Florida (UCF) on March 29, 2012. This workshop was organized as a result of the project "Building Disaster Resilience and Sustainability in Rural Communities in Central Florida: Applications of Social Network Analysis," funded by NIFA's Disaster Resilience for Rural Communities program. The three editors of this book, Naim Kapucu, Christopher Hawkins, and Fernando Rivera, are the project directors for this NIFA grant and have done tremendous work in organizing the presentations, discussion groups, and compiling a set of papers that fill gaps in the research on disasters and rural communities. Their three chapters emanating from this project focus on current hazard mitigation practices, emergency response networks, and disaster recovery methods used by rural counties and rural parts of urban counties in Central Florida.

This book sheds new light on how communities can increase their resiliency through policy interventions and governance mechanisms and re-emphasizes several lessons that we have learned in previous research. However, unlike most previous scholarly literature that has tended to approach resiliency from a case study perspective, many chapters in this book offer a systematic, empirical examination of the concept of resiliency. The book will also advance a scholarly understanding of what influences resiliency and will contribute to environmental, economic, and social sustainability of communities and regions.

In the context of why I started working with NSF on the Disaster Resilience in Rural Communities program, I am glad to note that this book includes a chapter on "Housing Recovery Patterns Following Hurricane

Katrina." The authors found Hurricane Katrina-affected rural counties experienced housing growth during their recovery period above their matched counterparts in terms of the net change of housing units, starting immediately after the hurricane. Ultimately, communities with higher in-migration saw the most growth in housing stock over time. Findings in several other chapters of this book re-emphasize the lessons we have learned from Hurricane Katrina, including:

a) It is critical to create networks involving multiple organizations, agencies, and government in a partnership model before a disaster occurs. The structure and knowledge about the communities that these organizations are serving has a critical influence on the success of the relationship.

b) Thc families who stay together during an event experience less emotional strain than the ones who separated, regardless of whether they evacuated or not. Communities with strong social health may be more resilient in times of strain.

c) One major barrier to resilience is the lack of access to assets that provide the ability to respond to a disaster or disaster threat.

However, I am also glad to note, as indicated in this book, most of these results are generalizable and more widely applicable to a variety of circumstances in the U.S. and abroad.

The Kapucu, Hawkins, Rivera book is rich in evidence and thinking on the concept of disaster resilience and can serve as a resource for policy development and future research.

Siva Sureshwaran
Division of Agricultural Systems
Institute of Food Production and Sustainability
United States Department of Agriculture (USDA)
May 2012
Washington, DC

Preface

This book is intended to provide empirical evidence of the factors contributing to disaster resiliency. The concept of resiliency represents a goal for communities and regions in the United States and in other countries. Extant research in the emergency management literature has generally conceptualized resiliency as "bouncing back" after a stressor event.

This book offers a systematic, empirical examination of the concept of resiliency. It contributes to the literature on emergency management, social sciences, policy and planning by positioning the concept of resiliency as a framework that provides guidance for undertaking empirical analyses. To date the literature has tended to approach resiliency from a case study perspective. Our book approaches research on disaster resiliency in diverse contexts and regions in a comprehensive and integrated manner. Thus the volume provides results that are generalizable and applicable to a variety of circumstances in the U.S. and abroad. This book sheds new light on how communities can increase their resiliency through policy interventions and governance mechanisms.

The literature on disaster mitigation, preparedness, and recovery casts the challenges for communities as creating an integrated disaster management effort. This refocuses attention away from a "silo" approach to a "network" approach in creating disaster-resilient communities. Thus, a collection of chapters utilize network perspectives to assess the vulnerabilities and resilience of communities. The results of the analyses point to a need for the development of strategies that improve management in the mitigation, preparation, response to, and recovery from hazards.

The relationships among land uses, housing decisions, and mitigation strategies to vulnerability of disasters are in some cases clearly evident, such as the recent floods in the Midwest and coastal communities. Chapters in this volume test the relations among these factors, which have implications for planning decisions and policy for both inland and coastal regions. In other cases, the implications are more widely diffuse and long term, but are certainly just as important. The difficulty of coping with the gradual event of sea level rise, for example, is addressed in this volume.

This volume also presents chapters that measure vulnerability and resilience at the individual level of analysis. The roles of culture, social capital, socio-economic vulnerabilities, and interpersonal social networks provide complementary evidence to the analyses conducted at the community and regional scale of disaster planning and management.

The following research questions are not intended to be exhaustive of the propositions examined in this volume, but indicate some basic questions that we seek to investigate with the contributing scholars. How can the concept of resiliency be operationalized in a way that is useful as a framework to investigate the conditions that lead to stronger, safer, and more sustainable communities? What explains the resiliency of communities and regions? In other words, what factors account for the variation across jurisdictions and geographic units in the ability to withstand and "bounce back" from a disaster or natural event? What are the various policy interventions and governance mechanisms that can be developed to improve the resiliency of communities and reduce their vulnerability to natural disasters?

This book is developed based on the papers presented at the Disaster Resiliency Workshop at the University of Central Florida (UCF) on March 29–30, 2012. The workshop was funded by NIFA's Disaster Resilience for Rural Communities program of the U.S. Department of Agriculture (USDA), as part of the project entitled "Building Disaster Resilience and Sustainability in Rural Communities in Central Florida: Applications of Social Network Analysis." Each chapter in the book has been reviewed by at least two workshop participants. The book contains three parts and each part is introduced by one of the discussants in the workshop.

Acknowledgements: We would like to thank the USDA for funding the workshop. We thank workshop participants for their careful review of the chapters included in the book. We are grateful for the feedback from anonymous reviewers to the proposal. We thank Fatih Demiroz and Marc Settembrino for their assistance in this project. We would also like to express our gratitude to Natalja Mortensen, Darcy Bullock, and the entire team at Routledge for their endless patience.

1 Disaster Resiliency

Interdisciplinary Perspectives

Naim Kapucu, Christopher V. Hawkins, and Fernando Rivera

Communities in the United States and abroad experience a variety of natural disasters including droughts, earthquakes, floods, hurricanes, and tornadoes among others. Disasters, such as Hurricane Katrina in 2005, the four hurricanes that damaged portions of Florida within a six-week span in 2004, and an earthquake and tsunami in Japan in 2011, are unfortunate reminders of the vulnerability of communities to natural disasters. These events, like many others, illustrate how disasters can impact communities and individuals, and disrupt socio-technical systems and community functions. Threat of disasters will continue, but their consequences can be minimized if communities and people reduce their vulnerabilities and increase their resilience.

There tends to be an imbalance between rural and urban communities with respect to their capacity to manage disasters. Vulnerable populations and infrastructure of rural areas, an undiversified economic base in many instances, and the geographic dissimilarities with urban areas produce artificial boundaries in communication and coordination among emergency magement organizations. Although conceptual models of resilient communities and regions have been developed, accurate measurement that informs practice and basic research and that provides a foundation for advancing the field of emergency management, particularly in rural areas, is needed. This book is a modest contribution to this end.

DISASTER RESILIENCY

The focal point of our interdisciplinary approach is the four elements common to traditional models of the emergency management process: (1) prevention/mitigation, (2) preparedness, (3) response, and (4) recovery. Mitigation represents activities that prevent a disaster, reduce the chance of it occurring, or reduce its damaging effects. Preparedness represents those actions taken before impact, including plans and preparations for disaster. Response refers to actions taken during the initial impact of a disaster to save lives and prevent further property damage. Recovery includes actions taken after the initial impact, including those directed toward a return to normality (McEntire, Myers, & Foss, 2003; McLoughlin, 1985; Tierney et al., 2001; Waugh, 2000; Tobin & Montz, 1997; Waugh & Tierney, 2007; Weichselgartner, 2001). Underlying each of these components is a "culture of preparedness"—a future-oriented approach towards planning for disasters and unexpected events (Kapucu, 2008b; Kapucu & Özerdem, 2012).

Our approach also emphasizes the multi-organizational and multi-jurisdictional planning processes in preparing for emergencies and crises. Although collaboration and coordination is one of the most important functions of managing crises and disasters, they create certain challenges, including developing effective channels of communication, working with different agencies that have multiple roles and responsibilities, and acquiring the necessary resources needed to properly prepare for a disaster.

In the following sections we discuss the components that comprise our interdisciplinary approach: public policy and management; planning and development; and behavioral and social networks.

PUBLIC POLICY AND MANAGEMENT

The public policy and management perspective focuses on the policy tools for effective management of planning and response networks and maintaining community partnerships. Public, private, and nonprofit sector organizations have always had a role in disaster management. Recently, the importance of non-governmental actors has been increasing in the disaster management system. Although in many cases the private sector supplies water, power, and communication in times of disasters, nonprofits often provide shelter to disaster victims, disaster relief in the form of food and water, man power for search and rescue efforts, and fundraising (Rubin, 2007).

The importance of collaboration with the private and nonprofit sector extends beyond disaster conditions. For instance, supply and production chains, service delivery networks, and contract relationships flow within collaborative relationships. Nonprofits and other community organizations work in a similar fashion and they maintain their activities within a network of public-private-nonprofit agencies. They receive contracts from private and public agencies and serve individuals or other organizations (mostly nonprofits); all these actions are generated as a part of a network system, which oftentimes necessitates collaboration between partners (Kapucu, 2008a; Rubin, 2007).

The collaborative nature of the emergency management system is well understood at different levels of government. The roles and responsibilities of public and non-governmental actors are framed by federal legislation in the U.S. The Stafford Act, Disaster Mitigation Act of 2000, Post Katrina Emergency Management Reform Act, Presidential Directives 5 and 8, Federal Response Plan, National Response Plan, and the National Response

Framework are the foremost legislation and federal initiatives towards regulating emergency management systems in this context (NRC, 2010).

A challenge in developing resilient communities is not only recognizing and anticipating the scope of damages, but integrating multiple agencies, jurisdictions and stakeholder groups in a response to a disaster (Pelling, 2003; Ronan & Johnson, 2005). Hazard preparedness should include strong coordination across federal/central, state/province, and local agencies with clear lines of responsibility. Coordination is particularly important across jurisdictions because disasters do not confine themselves to political and administrative boundaries. Thus regionalism and associated intergovernmental cooperation is another key component in assessing vulnerability (Caruson & MacManus, 2008; Kapucu, 2008a). Due to the catastrophic nature of disasters, it has been imperative that cross-sector collaboration occurs in emergency management. Because no single agency or organization has all the resources needed in a response/recovery phase, creating these networks before a disaster is vital (Kapucu, 2007). Building relationships and collaborating with other agencies will increase networking and allow agencies to deal with multiple problems related to emergency management. Community response to a disaster or emergency can be successful if capacity building has taken place and the collaborating agencies have developed a shared vision, a common understanding of the problem, leadership skills, and sustainable community involvement (Beatley, 1998). For instance, the National Research Council (NRC) (2010) recognizes four principles of resiliency, which are: identifying and creating incentives; adopting an appropriate planning perspective; using decentralized partnerships; and allowing for multiple levels of engagement.

First, community collaborative networks have to deliver benefits to participants. Each participant's expectations are different than the others. For instance a private company largely pursues incentives that will yield a higher return on investment, business continuity, and more accurate risk and benefit analysis based on enhanced trust. Nonprofits, on the other hand, are motivated by more efficient service delivery, increased income (and perhaps contracts), and capacity building. One critical challenge for community-based collaboration is that the benefits will not be visible for the partners in the near future, which may inhibit the willingness to participate (NRC, 2010).

Second, factors such as community priorities, vulnerabilities, and culture necessitate custom planning for each community. Because the needs of each community vary depending on the disaster cycle, NRC (2010) notes that a customized plan requires "systematic identification of the resources and strategies needed to accommodate land use planning, public preparedness education, and short- and long-term disaster recovery for likely scenarios" (43). Flexibility of the plans is critical because of the uncertainty created by disasters, yet flexibility should be built on strong and rooted relationships among partners; otherwise loosely connected organizations may not be able to manage a flexible plan because they cannot predict the behaviors of their peers.

Third, the structure of the collaboration has a critical influence on the success of the relationship. The nature of collaboration requires decentralized network relationships. For instance, the U.S. Coast Guard's successful response to Hurricane Katrina is directly associated with its decentralized structure and knowledge about the communities they were serving. Finally, collaboration should exist not only at a single level but at all levels in accordance with the appropriateness of the participating agency. In other words, organizations are more likely to collaborate with each other if their goals and resources are appropriate for each other. An organization will build a relationship with others that can share the necessary knowledge and resources that it needs; otherwise, the relationship inevitably fades because it does not generate resources.

The system for risk interpretation, communication, and action requires significant collaboration and cohesion among the partners in the system. The framework for collaboration towards disaster resilience developed by NRC identifies five key premises: "1) disaster resilience correlates strongly with community resilience; 2) public private collaboration is based on relationships in which two or more private and public entities coordinate resources toward common objectives; 3) effective collaboration depends on a community engagement approach and 4) principles of comprehensive emergency management ideally guide resilience-focused collaboration" (NRC, 2010: 5). The model developed based on this framework integrates community participants (public, private, nonprofit organizations, volunteers, faith-based initiatives) and operations and processes (e.g., collaborative management, interorganizational networking) for creating synergy toward building disaster resiliency.

PLANNING AND DEVELOPMENT

Planning and development represents the second perspective. An extensive literature in planning has emerged that focuses on changes in the built environment, particularly sprawl (e.g., the outward expansion of low-density commercial and residential development). Sprawl is especially relevant to hazards because development patterns consume land in areas that may not be suitable for development and alters environmental systems that increase the vulnerability of populations and property to hazards. Although sprawl remains an elusive concept and problems of accurate measurement persists, there is general agreement among scholars that land use patterns often used to describe sprawl produce negative externalities and are harmful to the environment. Fragmented open spaces, infringement of development on wetlands, and impervious surfaces contribute to the vulnerability of some hazards, such as flooding.

Communities can strengthen their resilience and reduce risks to their residents by using a suite of development management tools to mitigate hazards (Burby et al., 2000; Brody, Kang, & Bernhardt, 2010; Godschalk et al., 1999). Structural approaches have historically dominated flood mitigation in the U.S. (Birkland et al., 2003). They are oriented to modifying the built environment to mitigate flood damage, including levees, floodwalls, and fills (Alexander, 1993; Few, 2003). Structural measures, however, have been critiqued. Some suggest that flooding can exceed the capacity of a structure (Burby, French, & Cigler, 1985; Burby et al., 1999; Larson & Plasencia, 2001; Stein et al., 2000). Moreover, they are often neglected, leading to catastrophic events (Brody, Kang et al., 2010). In addition to having adverse environmental impacts on natural ecosystems and hydrology (Birkland et al., 2003; Abell, 2000), they may also bring a false sense of security (Dalton & Burby, 1994), are very expensive to construct and maintain (Stein et al., 2000), may also increase the velocity of water downstream by changing the natural floodplain (Birkland et al., 2003), and encourage development in inappropriate locations (Burby, 2006; Pielke, 1999; Tobin, 1995).

Non-structural measures provide alternatives to structural mechanisms. These approaches largely focus on land use planning, integrating mitigation into land development policies, and capital facilities policies (Burby et al., 1985, 1999; Godschalk, Brower, & Beatley, 1989; Brody et al., 2007; Burby, 1998; Burby et al., 2000; Brody, Groverb et al., 2010; Mileti, 1999). Non-structural measures seek to reduce vulnerability by directing land use away from hazardous areas, communicating mitigation information, protecting sensitive areas, and insurance programs that distribute risk. Non-structural measures are also seen as being able to achieve both mitigation and environmental and ecological goals (Birkland et al., 2003; Birkland & Waterman, 2008; Olschansky & Kartez, 1998). Alternative mechanisms, which are sometimes called low impact development, are seen as relatively cost effective and more suitable to land use that is increasingly diffuse in nature (Braden & Johnston, 2004; Marsalek & Chocat, 2002; Thurston, 2006; Thurston et al., 2003). Local governments have therefore turned to a variety of regulations and incentives to protect environmental resources and improve disaster resiliency (Bengston, Fletcher, & Nelson, 2003; Porter, 1997).

Studies have also shown that the risk of, and damage from, hazards are significantly reduced when appropriate land use planning and development oversight is performed, including monitoring efforts to ensure land use regulations are properly implemented (Burby, 2005, 2006; Burby, French, & Nelson, 1998; Nelson & French, 2002). From a preparedness and response perspective, planning can reduce vulnerability through targeted emergency management plans or though broader comprehensive plans that incorporate disaster preparation and response elements. Reviews of both approaches indicate that they have a positive effect in fostering the development of more rigorous hazard programs (Burby et al., 1999; Burby et al., 1998). Communities that invest the time to specify hazard mitigation goals and objectives are more likely to reduce loss through appropriate development and adoption of effective strategies (Burby et al., 1998; Nelson & French, 2002; Schwab et al., 1998).

Although many call for improving planning efforts that incorporate hazards planning, the extent to which communities can implement strategies is linked to their management capacity. Rural communities in particular may lack resources to undertake planning efforts. Local governments with poor fiscal conditions may be the least likely to be able to finance activities in order to prepare for emergency management situations. Furthermore, state and federal grants that are based on population size as a formula may be biased against rural communities, thus impacting their efforts to develop and implement long-term planning and hazard mitigation practices (Caruson & MacManus, 2007).

The resource constraints of rural communities are more pronounced when considering infrastructure vulnerabilities. Antiquated public infrastructure is a source of vulnerability in smaller and more rural communities because these areas may lack the resources to rebuild or undertake recovery efforts that do not result in a large financial burden (Chang & Shinozuka, 2004). These represent just a handful of challenges facing rural communities in their efforts to integrate planning with disaster resiliency.

BEHAVIORAL AND SOCIAL NETWORKS

Our interdisciplinary approach also builds on the literature of social, communication, and implementation networks. Social networks can facilitate access to information, resources, and opportunities (Burt, 1997; Coleman, 1988; Ronan & Johnson, 2005). This perspective has direct relevance to social and behavior responses to disasters, particularly the role of external organizations and forces that influences risk perceptions of local stakeholders. Studies of interorganizational networks generally describe patterns of coordination and the structural and institutional arrangements of collaborative governance as described in the first section of this chapter. This perspective focuses attention to both the formal institutions and the informal arrangements among actors and views policy making as decisions made by interdependent actors who engage in exchange relations. The preparation and response behavior of key governing units to disasters, we believe, is strongly tied to the patterns of relations and key organizations within a network that influence the perceptions of stakeholder groups.

It is important to understand that natural hazards do not occur in a vacuum. The vulnerability and resilience of individuals, groups, and communities happen within the social context (e.g., socio-demographic factors). Thus there is a very strong link between the factors that influence the vulnerability of a community to natural hazards and the extent of their resilience. Past research has provided two distinctive ways to analyze the role of culture in preparing and responding to natural hazards. One set of studies has examined culture at the organizational level, specifically on how communities and organizations develop a culture that might increase vulnerability, particularly in socio-technical disasters (Hopkins, 1999; Turner, 1978). Another set of studies has investigated culture as it relates to the processes of political, social, and cultural marginalization on disaster vulnerability and recovery (Bolin & Stanford, 1991; Phillips, 1993).

NRC (2009) defines networks as "the interactions between people and organizations, including who knows, works with, or communicated with whom, that can be mapped" (2). These networks can be affected by natural hazards. In essence, disaster situations usually disrupt social interactions and it is within the community that these social networks are re-established in order to continue normal daily activities. Thus knowing the vulnerabilities of the social networks is intrinsically related to the resilience of the community. In rural communities this progress might be complicated if they lack access to social capital and limited institutional and organizational assets (Green, Kleiner, & Montgomery, 2007; Ritchie & Gill, 2007). Disasters are the result of the intersection of environmental hazards and social systems that impact the ability of communities to anticipate, resist, and recover from these events. Allen and Katz (2009) argue that public health and emergency preparedness can benefit from sociological research. Indeed, a large body of sociological research has concentrated on the identification of social inequalities such as gender (Morrow & Phillips, 1999; Tierney, 2006), race (Elliott & Pais, 2006), socio-economic status (Bolin, 2006; Fordham, 1999), among other individual characteristics that render communities vulnerable to disasters.

From this perspective, race, class, and gender represent a Gordian knot of inequalities. Collins' (2009) theory of the matrix of domination explains how race, class, and gender intersect to create varying systems of oppression. Specifically, Collins argues that depending on one's race, class, and gender his or her life is qualitatively different, in terms of opportunities and experiences. It is not surprising then, that disaster researchers have identified race, social class, and gender as barriers to disaster resilience. Ursano, Fullerton, and Terhakopian (2008) identified low socio-economic position and mistrust of governmental agencies/officials as potential barriers to community resilience. Additionally, Ojeiro et al. (2011) found that one's poverty status significantly affected one's participation in federal wildfire risk mitigation programs in Arizona. This finding is an echo of Bolin's (2006) statement that "federal assistance programs consistently failed to meet the needs of the homeless, Latino farm workers, and low-income African Americans" (122). Furthermore, in their study of the aftermath of Hurricane Katrina, Green and his colleagues (2007) discovered that one major barrier to resilience was the lack of access to assets that provide the ability to respond to a disaster or disaster threat; particularly, the lack of access to environmental, physical, built, human, social, and financial assets.

Racial and gender inequality also have significant effects on disaster resilience. For instance, Elliott and Pais (2006) found that blacks were less likely to take hurricane warnings seriously and were almost seven times more likely to lose their jobs than whites in the aftermath of a storm. Bolton, Liebow, and Olson (1993) found that low-income Latinos in Los Angeles struggled to receive information about housing and assistance after being displaced by an earthquake. The authors also note that language was a significant barrier in obtaining such information.

Women may take on additional responsibilities such as caring for children and the elderly, before and after disasters (Morrow & Phillips, 1999). Due to gender roles and the gender division of labor, women experience and respond to disasters much differently than men (Enarson, Forthergill, & Peek, 2006). Social class also mediates women's experiences of disaster. Although women of lower socio-economic status face greater economic losses during disasters, more wealthy women are not immune to the emotional impacts of disasters (Enarson et al., 2006; Fordham, 1999).

Not surprisingly, low-income women of color may face the greatest barriers to disaster resilience. Morrow and Enarson (1996) found that immigrant women, especially Latinas, faced several obstacles after Hurricane Andrew in 1992. Most notably, they found that women disproportionately faced a lack of affordable housing and slow repairs to public housing in the storm's wake. Echoing these findings, Enarson and Fordham (2001) found that after the 1997 Grandforks Flood, disaster relief efforts excluded migrant workers, thus disproportionally affecting Latinas. The exclusion of Latinas in relief efforts after the 1997 flood may have led to greater resilience in future disasters. After the flood many women came together to organize a political presence for Latinos in the region (Enarson, 2001).

Although there are many barriers to disaster resilience, there are several factors which may build, or facilitate, disaster resilience. Quick response and replacement of community resources and knowledge of pre-disaster community vulnerabilities also build resilience. Some suggest the need for the development and maintenance of community networks in every aspect of disaster planning (Green et al., 2007; Kim & Kang, 2010). Varghese and her colleagues (2006) demonstrated that local ownership and engagement facilitated community resilience in logging and mill towns in Canada. Additionally, Kim and Kang (2010) found that integration into community-level communication resources including community organizations and interpersonal networks impact the likelihood of engaging in hurricane preparedness activities.

In their study of East Biloxi, Alabama after Hurricane Katrina, Green, Gill, and Kleiner (2006) recommended addressing vulnerability in planning for disaster preparedness, relief, recovery, and redevelopment. They also put forward the need to provide fast and dignified disaster relief in order to build trust with the community. Fordham (1999) also supports providing dignified disaster relief, noting that the relationships developed during post-disaster relief efforts may build future disaster resilience. Ultimately, in order for communities to be resilient during and after a disaster, steps must be taken to mitigate their vulnerabilities. Recent sociological research suggests that the inclusion of local civic and faith-based organizations is vital in the planning and response to disasters (Berke & Campanella, 2006; Brennan & Flint, 2007; Elliott & Pais, 2006; Green et al., 2006). The inclusion of local groups can ameliorate issues of mistrust and can accelerate and promote disaster resiliency.

In community disaster resilience research, its prominent components include interpersonal and group linkages, social support, and community bonds. Interpersonal and group connectedness constitutes a common social memory of community disasters experiences (Longstaff et al., 2010) which generates innovative learning and better behavioral adaptation towards future incidents (Pelling & High, 2005). Social support is based on social interactions among individuals that in turn generate a social environment of mutual assistance, caring, and support. Aldrich (2010) stresses the importance of social support, particularly in the aftermath of disasters. He notes that people who are better connected to others receive more assistance than less connected people. Also, interpersonal connections enhance communication among people, which allows individuals to share knowledge about how to follow formal disaster assistance procedures, deter looting, and "voice" their expectations from recovery actions rather than exiting their communities (Aldrich, 2010). Community bonds, as Norris et al. (2008) note, are the primary drivers of sense of community, place attachment, and citizen participation. These three factors enable individuals to be emotionally connected to their communities, voice their concerns, mitigate potential limitations of population diversity, and distribute fairly the roles and responsibilities within the community (Aldrich, 2010; Norris et al., 2008).

Community competency refers to the ability to have accurate and reliable information sharing during and after a disaster for solving emergent problems. It measures "how well the community functions pre- and post-disaster including sense of community and ideals as well as the attachment to place and the desire to preserve pre-disaster cultural norms and icons" (Cutter et al., 2008: 604). Norris et al. (2008) suggest community competence is a function of collective action and decision making, and collective efficiency and empowerment. Social capital and communication are the key components of community competence. These are particularly important in collective action and decision-making processes because satisfactory decisions about community-related issues require consideration of all the community stakeholders. Trusted and information sharing recommends people to take necessary mitigation actions in the face of disasters. Empowerment is another strongly related factor that influences community competence. Norris et al. (2008) identify empowerment as "a process through which people lacking an equal share of resources gain greater access to and control over resources" (142). An individual's personal competence, ability to participate in and influence decision-making processes, and ability to reach useful decisions are necessary components of empowerment for collective action.

CONCLUSION

In this introduction, disaster resiliency is explained through three perspectives: public policy and management, planning of development, and behavioral and social networks. The public policy and management perspective of community disaster resiliency focuses on roles and responsibilities of public organizations as well as their partnership efforts to involve private, nonprofit, and other community actors to take role in resiliency-building efforts. Interorganizational networks and collaboration are the epicenter of this perspective because managing disasters and emergencies requires significant common effort from various actors in all of its phases. Planning and development focuses on reducing physical vulnerabilities and enhancing community capacity. There are numerous factors impacting the built environment and shaping the environmental, social, and economic systems of communities. Behavioral and social network perspective focuses on the fabric of the community and identifies the social vulnerabilities embedded in that community. Ethnic and cultural diversity, socio-economic status, level of social capital, and the existence of social institutions determine the scope of social vulnerabilities in a community. A large volume of literature suggests that cultural and racial diversity increases vulnerability, whereas social capital and social institutions mitigate the effects of disasters on communities. This dimension emphasizes the community actions that are necessary to overcome these vulnerabilities.

ACKNOWLEDGEMENTS

We thank Fatih Demiroz, Marc Settembrino, and Rebecca Dodson for their assistance in this chapter.

REFERENCES

Abell, R.A. (2000). *Freshwater ecoregions of North America: A conservation assessment.* Washington, DC: Island Press.

Aldrich, D.P. (2010). Fixing recovery: Social capital in post-crisis resilience. *Journal of Homeland Security*, 6, 1–10.

Alexander, D. (1993). *Natural disasters.* New York: Chapman & Hall.

Allen, H., & Katz, R. (2009). Demography and public health emergency preparedness: Making the connection. *Population Research and Policy Review*, 29, 527–539.

Beatley, T. (1998). The vision of sustainable communities. In J.R. Burby (Ed.), *Cooperating with nature: Confronting natural hazards with land-use planning for sustainable communities*. Washington, DC: Joseph Henry Press.

Bengston, D.N., Fletcher, O.J., & Nelson, K.C. (2003). Public policies for managing urban growth and protecting open space: Policy instruments and lessons learned in the United States. *Landscape and Urban Planning*, 69, 271–286.

Berke, P.R., & Campanella, T.J. (2006). Planning for postdisaster resiliency. *Annals of the American Academy of Political and Social Science*, 604, 192–207.

Birkland, T.A., Burby, R.J., Conrad, D., Cortner, H., & Michener, W.K. (2003). River ecology and flood hazard mitigation. *Natural Hazards Review*, 4(1), 46–54.

Birkland, T., & Waterman, S. (2008). Is federalism the reason for policy failure in Hurricane Katina? *Publius: The Journal of Federalism*, 38, 692–714.

Bolin, B. (2006). Race, class, ethnicity, and disaster vulnerability. In H. Rodriguez, E.L. Quarantelli, & R. Dynes (Eds.), *Handbook of disaster research*. United States of America: Springer.

Bolin, R.C., & Stanford, L. (1991). Shelter, housing and recovery: A comparison of U.S. disasters. *Disasters*, 15, 24–34.

Bolton, P.A., Liebow, E.B., & Olson, J.L. (1993). Community conflict and uncertainty following a damaging earthquake: Low-income Latinos in Los Angeles, California. *Environmental Professional*, 15, 240–247.

Braden, J.B., & Johnston, D.M. (2004). Downstream economic benefits from storm-water management. *Journal of Water Resources Planning and Management*, 130(6), 498–502.

Brennan, M.A., & Flint, C.G. (2007). Uncovering the hidden dimensions of rural disaster mitigation: Capacity building through community emergency response teams. *Southern Rural Sociology*, 22, 111–126.

Brody, S.D., Groverb, H., Lindquist, E., & Vedlitz, A. (2010). Examining climate change mitigation and adaptation behaviours among public sector organisations in the USA. *Local Environment*, 15, 591–603.

Brody, S.D., Kang, J.E., & Bernhardt, S. (2010). Identifying factors influencing flood mitigation at the local level in Texas and Florida: The role of organizational capacity. *Natural Hazards*, 52, 167–184.

Brody, S.D., Zahran, S., Praveen, M., Grover, H., & Highfield, W.E. (2007). The rising costs of floods: Examining the impact of planning and development decisions on property damage in Florida. *Journal of the American Planning Association*, 73(3), 331–345.

Bullock, A., & Acreman, M. (2003). The role of wetlands in the hydrological cycle. *Hydrology and Earth System Sciences*, 7(3), 358–389.

Burby, R.J. (1998). Natural hazards and land use. In R.J. Burby (Ed.), *Cooperating with nature: Confronting natural hazards with land-use planning for sustainable communities*. Washington, DC: Joseph Henry Press.

Burby, R.J. (2005). Have state comprehensive planning mandates reduced insured losses from natural disasters? *Natural Hazards Review*, 6(2), 67–81.

Burby, R.J. (2006). Hurricane Katrina and the paradoxes of government disaster policy: Bringing about wise government decisions for hazardous areas. *Annals American Academy of Political and Social Sciences*, 604(1), 171–191.

Burby, R.J., Beatley, T., Berke, P.R., Deyle, R.E., French, S.P., Godschalk, D.R., Kaiser, E.J., . . . & Platt, R.H. (1999). Unleashing the power of planning to create disaster resistant communities. *Journal of the American Planning Association*, 65(3), 247–258.

Burby, R.J., Deyle, R.E., Godschalk, D.R., & Olshansky, R.B. (2000). Creating hazard resilient communities through land-use planning. *Natural Hazards Review*, 1(2), 99–106.

Burby, R.J., French, S.P., & Cigler, B.A. (1985). *Floodplain land use management: A national assessment*. Boulder, CO: Westview Press.

Burby, R.J., French, S.P., & Nelson, A.C. (1998). Plans, code enforcement, and damage reduction: Evidence from Northridge Earthquake. *Earthquake Spectra*, 14, 59–72.

Burt, R. (1997). The contingent value of social capital. *Administrative Science Quarterly*, 42, 339–365.

Caruson, K., & MacManus, S.A. (2007). Structuring homeland security grants: Florida's local finance officials evaluate the funding process. *Journal of Public Finance and Management*, 7, 191–226.

Caruson, K., & MacManus, S.A. (2008). Disaster vulnerabilities: How strong a push toward regionalism and intergovernmental cooperation? *The American Review of Public Administration*, 38, 286–306.

Chang, S.E., & Shinozuka, M. (2004). Measuring improvements in the disaster resilience of communities. *Earthquake Spectra*, 20, 739–755.

Coleman, J.S. (1988). Social capital in the creation of human capital. *The American Journal of Sociology*, 94, S95–S120.

Collins, P.H. (2009). *Black feminist thought*. 2nd edition. New York: Routledge.

Cutter, S.L., Barnes, L., Berry, M., Burton, C., Evans, E., Tate, E., & Webb, J. (2008). A place-based model for understanding community resilience to natural disasters. *Global Environmental Change*, 18(4), 598–606.

Dalton, L.C., & Burby, R.J. (1994). Mandates, plans and planners: Building local commitment to development management. *Journal of the American Planning Association*, 60, 444–461.

Elliott, J., & Pais, J. (2006). Race, class, and Hurricane Katrina: Social differences in human responses to disaster. *Social Science Research*, 35, 295–321.

Enarson, E. (2001). What women do: Gendered labor in the red river valley flood. *Environmental Hazards*, 3, 1–18.

Enarson, E., & Fordham, M. (2001). From women's needs to women's rights in disasters. *Environmental Hazards*, 3, 133–136.

Enarson, E., Fothergill, A., & Peek, L. (2006). Gender and disaster: Foundations and directions. In H. Rodriguez, E.L. Quarantelli, & R. Dynes (Eds.), *Handbook of disaster research*. United States of America: Springer.

Few, R. (2003). Flooding, vulnerability and coping strategies: Local responses to a global threat. *Program Developement Studies*, 3, 43–58.

Fordham, M. (1999). The intersection of gender and social class in disaster: Balancing resilience and vulnerability. *International Journal of Mass Emergencies and Disasters*, 17, 15–37.

Godschalk, D.R., Beatley, T., Berke, P., Brower, D., & Kaiser, E. (1999). *Natural hazard mitigation: Recasting disaster policy and planning*. Washington, DC: Island Press.

Godschalk, D.R., Brower, D.J., & Beatley, T. (1989). *Catastrophic coastal storms*. Durham, NC: Duke University Press.

Green, J.J., Gill, D.A., & Kleiner, A.M. (2006). From vulnerability to resiliency: Assessing impacts and responses to disaster. *Southern Rural Sociology*, 21, 89–99.

Green, J.J., Kleiner, A.M., & Montgomery, J.P. (2007). The texture of local disaster response: Service providers' views following Hurricane Katrina. *Southern Rural Sociology*, 22, 28–44.

Hopkins, A. (1999). Counteracting the cultural causes of disaster. *Journal of Contingencies and Crisis Management*, 37, 141–149.

Kapucu, N. (2007). Building community capacity to respond. *The Public Manager*, 36(3), 21–25.

Kapucu, N. (2008a). Collaborative emergency management: Better community organizing, better public preparedness and response. *Disasters*, 32(2), 239–262.

Kapucu, N. (2008b). Culture of preparedness: Household disaster preparedness. *Disaster Prevention and Management*, 17(4), 526–535.

Kapucu, N., & Özerdem, K. (2012). *Managing emergencies and crises*. Burlington, MA: Jones & Bartlett Learning.

Kim, Y.C., & Kang, J. (2010). Communication, neighborhood belonging and household hurricane preparedness. *Disasters*, 34, 470–488.

Larson, L., & Plasencia, D. (2001). No adverse impact: A new direction in floodplain management policy. *Natural Hazard Review*, 2(A), 167–181.

Longstaff, P.H., Armstrong, N.J., Perrin, K., Parker, W.M., & Hidek, M.A. (2010). Building resilient communities: A preliminary framework for assessment. *Homeland Security Affairs*, 6(3), 1–23.

Marsalek, J., & Chocat, B. (2002). International report: Stormwater management. *Water Science and Technology*, 46(6/7), 1–17.

McEntire, D.A., Myers, A., & Foss, T. (2003). Disaster preparedness. *IQ Report*, 35(11).

McLoughlin, D. (1985). A framework for integrated emergency management. *Public Administration Review*, special issue, 165–172.

Mileti, D.S. (1999). *Disasters by design: A reassessment of natural hazards in the United States*. Washington, DC: Joseph Henry Press.

Morrow, B.H., & Enarson, E. (1996). Hurricane Andrew through women's eyes: Issues and recommendations. *International Journal of Mass Emergencies and Disasters*, 14(1), 5–22.

Morrow, B.H., & Phillips, B. (1999). What's gender got to do with it? *International Journal of Mass Emergencies and Disasters*, 17, 5–11.

Nelson, A., & French, S. (2002). Plan quality and mitigating damage from natural disasters: A case study of the Northridge Earthquake with planning policy considerations. *Journal of the American Planning Association*, 68(2), 194–207.

Norris, F.H., Stevens, S.P., Pfefferbaum, B., Wyche, K.F., & Pfefferbaum, R.F. (2008). Community resilience as a metaphor, theory, set of capacities and strategy for disaster readiness. *American Journal of Community Psychology*, 41(1–2), 127–150.

NRC (National Research Council). (2009). *Applications of social network analysis for building community disaster resilience*. Washington, DC: The National Academies Press.

NRC (National Research Council). (2010). *Building community disaster resilience through private-public collaboration*. Washington, DC: The National Academies Press.

Ojerio, R., Moseley, C., Lynn, K., & Bania, N. (2011). Limited involvement of social vulnerable populations to mitigate wildfire risk in Arizona. *Natural Hazards Review*, 12(1), 28–36.

Olshansky, R.B., & Kartez, J.D. (1998). Managing land use to build resilience. In J.R. Burby (Ed.), *Cooperating with nature: Confronting natural hazards with land-use planning for sustainable communities*. Washington, DC: Joseph Henry Press.

Pelling, M. (2003). *The vulnerability of cities: Natural disasters and social resilience*. London: Earthscan.

Pelling, M., & High, C. (2005). Understanding adaptation: What can social capital offer assessment of adaptive capacity? *Global Environmental Change*, 15(4), 308–319.

Phillips, B. (1993). Cultural diversity in disaster: Shelter, housing, and long-term recovery. *International Journal of Mass Emergencies and Disasters*, 11(1), 99–110.

Pielke, R.A. (1999). Nine fallacies of floods. *Climatic Change*, 42(2), 413–438.

Porter, D.R. (1997). *Managing growth in America's communities*. Washington, DC: Island Press.

Ritchie, L.A., & Gill, D.A. (2007). Social capital theory as an integrating theoretical framework in technological disaster research. *Sociological Spectrum*, 27, 103–129.

Ronan, R.K., & Johnson, D.M. (2005). *Promoting community resiliency in disasters: The role for schools, youth, and families*. New York: Springer.

Rubin, C. (2007). *Emergency management: The American experience, 1900–2005*. Fairfax, VA: PERI.

Schwab, J., Topping, K.C., Eadie, C.C., Deyle, R.E., & Smith, R.A. (1998). *Planning for post disaster recovery and reconstruction*. Planning Advisory Service report no. 483/484. Chicago: American Planning Association Publication.

Stallings, R.A. (2002). Weberian political sociology and sociological disaster studies. *Sociological Forum*, 17(2), 281–305.

Stein, J., Moreno, P., Conrad, D., & Ellis, S. (2000). *Troubled waters: Congress, the Corps of Engineers, and wasteful water projects*. Washington, DC: Taxpayers for Common Sense and National Wildlife Federation.

Thurston, H.W. (2006). Opportunity costs of residential best management practices for stormwater runoff control. *Journal of Water Resources Planning and Management*, 132(2), 89–96.

Thurston, H.W., Goddard, H.C., Szlag, D., & Lemberg, B. (2003). Controlling storm-water runoff with tradable allowances for impervious surfaces. *Journal of Water Resources Planning and Management*, 129(5), 409–418.

Tierney, K. (2006). *Facing hazards and disasters: Understanding human dimensions*. Washington, DC: The National Academies Press.

Tierney, K.J., Lindell, M.K., & Perry, R.W. (2001). *Facing the unexpected: Disaster preparedness and response in the United States*. Washington, DC: Joseph Henry Press.

Tobin, G.A. (1995). The levee love-affair: A stormy relationship. *Water Resources Bulletin*, 31(3), 359–367.

Tobin, G.A., & Montz, B.E. (1997). *Natural hazards: Explanation and integration*. New York: The Guilford Press.

Turner, B. (1978). *Man-made disasters*. London: Wykeham.

Ursano, R., Fullerton, C.S., & Terhakopian, A. (2008). Disasters and health: Distress, disorders, and disaster behaviors in communities, neighborhoods, and nations. *Social Research*, 75, 1015–1028.

Varghese, J., Krogman, N.T., Beckley, T.M., & Nadeau, S. (2006). Critical analysis of the relationship between local ownership and community resiliency. *Rural Sociology*, 71, 505–527.

Waugh, W.L. (2000). *Living with hazards, dealing with disasters: An introduction to emergency management*. Armonk, NY: M.E. Sharpe.

Waugh, W.L., & Tierney, K. (Eds.) (2007). *Emergency management: Principles and practice for local government*. 2nd edition. Washington, DC: ICMA.

Weichselgartner, J. (2001). Disaster mitigation: The concept of vulnerability revisited. *Disaster Prevention and Management*, 10(2), 85–95.

Part I

Introduction to Conceptual Insights and Applications of Resilience

Christine A. Bevc

Thus far the twenty-first century has stood witness to some of the most devastating events in human history—the terrorist attacks of September 11 (2001), Indian Ocean tsunami (2004), the catastrophe of Hurricane Katrina (2005), and the earthquake-tsunami-nuclear disaster in Japan (2011). The loss of life, in addition to the combined billions of lost and damaged physical and economic capital have brought the concept of resilience to the forefront. It is important to remember that these events are not the first, nor the last, to mar the collective conscience of society. In reality, they contribute to an increasing list of disasters that serve as markers in an evolving history. These events will continue to inextricably alter and transform the physical and social landscape of countless communities. As long as communities continue to build along fault lines, floodplains, tropical coasts, windswept plains, and fertile shadows of volcanoes, hazards will inevitably continue to have an effect on society.

There is a general consensus that disasters may be defined as events concentrated in time and space, where social structure is disrupted and essential functions prevented (Fritz, 1961; Killian, 1955; Turner, 1978). Beyond any specific event, the common phases of a disaster (preparedness, mitigation, response, and recovery) are referred to as the *hazards cycle*. Regardless of the event, the response is constituted by the ongoing activities and behaviors that continue to evolve over the time following the impact of the event within and around the affected area. The efforts of one response may also serve to mitigate future effects, thereby creating feedback loops into the system. In an effort to restore routine behaviors, the activities and interactions of individuals and organizations must adapt and change as response demands evolve. Disasters thus demand adaptation through an assortment of shifts and adjustment in the patterns of interaction among social and ecological systems. It is during this shift from response to recovery that the "line" is often blurred as task and activities overlap in the transition (Neal, 1997). It is also during the immediate and, perhaps more so, long-term recovery phase where resilience is most likely to be observed.

Frequently, and succinctly, defined as the ability to "bounce back," resilience remains a relatively ambiguous and amorphous concept spoken of more generally than specifically and raising more questions than answers. What does it mean to be resilient? What contributes to resilience? How can communities become more resilient? Does increased resilience in one area improve resiliency in others? Does resilience to more frequent small-scale emergencies lead to increased resilience in the face of catastrophic events? These and many other questions remain central to the agenda of resilience research. What is certain is the inherent complexity associated with the study of resilience.

In the face of events, individuals, organizations, and communities confront external threats and opportunities, the response to which is dependent upon a range of resources and risk perceptions (Chattopadhyay, Glick, & Huber, 2001). While innovation and improvisation may occur during this time, there is also a tendency to "fall back on well-known routines and procedures" (Chattopadhyay et al., 2001: 941) when facing threatening events (Ocasio, 1995; Staw, Sandelands, & Dutton, 1981). Crisis and disasters create an uncertain environment of evolving needs and demands during which multi-level systems must also adapt. Individuals, organizations, and communities are able to adjust to small routine shifts in these stable systems; however, "jolts," including disasters, are unplanned and may fundamentally alter the state of the system and its elements to disrupt and transform the embedded structure of the system (Meyer, 1982; Meyer, Brooks, & Goes, 1990; Mileti, Cress, & Darlington, 2002). In principle, all disasters are local, the response to which originates from the local level through a series of mutual aid networks comprised of existing and emergent relationships among a diverse set of actors (Drabek, 1987; Waugh & Streib, 2006).

Owing to the complexity of the matter, researchers must adopt a more systems-level perspective, or wide-angle lens, to understand the concept of resilience and address the challenge of achieving it. It must also be recognized that resilience refers to a complex system interacting across multiple scales and dimensions. The adoption of systems-thinking or a systems-science approach may assist in responding to the challenge. Complex challenges, such as disasters, represent multiple overlapping, interconnected subsets of problems that cut across multiple policy domains and levels of governance. The cross-cutting characteristics of problems that emerge (or are revealed) during disasters are inescapably connected to other problems, such as environmental preservation and economic development (Weber, 1998, 2003). Such problems can engage conflicting values, such as homeland security and privacy protection (Khademian, 2005), and they often generate high degrees of uncertainty (Mason & Mitroff, 1973; Van Bueren, Klijn, & Koppenjan, 2003).

Stakeholders attempting to grapple with the complexities of disaster responses must contend with an assortment of perspectives, worldviews, political agendas, educational and professional backgrounds, programmatic responsibilities, and cultural traditions. Individual (microlevel) interactions contribute to the formation of complex networks (meso- and macrolevel) through the patterns of their interactions, because interactions can have effects on other interactions, as well as higher level structures (Mayhew, 1980). As a result, "individuals are embedded in thick webs of social relations and interactions" that structure and influence the actions of individuals and that are in turn affected by them (Borgatti et al., 2009: 892). Within the study of hazards and disasters, there is an increasing recognition of the need and integration of systems-based approaches into current research, as evident in this volume.

However, as a complex systems-level issue, the concept of resilience remains a challenge to operationalize and subsequently assess. One of the most successful and succinct operationalizations of the resilience originates with researchers at the University at Buffalo, the State University of New York's Multidisciplinary Center for Earthquake Engineering (MCEER). Bruneau and Reinhorn (2006) presented their conceptualization of resilience, now more commonly referred to as the "four 'R's of resilience" (robustness, redundancy, resourcefulness, rapidity). Originating from the field of seismic engineering, the MCEER resilience framework identifies four fundamental properties including *robustness* to withstand a given level of stress or demand without suffering degradation or loss of function; *redundancy* of functional requirements in the event of disruption, degradation, or loss; *resourcefulness* to supply material and human resources to meet established priorities and goals; and *rapidity* to contain losses and avoid future disruption (Bruneau & Reinhorn, 2006). These properties extend across four defined dimensions—technical, organizational, societal, and economic—to further provide a means of operationalizing physical and organizational systems. In combination, this framework provides a foundation for research on the resilience of organizations and communities (Bruneau & Reinhorn, 2006).

The challenges of conceptualizing and operationalizing the concept of resilience as well as the opportunities in applying the concept to empirical research will continue to provide a goal towards which the research may endeavor to achieve. In the following chapters, the authors present their own conceptual insights and applications towards understanding resilience. As a collection, these chapters contribute to an observed shift in research from the retrospective to the prospective, from the autonomous individual to the interdependent community, and from a sense of fear to empowerment across multiple levels of society. Here, the topic of resilience is demonstrated in this research with regard to the context of vulnerability and disaster recovery as well as in the development of community assessment and capacity-building tools that can enhance community resilience. Lastly, it is important to keep in mind that resilience is not only conceived as multi-dimensional, but also relative to the context of concern.

To begin, Edwards uses the concept of "whole community" in her chapter, which recognizes the underlying dependencies among individuals, organizations, and communities responding to disasters. She argues that in order for a community to be resilient—to recover socially and economically and develop a "new normal" after a disaster—it must have a pre-disaster decision-making ability that encompasses the "whole community." As Edwards emphasizes the role of social capital in whole community, Rivera and Settembrino propose a framework that incorporates the importance of social capital in facilitating disaster resiliency and eliminating barriers.

Kusenbach and Christmann present findings from work on vulnerable populations in mobile and manufactured housing communities. They discuss perceptions of risk and vulnerability among the residents, highlighting the ongoing challenges of preparedness and specific needs. To address this population of interest, the fundamental issues must be addressed as well as the unique needs of the locals. This work highlights the potential of empowering communities and neighborhoods by addressing subjective perspectives. It is in these local concerns and interests that lessons might be learned to contribute to the effort to build resilient communities. They reveal, in a qualitative light, the complex issues associated with vulnerability and the subsequent nuances of resilience to be considered when attempting to increase the capacity of communities to prepare for, respond to, and recover from disasters. As risk remains a function of the probability of a hazard relative to the level of the vulnerability, so does resilience. It is through the shared accounts by which the social dimensions of resilience are illustrated in this Florida community.

Gazley addresses the relationship between nonprofit agencies, specifically Community Organizations Active in Disasters (COADs), and the collaborative planning process in emergency management. She discusses the collaborative capacity of COADs and their potential to increase the resilience of communities through community-based organizations joint action. This chapter highlights the recognition that while many organizations prepare continuity of operations plans, these plans often remain largely individualistic and independent of the larger systemic dependencies. To promote collaborative action, Gazley identifies a series of factors, including stable, but adaptable forms of coalitions; the ability to recognize shared goals and risks; the determination of links between capacity and performance; and the need to reconcile challenges between balance, power, and participation. More generally, this chapter addresses the ability of organizational systems to respond to the rapidly shifting needs and demands following an event.

Resilience is not only limited to the capacity to respond, but also the ability to recover from an event. Using a prospective quasi-experimental approach, Ganapati and her colleagues examine levels of housing recovery among rural counties affected by Hurricane Katrina. As an event inherently alters the system in which the housing units resided, pre-event points of comparison may no longer be appropriate as markers of recovery. In a novel application of counterfactuals, Ganapati and her colleagues demonstrate the utility of a quasi-experimental design with the statistical robustness of optimal matching algorithms to identify the best case-controls (unaffected counties) for a comparison. This innovative application of econometrics enables the ability to assess net change in housing stocks, in other words, the rapidity in which housing in these communities has recovered. This research helps to reiterate the importance of prospective research to help assess community recovery over time.

Together, these chapters highlight the diversity of research addressing the multiple dimensions of resilience. Importantly, this work emphasizes the point that resilience is relative to the context of the concern. As current public policy adopts and integrates the concept of resilience, such as FEMA's objective towards "whole community" planning and the President's National Preparedness Goal, the burden will rest at the local level to address issues of community resilience (also addressed in Chapter 2 by Edwards), the understanding of which extends to a series of relevant policy implications. The work in this section, as well as the larger collection, demonstrates how perceptions, practices, and efforts (or impediments) throughout the hazards cycle may mitigate (or magnify) potential effects of natural, technological, and human-induced hazards. The overlapping and interdependent nature of complex problems, such as those addressed in these chapters, requires the expansion of existing knowledge to better understand the relations that manifest in dynamic social, technological, socio-technological, and ecological systems. The examination and understanding of the mechanisms and processes that contribute to these complex systems further reveals the underlying tendencies and dependencies that lead to persistence and, ultimately, resilience of social and ecological systems.

REFERENCES

Borgatti, S.P., Mehra, A., Brass, D.J., & Labianca, G. (2009). Network analysis in the social sciences. *Science*, 323(892), 892–895.

Bruneau, M., & Reinhorn, A. (2006). Overview of the concept of resilience. Paper no. 2040. Proceedings of the 8th U.S. National Conference on Earthquake Engineering, San Francisco, April 18–22.

Chattopadhyay, P., Glick, W.H., & Huber, G.P. (2001). Organizational actors in response to threats and opportunities. *The Academy of Management Journal*, 44(5), 937–955.

Drabek, T.E. (1987). Emergent structures. R.R. Dynes, B. DeMarchi, & C. Pelanda (Eds.), *Sociology of disasters: Contributions of sociology to disaster research*. Milan, Italy: Franco Angeli.

Fritz, C. (1961). Disaster. In R.K. Merton & R.A. Nisbet (Eds.), *Contemporary social problems*. New York: Harcourt.

Khademian, A.M. (2005). The politics of homeland security. In D.G. Kamian (Ed.), *The McGraw-Hill homeland security handbook*. Columbus, OH: McGraw-Hill.

Killian, L.M. (1955). The significance of multiple-group membership in disaster. *The American Journal of Sociology*, 57(4), 309–314.

Mason, R.O., & Mitroff, I.I. (1973). A program for research on management information systems. *Management Science*, 19(5), 475–487.

Mayhew, B.H. (1980). Structuralism versus individualism; Part I: Shadowboxing in the dark. *Social Forces*, 9, 305–339.

Meyer, A.D. (1982). Adapting to environmental jolts. *Administrative Science Quarterly*, 27, 515–537.

Meyer, A.D., Brooks G.R., & Goes, J.B. (1990). Environmental jolts and industry revolutions: Organizational responses to discontinuous change. *Strategic Management Journal*, 11, 93–110.

Mileti, D.S., Cress, D.M., & Darlington, J.D. (2002). Earthquake culture and corporate action. *Sociological Forum*, 17(1), 161–180.

Neal, D.M. (1997). Reconsidering the phases of disaster. *International Journal of Mass Emergencies and Disasters*, 15(2), 239–264.

Oscasio, W. (1995). The enactment of economic diversity: A reconciliation of theories of failure-induced change and threat rigidity. In L.L. Cummings & B.M. Staw (Eds.), *Research in organizational behavior*, vol. 17. Greenwich, CT: JAI Press.

Staw, B., Sandelands, L., & Dutton, J. (1981). Threat rigidity effects in organizational behavior: A multilevel analysis. *Administrative Science Quarterly*, 26, 501–524.

Turner, R.H. (1978). Earthquake prediction volunteers: What can the United States learn from China? *Mass Emergencies*, 3(2), 143–160.

Van Bueren, E.M., Klijn, E., & Koppenjan, J. (2003). Dealing with wicked wicked problems, knowledge challenges, and collaborative capacity problems in networks: Analyzing an environmental debate from a network perspective. *Journal of Public Administration Research and Theory*, 13(2), 193–212.

Waugh, W.L., & Streib, G. 2006. Collaboration and leadership for effective emergency management. Special issue, *Public Administration Review*, December, 131–140.

Weber, E.P. (1998). *Pluralism by the rules: Conflict and cooperation in environmental regulation*. Washington, DC: Georgetown University Press.

Weber, E.P. (2003). *Bringing society back in: Grassroots ecosystem management, accountability, and sustainable communities*. Cambridge, MA: MIT Press.

2 All Hazards, Whole Community

Creating Resiliency

Frances L. Edwards

INTRODUCTION

Catastrophic disasters are becoming more common as people migrate to the hurricane-prone coasts and earthquake-prone areas, sea levels rise, and critical infrastructure becomes denser, as noted in Mileti's groundbreaking book, *Disasters by Design* (1999). The United Nations' *2011 Revision of the World Urbanization Prospects* noted that, of the 450 urban areas worldwide with populations over 1 million, 60% are exposed to hazards such as flood, drought, cyclones (including hurricanes), and earthquakes. Floods alone affect 233 major cities worldwide with a combined population of 663 million, while 890 million people are at risk of experiencing one of the natural hazard-driven disasters (UN, 2012). A NOAA study (2004) found that although only 17% of the land area of the United States is in coastal counties, 53% of the U.S. population lived on the coast in 2003, representing a 20% increase in just over twenty years. The population density of coastal counties is 300 persons per square mile, compared to 98 persons per square mile for the nation as a whole (NOAA, 2004).

In addition, the earthquake-producing Pacific Ring of Fire appears to be entering a more active state, with notable recent earthquakes in China (2008), Chile (2010), and Japan (2011). The U.S. Geological Survey (USGS) notes that seventy-five million people are at significant risk of earthquake shaking in thirty-nine states in the U.S., including twenty-six urban areas (USGS, 2006). Of the nation's one hundred largest cities, twenty-three were located in Alaska, California, Hawaii, Nevada, and Washington, the nation's most seismically active states (USGS, 2009) and twenty-one of these cities gained population between 2000 and 2003 (City Mayors, 2010).

Unique and unexpected aspects seem to be becoming more frequent elements of these catastrophes, as Haiti and China had earthquakes in areas not considered to have high seismic activity, and Japan's well-developed earthquake and tsunami mitigation plans were overwhelmed by the Great East Japan Earthquake-generated tsunami that drowned a nuclear power plant, resulting in ongoing harm (Nakamura & Kikuchi, 2011).

The northeastern United States and Canada lost power when a maintenance error in Ohio brought down the aging power grid (Minkel, 2008), and Hurricane Irene caused hurricane-related storm damage in Vermont (Cooper, 2011), which is not a coastal state. The lasting image of catastrophic disaster is the residents of New Orleans stranded on the roofs of their homes from the secondary disaster of levee failure related to poor maintenance (Cooper & Block, 2006).

ALL HAZARDS AND DISASTER MANAGEMENT

Emergency planning began as a response to specific, repetitive events. Communities had plans for floods, hurricanes, earthquakes, and wars that were developed as discrete activities. During the Cold War civil defense agencies in the U.S. conducted community-wide risk assessments to determine commonalities among their various disaster plans, leading to single functional plans. For example, most communities had a single plan for sheltering, regardless of whether the impetus was flood or war.

By the end of the Cold War, FEMA encouraged communities to evaluate the similarities of response to hazards, and create emergency plans that addressed all hazards, focused on "multiple-use opportunities for personnel, procedures, facilities, and equipment" (Lindell & Perry, 2007: 117). This led to the creation of integrated emergency operations plans, and multi-hazard oriented training and exercises. After September 11 there was a brief period of an "all terrorism, all the time" focus for emergency planning, as the new Department of Homeland Security was created and subsumed FEMA (Edwards & Goodrich, 2011). However, after Hurricane Katrina the National Response Framework (NRF) once again emphasized the commonalities among responses to disasters, regardless of their impetus. FEMA's latest emergency planning guidance is focused on "all levels of government in their efforts to develop and maintain viable, all-hazards, all-threats emergency plans" (FEMA, 2010).

RISK AND VULNERABILITY

Mileti (1999) noted that America was becoming prey to ever worsening disasters because of a failure to properly assess risk before permitting development. The elements of risk are the presence of natural hazards, (but also technological and human caused), a population that is vulnerable to the hazards, and the likelihood that the population would be affected by the hazard. Phillips (2009) notes that vulnerabilities may be created or enhanced by humans' efforts to "engineer" three "major misfits" in the environment: the natural systems, the built environment, and the location of human habitation.

Mileti (1999) also noted that the placement of the built environment contributes directly to the creation of natural disasters and the vulnerability of inhabitants. Examples are construction of homes, industries, and infrastructure in floodplains, below dams, or near seismic faults. Furthermore, people try to engineer their way out of disasters, sometimes inadvertently enhancing them. For example, levees give people a false sense of security from flooding, as the people in New Orleans learned in 2005 (Cooper & Block, 2006). Likewise seawalls designed to prevent coastal erosion may simply deflect damage to another part of the coast (Donner & Rodriguez, 2008).

The social system also enhances the vulnerability to disaster of some groups: "Gender, disability, income level and development status means that some population segments bear disproportionate losses" (Phillips, 2009: 38). Given that vulnerable populations are especially sensitive to disasters, communities must consider how to manage that risk. Phillips (2009) notes that "vulnerable populations bring resources to the table" (39). They have social capital in the form of relationships and organizations that meet their special needs in non-disaster times. For example, organizations that regularly provide services to people with hearing limitations will have staff members who know American Sign Language. If this organization were included in developing the community's disaster plan during the preparedness phase, it could assist people with hearing limitations during disaster response and recovery. Likewise, community-based nonprofit organizations providing paratransit services, language interpretation services, or medical procedures like dialysis can contribute knowledge and resources to the community disaster plan, as well as during the response and recovery phases. How can American communities better align their social capital to understand their hazards, anticipate response needs, and plan for recovery to create community resilience?

RECOVERY, RESILIENCE, AND SOCIAL CAPITAL

Phillips (2009) offers four lenses through which to view disaster recovery. First, in systems theory the problem is viewed through built, natural, and human environments and disruption of their relationships. Second, in vulnerability theory the focus is on the unequal sharing of risk and an opportunity during recovery to "alter differential risks [such as] income, development status, gender, race, and disability" (45). Third, emergent-norm theory points out that new threats like terrorism and pandemics can create new needs and new groups that meet those needs. Finally, resilience seems most closely related to her concept of socio-political-ecology theory. Acknowledging that "disasters disrupt social interactions, [which] influences how people recover" (45), she recognized that "participation of all people affected by disaster in the recovery process is essential" (45). As

Phillips notes, "Communities must . . . interact, [and] equitable participation . . . should lessen the uneven chance of returning home" (41).

However, Phillips acknowledges that powerful groups may unduly influence recovery. For example, economist Chamlee-Wright (2010) interviewed New Orleans residents who had lost their homes to post-Katrina flooding. FEMA bureaucrats, who controlled the creation and issuance of Flood Insurance Rate Maps (FIRMs), were seen as having disproportionate power in decisions related to community recovery. The FIRM definition of the boundaries of a floodplain would determine whether the homeowner would have to elevate the home in order to be eligible for flood insurance (Figure 2.1). In some areas homes would have to be elevated as much as 25 feet above ground level, adding cost and aesthetic challenges. Homeowners were often reluctant to invest in repairs until they knew what their new status would be under the post-Katrina FIRMs (Chamlee-Wright, 2010). Although the creation of new FIRMs includes a community meeting where comments are received from the public (FEMA, 2005a), residents from New Orleans had been relocated to other communities, making participation difficult (Chamlee-Wright, 2010).

Cox and Perry assert, "Resilience is generally understood as the capability of a community to face a threat, survive and bounce back or, perhaps more accurately, bounce forward into a normalcy newly defined by the disaster related losses and changes. Community resilience is, in effect, a

Figure 2.1 An elevated house in New Orleans.

reflection of people's shared and unique capacities to manage and adaptively respond to the extraordinary demands on resources and the losses associated with disasters" (Cox & Perry, 2011: 395–396). Erickson (2003) found that rural communities have greater social capital based on diverse networks and smaller community size. Woolcock (1998: 173) finds that these "high intra-community connections" create the social capital that leads to more successful disaster recovery.

SOCIAL CAPITAL, MITIGATION, AND RESILIENCY

Following Hurricane Katrina scholars asked why this storm had so overwhelmed the emergency management capability of the world's largest economy. Cooper and Block (2006) noted local political decision making that weakened mitigation measures through poor levee maintenance in advance of the storm, and failed to respond with adequate leadership as the storm bore down on the city. The congressional study of the catastrophe called it a "Failure of Initiative," as federal agencies became bound by regulations instead of focusing on life safety. The signal event was the insistence of the secretary of DHS, Michael Chertoff, that he had to know whether the levees had been overtopped or breached before making a decision about the level of federal aid that could be mobilized without a governor's request for a Presidential Disaster Declaration (Cooper & Block, 2006).

Community leaders in the disaster area told social science researchers about their frustration with bureaucrats who focused more on historic preservation and endangered species regulations than on helping the community reopen schools and medical facilities (Chamlee-Wright, 2010). Both Cooper and Block (2006) and Chamlee-Wright (2010) note the dynamic and motivating role played by community groups in New Orleans East's Mary Queen of Vietnam parish in moving forward with their own recovery plans, demonstrating the ability to recreate their community with little or no outside assistance.

Mileti's (1999) far-reaching analysis of disaster mitigation suggested that communities recover from small events quickly due to the availability of insurance—a form of financial mitigation—while the outcomes from larger disasters seem to last longer, especially if they include significant damage to public infrastructure. Mileti characterized resilience as "economies returning to pre-disaster conditions in a few months to a year" (214), and pointed to the value of the community anticipating the event and preparing for it.

Cox and Perry (2011) studied the 2003 McLure fire in British Columbia and concluded that the sense of place was key to recovery. The McLure fire resulted in the loss of houses and businesses, trees and birds and animals (Cox & Perry, 2011). They saw that the "sense of belonging, whether that is experienced as belonging or not belonging, in place, relationships, routines, and intentional engagement with the world through activities in the public and/or private spheres" (399) creates the social capital on which community recovery is built.

Some communities, like Mary Queen of Vietnam Parish in New Orleans East and Biloxi, Mississippi, have bounced back from Hurricane Katrina while others, like parts of the Lower Ninth Ward and sections of the coastal Louisiana parishes, still remain in ruins six years later. Scholars have investigated social, economic, and physical differences to understand how to engender a better recovery for a community to be resilient. In 1990 Rubin and Popkin studied the rural areas damaged by Hurricane Hugo, the costliest natural disaster up to that time. They noted that previous models of recovery that suggested a series of waves of recovery did not represent the actual pattern, which was better depicted as mountain peaks. Physical recovery was the first peak, while peak two was an attempt to incorporate mitigation for future events into the recovery, and the third peak encompassed an understanding of all the hazards facing the community and the multiple means of lessening future impacts, incorporating recovery and mitigation into emergency management.

Sylves (2008) noted that after the Great Midwestern Flood of 1993, FEMA took a proactive approach to disaster damage assessment to speed recovery funds to inundated communities. The voluntary organizations, a key element of social capital, provided assistance to victims with shelter, food, and medical supplies external to the government's responses. The Voluntary Organizations Active in Disasters (VOADs) that formed after Hurricane Camille in 1969 provided a central point of coordination for community-based organizations in subsequent disaster response and recovery operations. The VOADs also advocated for mitigation against future disasters (also see Chapter 5).

A consistent theme in post-Katrina recovery analysis has been the role of mitigation in the slowing of recovery. Phillips (2009) offers four paradigms for recovery, each pointing to the need to evaluate relationships among the built, natural, and human aspects of the community, and consider realignments to prevent repetition of losses. This process often includes "new codes, ordinances and policies that govern the rebuilding process, . . . set aside flood plains and reduce wildfire threats" (44). She points out that community members may not understand the delays as they strive to cope with their displaced lives.

In Chamlee-Wright's (2010) interviews with Katrina survivors the theme of recovery planning and mitigation as a road block recurs. People complain of being unable to make a decision about reinvesting in their homes because FEMA flood maps are pending and city-driven community plans are evolving. While most people wait in uncertainty, some community members drive recovery and come back, such as the nurse who restarted the clinic in the Lower Ninth Ward and the priest who led the return and repairs in New Orleans East. Although working in the context of her professional role, the superintendent of schools in Saint Bernard Parish who spearheaded the reopening of the local schools as an anchor for overall community recovery was also a community member (Chamlee-Wright, 2010).

These successful community recoveries are based in the social capital of the community. The importance of social capital in disaster recovery has also been studied by international scholars. For example, Nakagawa and Shaw (2004) noted that after the Kobe earthquake of 1995 and the Gujarat, India earthquake of 2001, "the community's social capital and leadership are found to be the most effective elements in both cases in enhancing collective actions and disaster recovery" (1). Cox and Perry (2011) studied the 2003 McLure fire in Canada which displaced residents from two small rural communities and created many barriers to their return. The smallest town lost its saw mill and all its trees, the loss of its major source of employment. They found that "social capital . . . is expressed through a sense of belonging, a sense of community, place attachment, and participation in civil society" (396).

WHOLE COMMUNITY PERSPECTIVE

In the face of the uneven outcomes of recovery from Hurricane Katrina, the federal government undertook a re-evaluation of how communities recover from catastrophic disasters. The report of this study, *A Whole Community Approach to Emergency Management: Principles, Themes and Pathways for Action*, acknowledges, "In large-scale disasters or catastrophes, government resources and capabilities can be overwhelmed" (FEMA, 2011: 1). As population grows and public budgets shrink it is imperative to understand how those available resources can be leveraged to best advantage in restoring communities. The report noted that "the conversations with various stakeholders focused on how communities get motivated and engaged, how they understand risk, and what their experiences are with resilience following a disaster" (FEMA, 2011: 2). As a result of the conversations, FEMA administrator Craig Fugate asserted, "A government-centric approach to disaster management will not be enough to meet the challenges posed by a catastrophic incident. That is why we must fully engage our entire societal capacity," the "*whole community*" (FEMA, 2011: 2).

The "whole community" incorporates "private and nonprofit sectors, including businesses, faith-based and disability organizations, and the general public" and government partners (FEMA, 2011: 3). The goal is "greater integration and empowerment across the community, stronger social infrastructure, and increased individual and collective preparedness" (FEMA, 2011: 3). This community engagement is intended to make the needs, resources, and existing networks a part of the entire emergency management enterprise, following the vision of Rubin and Popkin (1990) in incorporating recovery and mitigation into emergency management.

The "whole community" approach to developing resilience addresses the challenges of politics, power, and economic condition by creating a safe space for all the sectors of the community to work together on their common hazards, anticipate the uneven workings of the market place, and offer everyone in the community the common tools that will allow them to recover the sense of place that they value.

Community

"Community" means different things to different members of the community. It may be the physical space that represents "home" or a unit of social environment that a person identifies with, which may be spatial or psychological. It is also a place where social capital is developed. Social capital may reside in community groups like a homeowner's association, school organizations like the PTA, volunteer fire departments, police or sheriffs reserves, and amateur radio operator organizations (ARES and RACES). It encompasses civic groups like the Grange, the Chamber of Commerce, Rotary, Lions Club, and Kiwanis; community-based organizations like American Red Cross, Salvation Army, Planned Parenthood, Goodwill, hospitals, and nonprofit child care groups; and faith-based groups like churches, synagogues, temples, mosques, and other places of worship and meditation. All of these groups represent the social capital of the community, and participation in them often overlaps, creating extended social networks that create the mesh of community life. Especially in rural communities where people do not live in close proximity, these groups form the backbone of community life, giving a sense of place and belonging to an extended area.

Community Sectors

"Whole community" has different meanings for different sectors of a community. An urban planner might envision the "whole community" as the physical spaces, land uses, real estate sector, economic development, natural resources, and natural hazards, encompassing Phillips' (2009) conceptualization of built, environmental, and human elements. A government official might see the "whole community" as the voters, taxpayers, residents, business community, community-based charitable organizations, political supporters, public information networks, and the media. Business people tend to see the community as customers, suppliers, financiers, employees, and contractors, and the supporting infrastructure like telecommunications, roads, and utilities. Community-based charitable organizations and faith-based leaders may envision the "whole community" as members, volunteers, fund-raisers, donors, clients, homeless, disability services, cultural community, and "household of faith." In considering the involvement of various constituencies it is important to remember that 90% of the workforce is in the business/nonprofit sector, and 85% of infrastructure is privately owned. This includes water companies, telecommunications, and Internet companies and electricity providers.

Clearly a "whole community" approach must incorporate the government sector, including local and state government entities, as well as special-purpose governments like schools, cemetery districts, and sewer authorities. County functions like the agricultural agent, public health officer, county hospital, and mental health department are all key stakeholders in disaster response and recovery. Public safety entities like paid and volunteer fire departments and sheriff's departments are key first responders.

DEFINING RESILIENCY

After Katrina the federal government's goal was to develop communities that could not only recover in the long run, but that could conform to Mileti's (1999) concept of a rapid economic rebound. Different entities have defined resiliency in slightly different ways. The NRC sees resilience as a positive response to stress. In its report NRC (2011) defines resilience as: "To adapt to stress . . . so that it may continue to function, or quickly recover its ability to function, during and after stress . . . The ability of groups, such as communities and cities, to withstand hazards or to recover from such disruptions as natural disasters. Building and maintaining resilience depend on the ability of a group to monitor changes and to modify its plans to deal with adversity appropriately" (14).

Social science researchers Coles and Buckle (2004), speaking at a workshop on recovery in New Zealand, expanded on the concept of a whole community by including the value of an informed community in developing resilience. "Community participates fully in the recovery process and . . . it has the capacity, skills and knowledge to make its participation meaningful" (92). This capacity for meaningful participation may be the difference between the frustration voiced by Chamlee-Wright's interviewees and the idealized "whole community" envisioned by FEMA.

INDICATORS OF RESILIENCY

Research in the social science and public policy literature, including the works of Mileti (1999), Coles and Buckle (2004), Phillips (2004), the United Kingdom's Cabinet Office (2011), Hays (2012), and the California Resiliency Alliance (2011), has suggested that there are four indicators of resiliency that lead to a successful recovery from a major or catastrophic disaster event. These are: understanding the risks to the community, mitigation planning and implementation in advance of the disaster event, preparation for the disaster event, and engagement of the "whole community" in advance of the event.

Understanding the Hazards

All communities are subject to some hazards. Most of these only generate small events that can be managed within a context of personal preparedness. For example, car accidents, house fires, and hazardous materials accidents like gasoline or antifreeze spills may occur anywhere, but foresight in purchasing insurance, installing smoke detectors, and using storm drain protection when working on a car can minimize the impact of such events.

Some hazards are based in the geology or topography of the location, or its proximity to bodies of warm water or floodplains. Local government leaders must understand the natural hazards posed by the environment, and be prepared to respond to them. USGS has studies that detail the natural threats related to geology and topography. Earthquake zones are well mapped, and return rates and likely intensities have been analyzed. Volcanoes are mapped and their activity is tracked. Maps of slopes, soil types, ancient landslides, and old creek beds are available from USGS or the state geologist. Hydrology maps include the location of springs, creeks, rivers, lakes, and aquifers, and the likely drainage based on topography. The National Weather Service provides seasonal and periodic estimates of storms, tornadoes, hurricanes, rainfall, and snow. High-heat periods may lead to forest fires. Rising ocean temperatures create stronger hurricanes. Natural disasters impact every element of community life, damaging infrastructure, making travel unsafe, and requiring people to become self-reliant for days to a week.

Technology creates the potential for accidents and disasters. Power outages caused by overused power lines and an aging electrical grid coupled with high demand have become common during hot weather. Wind and ice storms can damage power lines, causing outages. Transportation systems can experience accidents and failure of infrastructure, like bridge collapses. Hazardous materials users and hazardous waste generators may have on-site or transportation-related accidents. Power plants may catch fire or release radiation. Coal mines cave in, gas and oil wells catch fire. Silos explode and petroleum storage tanks can rupture or catch fire. Leaking underground storage tanks create environmental problems, especially in areas where well water is used for human consumption, animal watering, and irrigation. Many of these events create off-site consequences for properties immediately adjacent to the locations, and even to sensitive receptors at some distance.

Some technological disasters create cascading events. A bridge collapse may close the shipping channel and destroy power lines and telephone lines that travel over it, denying service to utility users and closing a port, resulting in economic impacts in the just-in-time delivery system of industrial supply chains across the country and around the world. A widespread power outage will close banks, force stores to refuse checks and credit cards, and even to close if the security systems are lost. The loss of power will impact water processing and sewer treatment. Hospitals operating on emergency power will have to cancel elective surgery, will lose air handling capabilities (heating and cooling), and be forced to limit outpatient services to emergencies. Traffic lights and electronic warning signs go dark, and Intelligent Transportation Systems controls are lost. Electrified mass transit stops, and heavy rail signals and communications systems fail. Schools close, forcing parents to leave work. Airports close because the terminals have no ticketing services, although the tower and runway have emergency lighting for the safety of incoming aircraft that cannot be diverted. This list of impacts was actually experienced in the Northeast Power Outage of 2003 (Minkel, 2008).

Some disasters are caused by human action. Sometimes the disaster is inadvertent, as when someone returned from a trip with an exotic disease that becomes an epidemic or even pandemic. Other disasters like terrorist attacks or criminal vandalism of infrastructure are intentional anti-social acts against a community or a government. Well-known attacks on the World Trade Center in 1993 and 2001 and the Murrah Building in Oklahoma City in 1995 provide examples of terrorist acts in the U.S., but industrial sabotage and vandalism are more common and often highly disruptive. Train derailments and destruction of power substations are examples of human-caused disasters that impact thousands of people, often creating the cascading disasters related to the loss of power.

Mitigation

A community that wishes to be resilient must not only be aware of existing hazards, but must also consider means to prevent as many consequences of the events as possible. Mitigation may be physical or knowledge-based. In the "Principles of Emergency Management" course FEMA defines mitigation as "taking sustained actions to reduce or eliminate long-term risk to people and property from hazards and their effects" (FEMA, 2005b: 3.1–3.2). "Examples of mitigation actions could include the use of engineering, construction, urban planning and population training to lessen the impacts of future disasters on a community" (Edwards, 2006: 1). Levees, pumps, seawalls, drainage systems, building elevation, safe rooms, storm cellars, and wind and seismic building codes represent physical mitigation.

Community education about the identified hazards and personal preparedness for these hazards offer other types of mitigation, known as "non-structural hazard mitigation" because they have no direct relationship to the physical structure of the building. Non-structural hazard mitigation includes securing a building's contents against the effects of disasters, such as shaking from hurricane winds or seismic forces. Non-structural elements of buildings, such as light fixtures, elevators, and piping, when fitted with the proper restraints, can withstand stresses from natural hazards with little damage. Laboratory equipment and furnishings are other types of building contents that are protected from disaster-induced damage by restraints (Masek & Ridge, 2009).

Community education can encourage the formation of Community Emergency Response Teams (CERTs) that harness the capabilities of community members to help each other during a disaster response. This is especially important in rural areas where people live at a distance from each other and professional first responders are few. Community members can be trained in light search and rescue, first aid, fire suppression, hazardous materials management, and home and personal preparedness. These skills will make them assets to the community's disaster response. They are especially important in schools and business locations.

Communities can construct emergency operations centers (EOCs) with backup power and communications systems to ensure coordination with other levels of government and better management of field response activities. They can develop emergency operation plans for the specific threats to the community, including evacuation plans and vulnerable population assistance plans, where appropriate. They can work with local utilities to devise continuity of operations plans and rapid restoration strategies to support the community. They can create volunteer management plans and donations management plans in conjunction with the community-based organizations and VOAD members. First responders should receive training in all hazards disaster response, including annual exercises led by emergency management, bringing together law enforcement, fire, emergency medical services and public works resources, VOAD members, utility providers, and business community representatives.

Land use limitation is another effective mitigation measure. Zoning can prevent construction of high-occupancy buildings in inherently dangerous areas, like in the Alquist-Priolo Zones in earthquake-prone areas of California, and on beaches on barrier islands and other beach fronts, and eroding cliffs. Building code enforcement can result in stronger buildings designed to withstand expected winds or seismic activity.

The business community will benefit economically by becoming a partner with the local government in emergency mitigation. Key employment centers and businesses should have business continuity plans to ensure that they recover rapidly, for the benefit of their owners, stockholders, and employees. Jobs for the community members and tax generation are crucial for community recovery following a disaster. Companies like Wal-Mart with robust logistics capabilities and a forty-position EOC that operates 24/7 can become pre-disaster partners in preparing for distribution points for needed goods. Pre-plans can be made for notification of corporate EOC staff members to begin their support of the disaster area. In California the business community has a representative in the state's emergency operations center to receive and coordinate requests for logistics support (California Resiliency Alliance, 2011).

Waffle House is another community business partner that sees rapid reopening as a priority. The CEO, Walt Elmer, says their disaster philosophy is "show up." He notes that it gives employees their jobs and income back, but also provides hot meals for community disaster victims and first responders. Their philosophy is based on lessons learned from numerous disasters. "Get momentum going quickly after a disaster; think of the business as a member of the local community and get to know their customers; prepare employees before a disaster happens with information like what to do during and after the storm; and do a better job of building relationships with government" (Pittman, 2012).

Insurance is a form of financial mitigation that can enhance recovery at the individual, business, or community level. Biloxi, Mississippi was able to recover more quickly from Hurricane Katrina's damage because the mayor had taken out a business interruption insurance policy on the gaming casinos. He was concerned that they might suddenly be shut down by the federal government and the town would be left with a reduced income but ongoing expenses. He was ridiculed at the time and threatened with recall, but he became a hero in the post-storm environment when the insurance payment came each month and the payroll was met for the public employees facilitating community recovery through debris removal, road repair, and building permit issuance. They also had the cash flow to make contracts for major repairs to public works, and were able to receive the FEMA reimbursement on successful completion of the work (Sigo, 2005).

On a personal level insurance has become problematic for some residents. Chamlee-Wright's interviewees told stories of drawn-out fights between the FEMA flood insurance adjuster and the homeowner's insurance company over whether the damage was from wind or flooding, with each saying the damage was caused by the other and refusing to pay. One particularly distraught shopkeeper noted that his cash register was looted, but because it was below the 8-foot water line his normal property insurance company refused to pay. In disgust he said that the cash was not wet, it was gone! Homeowners with soaked interiors fought with insurers over whether the wind blew off the roof, letting the water in, or the flood water pushed the roof off (Chamlee-Wright, 2010). But for those who finally got a settlement the funding permitted them to make decisions about whether and how to repair their properties.

Preparation

While mitigation steps are effective there is still a need for preparation for the inevitable disaster event. Simple steps include preparing a personal support kit for every car in the family, and a family emergency kit for the home. Water, medications, cash, and sturdy shoes are crucial elements for all car kits, as well as keeping the gas tank at least half full all the time.

Guidance on the contents of these kits and other simple and inexpensive household preparedness steps can be found at FEMA and ARC sites.

Parents should work with schools to ensure that they have plans for caring for the children during disasters, and determine whether they need to provide a disaster kit for the child at school. Similar conversations should be held with day care, day camp, and after-school care providers who do not have the same legal responsibilities as the public schools.

Hospitals and nursing homes have special needs during and after a disaster. Their residents and patients are inherently vulnerable due to short-term or long-term disabilities. Loss of power and water can have critical impacts on their care. Hospitals have accreditation requirements for emergency plans and exercises. The first responders and emergency managers should provide support for the medical sector in updating their plans and holding their required exercises. Disaster-oriented and medical nonprofit organizations could work with nursing homes as they plan for post-disaster patient care, when workers may need to go home to manage their families' emergencies. Nursing home patients were dumped in the Superdome during the hours before Hurricane Katrina as workers fled the city or went home to their own family members (Cooper & Block, 2006). The community should have a plan for sustaining care of the nursing home patients through family or community-based support to avoid the demands on public safety personnel created by rescuing stranded patients. Such plans should include sheltering in place when possible, because moving the patients elsewhere may be precluded by road conditions.

Hazards education should be an integral part of the K–12 curriculum. Children should learn about the hazards in the area, as well as safety precautions to take during events. Older students should be taught about mitigation measures and disaster avoidance strategies.

The federal government requires that all multi-community disasters be managed using the Incident Command System (ICS) in the field. The National Incident Management System (NIMS) is required as the framework of preparedness in order to qualify for post-disaster funding. Communities should ensure that their plans are compliant with these requirements, that all required employees have received the training appropriate to their ranks and roles, and that the exercise requirements are fulfilled annually as required. Details on ICS/NIMS and the Homeland Security Exercise and Evaluation Program (HSEEP) are available at the FEMA and DHS websites. Many of the courses are available online, or at no cost to local governments through the Emergency Management Institute (EMI) at Emmitsburg, Maryland.

Engagement

Resiliency demands the participation of the "whole community," starting with the individual. FEMA's guidance for developing a "Basic Disaster Supplies Kit" (n.d.) says to stockpile "at least three days" of food and water. Each person must be responsible for his own support while sheltering from an impending disaster at home, and for the first days after a disaster. Recent American disaster response experience suggests that it takes more than seventy-two hours to mount a response to a catastrophic event.

For example, in 2011 Hurricane Irene left Vermont towns without potable water and electricity for four days. The storm struck mid-day on Sunday, yet expedient access roads for emergency response vehicles could not be completed by the National Guard until midnight on Wednesday, eighty-two hours after the storm hit, and caravans of supplies could not begin deliveries over the makeshift roads until daylight, another seven hours (VanSusteren, 2011).

Disaster Medical Assistance Teams (DMATs), Urban Search and Rescue Teams (USARs), and ARC shelter nurses are all volunteers until they are called up. They have jobs they have to leave. They have to assemble their personnel and equipment at debarkation points, travel to the disaster area, and then find a way into the area of need. National Guard members have other jobs that they have to notify of their absence, and then join their units for deployment. Three days may not be enough time for assembly, travel, and distribution of rescue supplies in the disaster area.

Families and Individuals

Families and individuals should prepare for five days of independence from infrastructure support. Car and family kits and personal emergency plans should be made in advance and discussed with the whole family every year, ideally in concert with an exercise. An out-of-area contact should be established who can serve as the family bulletin board if family members are separated and phone service is limited. Generally speaking there will be more lines opened for calling out of the area than for calling within the area. Each adult should have access to a disaster credit card that is not used for other purposes. This ensures that the entire line of credit is available for the costs of evacuation to a motel and meals. Evacuation kits should include small bills and coins for small purchases when credit cards and checks are not accepted.

Be sure to plan for pets and economic animals like cattle, sheep, and goats. ARC shelters do not permit pets, so owners must make provisions for their animals to be with family members in the daytime and shelter in the car at night. Pet-specific equipment like leash and muzzle for a dog and a carrier for a cat or other small pets are essential. Pet needs must be considered when developing the family's food and water supplies for disaster use. Families must also plan for infants and elderly relatives who may need diapers and special food.

Local Government

The role of local government is to coordinate with other levels of government to obtain the disaster support needed by the community. The mayor and council will declare a local disaster and request a governor's declaration of a state of emergency. This opens the door for the governor to request assistance from FEMA under the Stafford Act.

The local government is responsible for the disaster response. They manage the first responders, issue evacuation orders, and open shelters (usually in concert with the VOAD community). They work to prevent looting and price gouging. They designate locations for the establishment of FEMA trailer parks, survey the availability of replacement housing for disaster victims, and coordinate with utility companies for the restoration of services, or the provision of emergency services.

The local government works closely with the nonprofit organizations sector to meet the disaster needs of the community. This starts with public education of the community on individual and family preparedness. They coordinate with VOAD members and the business community on the management of post-disaster untrained volunteers, warehousing and sorting of donations, and provision of services for the special needs community. This includes services for people living with disabilities, the very young and the very old, and people with literacy issues including English learners and people for whom English is a second language.

Local government emergency managers are specialists in dealing with other levels of government. In addition to managing interactions to trigger Stafford Act assistance for individuals and families, they work with FEMA to obtain financial support for the repair of public property that has been damaged. They coordinate with the state transportation agency to get funding for damaged roads that are part of the State Highway System program. They work with the Department of Housing and Urban Development to get disaster housing assistance for community members. They coordinate emergency work contracts and permanent repair contracts to ensure maximum FEMA reimbursement.

Businesses

The business sector has had a key role in recent disaster response and recovery. Large local retailers like Wal-Mart, Home Depot, and Lowe's have plans for providing assistance to communities after disasters, which were used during the response to Hurricane Katrina and San Diego's 2007 wildland fires. In addition to providing donations to community-based organizations they change their just-in-time supply chain to bring in the materials needed by the community for cleaning up and repairing damage, a crucial part of disaster recovery. Wal-Mart has a forty-seat emergency operations center that is staffed 24/7 to respond to the needs of its delivery fleet (Raths, 2010). These resources can be used to provide logistics assistance and coordination for post-disaster needs.

Some states have organized groups to represent the business community to ensure coordination after the disaster. For example, the Business

Executives for National Security (BENS) has chapters in many states. They also have a guide for "public private disaster response coordination" (BENS 2007) that is available online. In California they host the California Resiliency Alliance and its database of pre-identified private sector resources that are available to emergency management officials in time of need. Raths (2010) has written about the success of this coordination during the 2007 wildland fires in San Diego County. Such pre-disaster planning can be beneficial when local governments need resources in the aftermath of disaster.

Small local businesses are also a key to economic recovery in a community hit by disaster. These jobs are critical for local residents to have the income to pay for repairs. They are also a key source of local tax revenues that allow the local government to pay for its share of the disaster response and related repairs. But Main Street serves another crucial role in the post-disaster community. As Mileti (1999) noted, it provides a sense of place, of familiarity in a sea of damage. Being able to go to a familiar store and buy familiar goods is an important part of psychological recovery for the community members.

After the Loma Prieta Earthquake the Garden Mall in Santa Cruz was destroyed. The buildings were unreinforced masonry that gave the downtown a sense of history and permanence, but the brick walls crumbled under the seismic forces. That space had been not just shops, but also a community gathering place for generations of college students, anti-war protestors, and outdoor musicians. Its loss caused grieving throughout the community. People made videotapes of their sense of loss, with a song written and performed by local musicians. They sold the tapes all over the nation to raise money for a memorial plaque so the old mall would never be forgotten. Shops are not just places to spend money, but places where memories are made.

Businesses benefit from being part of the recovery. Large corporations with nationwide reach and robust capital build customer loyalty by meeting their needs in a time of trouble. Wal-Mart's plan includes providing survival supplies in the earliest shipments—bottled water, diapers, cleaning supplies. These low-profit-margin items do not cover their costs of doing business, but they establish the relationship with the customer and community. Later shipments include the household goods replacements and discretionary purchases that restore company profitability. Small local businesses have a more direct stake in the community. Their own survival is enhanced by a quick reopening, providing income for their own business restoration costs (Chu, 2007).

Schools

Schools and colleges are resources to a community. In addition to teaching children about disasters as part of the science curriculum, they teach disaster-related safety. Most states mandate disaster-specific emergency response training for their students. For example, in California students learn to "duck and cover" for earthquakes in school, practicing with regular drills. That learned behavior carries over into adulthood, and the "duck and cover" reflex prevented injuries for many residents during the 1989 Loma Prieta Earthquake in California. The March 2012 tornadoes in southern Indiana did significant damage to a middle school, but the teachers and students were unharmed because they followed their safety training (AP, 2012).

Schools and colleges should be invited into the "whole community" development of disaster recovery plans, where their unique skills and resources could be used. The schools will benefit from being able to ensure that community plans to use their facilities are in concert with educational needs and liability concerns. The community will benefit from being able to access the facilities and personnel resources of the school.

Children and their safety must come first. Organized faculty members should receive CERT training to perform the search and rescue and medical care functions on the campus until first responders are available. In rural areas where there may be few first responders the CERT role is crucial.

Parent Teacher Associations have organized CERT programs for community members, using the school's auditorium or lunchroom as a training center. For example, in the Alviso neighborhood of San Jose, California the elementary school became the focus for not only CERT classes but also more general community education on local hazards and community preparedness. The Santa Clara Valley Water District and the City of San Jose's Office of Emergency Services organized joint community outreach events that were held in the lunchroom and attended by community members, including those with no children at the school. Information on flooding potential, evacuation plans, and mitigation projects was displayed, and public education materials on disaster preparedness for the home and family were distributed in English, Spanish, Vietnamese, Portuguese, and Tagalog, the languages of that community.

Schools are often the community's social focus, especially in small towns where the K–8 school feeds into a larger regional high school in another community. The school gym or cafeteria may be the Red Cross shelter. Parents will gather at the school to get their children and exchange news with neighbors. This is a good place to organize community response and self-help efforts. Teachers and school staff members may be a source of translators for community members seeking information or disaster assistance. However, a child under eighteen should not serve as a translator in a medical situation unless there is no other option for life safety.

Colleges are a good source of professional assistance. Schools with nursing programs or engineering programs may have faculty and students who could be registered as disaster service workers and volunteer their skills in the community response. Staff members at community colleges may be retired first responders or emergency managers, or have practical skills like electrical and plumbing repairs that could help get a shelter open or a hospital functioning.

Community-Based Nonprofit Organizations

Nonprofits are where the social capital of a community is most visible. Their members volunteer their time and expertise to help the broader community on a daily basis, and often have planned roles in disaster response. Nationally the organizations focused on disaster relief are generally part of VOAD, which also has regional chapters. In many communities these organizations have also formed a disaster planning and response organization, often called Collaborative Agencies' Disaster Relief Effort (CADRE) or Community Agencies Responding to Disasters (CARD). These are umbrella organizations for all of the community-based organizations that respond to disasters. The structure is often depicted as an umbrella, with each section representing a different type of disaster work, such as feeding, sheltering, mental health services, physical health services, transportation, translation, and donations management.

Pre-disaster coordination of roles prevents overlapping services while other needs go unmet. It also allows for those with the best pre-disaster capabilities to handle various tasks. For example, Goodwill has a warehouse, loading dock, and trained goods-sorting staff members, making it the best organization to manage disaster donations. The food bank has large refrigerators and shelving to manage food donations.

It should be noted that while most members of VOAD have an ongoing commitment to emergency preparedness, response, and recovery, there are other nonprofit sector organizations that only participate in disaster response for specific segments of the community, or under catastrophic circumstances. For example, a dialysis center may be run as a nonprofit entity, but also function as a business with a specific clientele. The dialysis center could still become part of CADRE for the purpose of coordinating its services for the post-disaster benefit of the entire community, not just their existing clientele. Similarly, a community clinic might be a nonprofit, but might be willing to become part of the community's emergency medical response resource base in a large-scale emergency.

CARD or CADRE groups are organized at the community or county level, and meet several times a year to maintain and develop their disaster response skills. The leaders develop and maintain the disaster response plan. The organization offers training to new nonprofits as they join, including how to make a continuity of operations plan within the organization. This includes having each agency train its volunteers in its expectations of them during a disaster, and home and personal preparedness that will allow them to serve the community.

A major challenge after disasters is donations management. In 1999 FEMA developed an annex to the Federal Response Plan (FRP) on the management of donations and untrained post-disaster volunteers. They also have a course, E288, "State Volunteer and Donations Management" (FEMA, 2004: 47), that provides guidance on developing a system for managing unexpected donations and personnel during a disaster.

Although the disaster professionals encourage people to donate money to experienced disaster relief agencies, such as the ARC, many people prefer to make a donation of goods. Often they are used items that have to be sorted and warehoused, creating a "disaster within a disaster" for communities struggling with delivering basic services. Community emergency managers have created donations management plans in concert with nonprofits that have expertise in this field. For example, in Santa Clara County, California after the Loma Prieta Earthquake, the Salvation Army was the lead agency for donations management. When people wanted to donate clothing, blankets, toys, and household goods they were directed to existing Salvation Army locations where warehouses are already established for sorting and organizing goods. Disaster victims were given vouchers by their case workers for specified goods that could be obtained from the warehouse. This system matched goods and victims efficiently while limiting waste. Leftover items could be absorbed into the normal donated goods supply chain.

Volunteer management is another challenge during disasters. Pre-disaster volunteers can be trained and organized into specific work areas to augment the professional first responders. CERT, RACES, and ARC volunteer shelter managers are good examples of these organized volunteers. After a disaster there are always well-meaning people who want to help, and many claim to have useful skills in medical care and other professions. Without a way to screen personnel for credentials and licenses, these people are often turned away. So, for example, if a paramedic offers his services after a disaster he might be placed in a hospital with oversight by medical personnel so he can be evaluated as he delivers basic services. Because most VOAD members deal with volunteers every day they have volunteer managers who are skilled at matching offered services with appropriate oversight. For example, in Santa Clara County the ARC chapter, whose day-to-day disaster services section is volunteer-based, is the CADRE member that coordinates post-disaster, unregistered volunteers with demand for workers.

Another area where nonprofits are active is in the provision of services to people with special needs, including languages other than English, age-related services for children and senior citizens, and people with access and functional needs. Programs like Guardian Angels for Partners with Disabilities match community members with access and functional needs with others who can assist them before and during a disaster. Before the disaster the two partners work on non-structural hazard mitigation in the home, ensuring that the home's contents do not injure the residents and that access to medication and special equipment is maintained even in disaster conditions. They develop a home preparedness supplies kit with appropriate water, medications, and communications capabilities. During the disaster, the partner ensures that the person with access and functional needs has access to services, including evacuation assistance.

Senior centers, child care facilities, and after-school activities fulfill age-related special needs routinely. During a disaster these same facilities and personnel would be a community asset, as working adults need to ensure the safety of their dependents, but also need to provide services to the community. Adult day care and child care are especially critical for first-responder and medical personnel who may work extended shifts to provide key disaster response services.

Some members of the community live with disabilities that make them more vulnerable during a disaster. People with visual, auditory, mobility, and mental health limitations routinely receive services from nonprofits that focus on specific disability communities. If these nonprofits have a plan for disaster response they will be able to continue to provide these services if their clients have to go to a shelter or a hospital. Critical post-disaster capabilities include sign language interpreters, Braille writers, and counselors familiar with the special needs of people living with mental health challenges, including autism, Alzheimer's disease, and developmental disabilities. Many of these organizations are licensed or accredited, and required to have an emergency response plan. Such organizations make excellent disaster response partners because they are likely to be able to continue to operate, based on their existing emergency plans.

Faith-Based Community

Faith-based community organizations are another rich source of social capital. These organizations provide a network of social and cultural activities for their members, and often include charitable works as part of their ethos. The routine services provided to the grieving, distraught, and worried members can be extended to provide assistance to members and the broader community after a disaster.

The focus of the faith-based community is generally worship. Many people confronted with a disaster experience a desire to pray, worship, or meditate as a way to sort out the feelings of anger, fear, and grief that they are experiencing. Houses of worship, in keeping with their normal mission, can provide a community service by opening their doors more often and providing more frequent opportunities for corporate worship and private religious observances. For example, after Hurricane Hugo many people in the Caribbean experienced the storm as God's judgment. They sought forgiveness and a sense of healing through their faith communities.

After September 11 the first responders at the World Trade Center used Trinity Chapel as an interdenominational resource for prayer and corporate worship. The graveyard of the church faces the World Trade Center site, and was covered with ash from the exploding buildings that included human remains. For the first responders and family members of those who died it became "holy ground," a quiet place to pray, meditate, and rest. As police and fire personnel from all over the nation arrived on mutual aid to assist at the site they used the graveyard's fence as a memorial, placing caps and t-shirts from their departments along its spiked length.

Many religious leaders are trained in personal counseling for those in grief or distress. These skills are needed after a disaster for the benefit of the community. For example, after the Cerritos Air Crash in California in 1988, Pastor Swan of the local Lutheran Church and Dr. John O'Connor of the local Los Angeles County mental health clinic created the concept of "community diffusing." They used the church's social hall as a community gathering place, and conducted corporate grief counseling for the community where the plane had landed, killing their neighbors and family members. They combined their skills in faith-based and secular counseling to bring peace and closure to the community members who chose to gather to process their disaster experience.

Many faith-based organizations have a strong tradition of community service. "Good works" are part of the culture of many of the world's religions. After a disaster, many members express their concern for their neighbors through organized service provision. Some are organized on a national level. For example, the Southern Baptist Men have a disaster service based around a barbeque trailer that they use to feed first responders and disaster victims. The Church of the Brethren Women have a child care service that goes to disaster assistance centers and provides respite care for parents so they can go through the FEMA registration process while their children are playing in a corner of the same room. A coalition of United Methodist Commission on Relief, Lutheran Social Services, and ARC social workers provided case work services for the victims of Hurricane Katrina who were scattered to Texas, Arkansas, and Georgia.

Local faith-based organizations could plan to use their existing facilities and volunteers to assist the community with feeding, shelter, child care, and other needed services. When Mayor Ray Nagin announced in July 2005 that the city did not have the resources to evacuate everyone without a car, the Red Cross and the pastors of New Orleans—members of Total Community Action—organized a project called Operation Brother's Keeper that matched people with cars and extra space with people who had no vehicle. "Operation Brother's Keeper . . . enlist[s] churches in a vast, decentralized effort to make space for the poor and the infirm in church members' cars when they evacuate" (Nolan, 2005). This effort is credited with evacuating thousands of those who would otherwise have been left behind.

The Latter Day Saints churches had already begun a member evacuation matching program. Their plan included arrangements with out-of-area congregations in Mississippi and Louisiana to receive the carloads of evacuees. Members had also prepared three-day support car kits for use during the evacuation journey. The church maintained a toll-free line for regular communications that was used as a message drop box that church leaders monitored during the evacuation (Nolan, 2005).

Churches have also been organizers for community recovery. Chamlee-Wright (2010) and Cooper and Block (2006) both chronicle the work of the community of Mary Queen of Vietnam Church in bringing New Orleans East back to life after Hurricane Katrina. Chalmee-Wright's ethnographic study chronicled the work of the social networks based in the church. Her interview subjects described how the pastor and assistant pastor contacted the church council members who were spread throughout the shelters in other states, and used this network to get people to come back soon after the water receded. They offered land they had set aside for a senior citizens' housing development to FEMA for the trailers, knowing that the water, sewer, and electrical infrastructure created by FEMA to support the trailers would provided the basis for the future housing development's infrastructure. The pastor collected enough commitments from residents to return to their homes to get the electric company to repair the lines and restore service. The church served as a shelter in the early days as community members helped each other repair damaged homes and stores. In this community the church was the facilitator of recovery while the formal community organizations were debating priorities and plans.

WHOLE COMMUNITY AND RESILIENCE

Social science researchers have recognized the central role of the "whole community" in disaster recovery. In New Zealand social scientists recognized the crucial role of psychosocial impacts on people from the loss of livelihood and property, but they noted the significant role of "community cohesiveness and structures" (Angus, 2004: 28). At a New Zealand symposium on recovery Susan Norman noted, "Recovery is a highly political process. Leaders (both political and organisational) will be required to make hard decisions that may range from placing a caveat on an individual property title, to closing a school or business in a hazardous location, through to relocating entire suburbs. The leader's decisions may alienate or reassure members of their community as they grapple with conflicting recovery needs, reducing future risks, and introducing development goals" (Norman, 2004: 40).

It is exactly these dilemmas that are best managed through the "whole community" approach to disaster recovery. Resilience is developed when communities have confronted these trade-offs in advance of the disaster, at a time when corporate community decisions can be made and planned for, when resources to solve problems are available. As noted earlier, Coles and Buckle (2004) have defined resilience as being when the "community participates fully in the recovery process and . . . has the capacity, skills and knowledge to make its participation meaningful" (92).

Morse (2008) noted the existing disconnection between the community and outside bureaucracy that impedes community recovery, restoration, and rebuilding, which must be overcome to create resilience. "Community organizations [are] aware of the very local needs and focused on solving problems in areas where the private market will not go" (37). Cooper and Block (2006) noted how the slow decisions on the adoption of new FEMA flood maps impeded rebuilding in the Lower Ninth Ward, and Chamlee-Wright's (2010) interviewees described the dilemma they faced about whether to reinvest in the reconstruction of their homes if they could never again get flood insurance.

An overarching distrust of bureaucracy led to statements like, "They are going to say what they know you want to hear, and I don't believe them anymore, I don't have faith or trust in them" (Chamlee-Wright, 2010: 139). Reflecting on the fact that her property still was not rehabilitated, one resident said that she thought the planning effort was a scam to deprive her of her home and make money for the city. "And [the city says], 'you didn't take care of your property, so we are going to take it.' . . . and sell it for three times what I paid for it" (Chamlee-Wright, 2010: 137).

Such outcomes could be minimized by pre-disaster discussions of post-disaster community goals. General plans and zoning can be modified to prevent new construction in inherently dangerous areas like ancient landslides or steep slopes so that there is less damage after an earthquake. Community discussions of emergency response plans and organizational participation can lead to the creation of entities like CADRE or the California Resiliency Alliance through which groups may choose to make specific commitments for being part of the response and recovery.

CONCLUSION

The frequency and impact of disaster seems to be increasing in the U.S. Budgets at the federal, state, and local levels are not increasing. Communities need to harness all the capabilities, knowledge, and resources at their disposal to assess their risks, determine what can be prevented or mitigated, and prepare for those disasters that cannot be avoided. By bringing the whole community to the emergency planning table, each sector can become better educated about both its own self-interest and what it can do for the rest of the community in the aftermath to speed recovery and a return to the "new normal." Whether the motivation is profit, altruism, or professional responsibility, the contribution of each sector will enhance the planning process and the likelihood of community-wide resilience.

REFERENCES

Angus, L. (2004). Our direction for recovery in New Zealand. In S. Norman (Ed.), *New Zealand Recovery Symposium 2004*. Wellington, New Zealand: Ministry of Civil Defense and Emergency Management. Retrieved March 1, 2012 from http://www.civildefence.govt.nz/memwebsite.NSF/Files/Recovery/$file/Recovery_symposium2004.pdf#page=91.

AP (Associated Press). (2012). Tornadoes kill at least 14 across southern Indiana. March 2. Retrieved April 16, 2012 from http://www.wthitv.com/dpps/weather/us_wx_news/authorities-report-extreme-tornado-damage-in-indiana-nt12-tvw_4091754.

BENS. (2007). *Getting down to business: An action plan for public-private disaster response coordination; The report of the business response task force.* Washington, DC: Business Executives for National Security.

Cabinet Office of the United Kingdom. (2011). Strategic national framework on community resilience. Retrieved March 5, 2012 from http://www.cabinetoffice.gov.uk/sites/default/files/resources/Strategic-National-Framework-on-Community-Resilience_0.pdf.

California Resiliency Alliance. (2011). About us. Retrieved March 1, 2012 from http://www.caresiliency.org/page/about-us.

Chamlee-Wright, E. (2010). *The cultural and political economy of recovery.* London: Routledge.

Chu, K. (2007). Small businesses struggle to recover from Katrina. *USA Today*, August 28. Retrieved August 29, 2007 from http://www.usatoday.com/money/smallbusiness/2007–08–28-katrina-finances_N.htm.

City Mayors. (2010). *The largest US cities: Cities ranked 1 to 100.* Retrieved April 27, 2012 from http://www.citymayors.com/gratis/uscities_100.html.

Coles, E., & Buckle, P. (2004). Developing community resilience as a foundation of effective disaster recovery. In S. Norman (Ed.), *New Zealand Recovery Symposium 2004.* Wellington, New Zealand: Ministry of Civil Defense and Emergency Management.

Cooper, M. (2011). Battered Vermont looks first to its roads. *New York Times*, September 2. A13.

Cooper, C., & Block, R. (2006). *Disaster! Hurricane Katrina and the failure of homeland security.* New York: Henry Holt and Company.

Cox, R.S., & Perry, K.M.E. (2011). Like a fish out of water: Reconsidering disaster recovery and the role of place and social capital in community disaster resilience. *American Journal of Community Psychology*, 48, 395–411.

Donner, W., & Rodriguez, H. (2008). Population composition, migration and inequality: The influence of demographic changes on disaster risk and vulnerability. *Social Forces*, 87(2), 1089–1114.

Edwards, F.L. (2006). Mitigation versus prevention: A view from the local government level. Paper presented at the ASPA National Conference, Washington, DC.

Edwards, F.L., & Goodrich, D.C. (2011). *Continuity of operations/continuity of government for state-level transportation organizations.* San Jose, CA: Mineta Transportation Institute.

Erickson, B. (2003). Social networks: The value of variety. *Contexts*, 2(1), 25–31.

FEMA (Federal Emergency Management Agency). (2004). Emergency Management Institute catalog of activities. Retrieved March 10, 2012 from http://training.fema.gov/EMIWeb/downloads/EMICatalog0304PartI.pdf.

FEMA (Federal Emergency Management Agency). (2005a). *Adoption of flood insurance rate maps by participating communities.* Publication no. 495. September. Retrieved April 16, 2012 from www.fema.gov/library/file?type=publishedFile&file=fema.

FEMA (Federal Emergency Management Agency). (2005b). Principles of Emergency Management. 3.1–3.2. Retrieved March 15, 2005 from http://training.fema.gov/EMIWeb/Downloads/is230.doc.

FEMA (Federal Emergency Management Agency). (2010). *Developing and maintaining emergency operations plans: Community preparedness guide (CPG) 101.* November. Washington, DC: FEMA.

FEMA (Federal Emergency Management Agency). (2011). *A whole community approach to emergency management: Principles, themes, and pathways for action*. FDOC 104–008–1. January. Washington, DC: FEMA. Retrieved January 12, 2012 from http://www.pascenter.org/publications/publication_home.php?id=1256&focus=PAS%20Library.

FEMA (Federal Emergency Management Agency). (n.d.). Basic disaster supplies kit. Retrieved April 3, 2012 from http://www.ready.gov/basic-disaster-supplies-kit.

Hays, J. (2012). *The community resilience framework: A pathway toward resilience*. Oakland, CA: FEMA Region IX.

Lindell, M.K., & Perry, R.W. (2007). Planning and preparedness. In W. Waugh and K. Tierney (Eds.), *Emergency management: Principles and practice for local government*. Washington, DC: ICMA.

Masek, J., & Ridge, R. (2009). *Identification of methods to achieve successful implementation of nonstructural and equipment seismic restraints*. Oakland, CA: Earthquake Engineering Research Institute.

Mileti, D.S. (1999). *Disaster by design*. Washington, DC: John Henry Press.

Minkel, J.R. (2008). The 2003 Northeast blackout—five years later. *Scientific American*, August 13. Retrieved January 31, 2012 from http://www.scientificamerican.com/article.cfm?id=2003-blackout-five-years-later.

Morse, R. (2008). Federal recovery is off the rails. *Social Policy*, Spring, 34–37.

Nakagawa, Y., & Shaw, R. (2004). Social capital: A missing link to disaster recovery. *International Journal of Mass Emergencies and Disasters*, 22(1), 5–34.

Nakamura, A., & Kikuchi, M. (2011). What we know, and what we have not yet learned: Triple disasters and the Fukushima Nuclear fiasco in Japan. *Public Administration Review*, 71(6), 893–899.

NOAA (National Oceanic and Atmospheric Administration). (2004). *Population trends along the coastal United States: 1980–2008*. Retrieved April 27, 2012 from http://oceanservice.noaa.gov/programs/mb/pdfs/coastal_pop_trends_complete.pdf.

Nolan, B. (2005). In storm, N.O. wants no one left behind; Number of people without cars makes evacuation difficult. *New Orleans Times-Picayne*, July 24.

Norman, S. (2004). Focus on recovery: A holistic framework for recovery. In S. Norman (Ed.), *New Zealand Recovery Symposium 2004*. Wellington, New Zealand: Ministry of Civil Defense and Emergency Management. Retrieved March 1, 2012 from http://www.civildefence.govt.nz/memwebsite.NSF/Files/Recovery/$file/Recovery_symposium2004.pdf#page=91.

NRC (National Research Council), Committee on Private-Public Sector Collaboration to Enhance Community Disaster Resilience, & Geographical Science Committee. (2011). *Building community disaster resilience through private-public collaboration*. Washington, DC: The National Academies Press. Retrieved March 1, 2012 from http://www.nap.edu/catalog/13028.html.

Phillips, B.D. (2004). Using Online Tools to Foster Holistic, Participatory Recovery: An Educational Approach. *Australian Journal of Emergency Management*, 19(4), 32–36.

Phillips, B.D. (2009). *Disaster recovery*. Boca Raton, FL: CRC Press.

Pittman, E. (2012). How recovery is ingrained in Waffle House's culture. *Emergency Management*, January 19. Retrieved March 1, 2012 from http://www.emergencymgmt.com/disaster/How-Recovery-Is-Ingrained-in-Waffle-Houses-Culture.html.

Raths, D. (2010). Working together. *Emergency Management*, May–June. http://api.ning.com/files/OxOZ*R4YvzBlxE3bhI-FYyoERYfCQoLoJ9yAedwP-0PVPpA8ea0vobBQpPhgNVhd06460T9RSmxE8WF89x4*IkxoV1n9HFLyg/EmergMgtMagWorkingTogetherMay2010.pdf.

Rubin, C.B., & Popkin, R. (1990). *Disaster recovery after Hurricane Hugo in South Carolina*. Working paper no. 69. January. Boulder, CO: Natural Hazards Center.

Sigo, S. (2005). Mississippi: Casino insurance. *The Bond Buyer*, October 6.

Sylves, R. (2008). *Disaster policy and politics: Emergency management and homeland security*. Washington, DC: CQ Press.

UN (United Nations). (2012). *2011 revision of the world urbanization prospects*. New York: United Nations, Department of Economic and Social Affairs, Population Division. Retrieved April 23, 2012 from http://esa.un.org/unpd/wup/index.htm.

USGS (United States Geological Survey). (2009). Top earthquake states. October 21. Retrieved April 23, 2012 from http://earthquake.usgs.gov/earthquakes/states/top_states.php.

USGS (United States Geological Survey) & Department of the Interior. (2006). *Earthquake hazards: A national threat*. Fact sheet 2006–3016. Retrieved April 26, 2012 from http://pubs.usgs.gov/fs/2006/3016/2006–3016.pdf.

VanSusteren, D. (2011). Vermont officials assess the risks in towns lacking water and power. *New York Times*, August 31. Retrieved September 1, 2011 from http://www.nytimes.com/2011/09/01/us/01flood.html?nl=todaysheadlines&emc=tha23.

Woolcock, M. (1998). Social capital and economic development: Toward a theoretical synthesis and policy framework. *Theory and Society*, 27(2), 151–208.

3 Sociological Insights on the Role of Social Capital in Disaster Resilience

Fernando I. Rivera and Marc R. Settembrino

INTRODUCTION

The concept of resilience has received increasing attention in the study of disasters, particularly after the terrorist acts of September 11, 2001 and Hurricane Katrina in 2005. Indeed, there is an ongoing effort to study effective ways to establish resilient communities that are not only prepared to survive a disaster, but that are able to "bounce back" from any particular natural or social threat. While the concept has been the subject of inquiry in various disciplines, including psychology (Norris, 2009), ecology (Donoghue, 2007), geography (Cutter, Boruff, & Shirley, 2003; Cutter, Burton, & Emrich, 2010), organization and management sciences (McManus et al., 2008), and safety management (Dynes, 2003; O'Brien et al., 2006), our interest is to investigate the sociological views on disaster resilience. In this chapter, we address how sociological tenets can help shape a better understanding on ways that may facilitate or impede disaster resilience. Based on our review of the research literature we create a sociological framework of disaster resilience. Finally, we examine the social capital component of our framework utilizing social capital data from seven counties in Central Florida.

SOCIOLOGICAL PERSPECTIVES

A recent report by the NRC (2010) defines resilience as "the continued ability of a person, group, or system to function during and after any sort of stress" (3). This definition should resonate within a sociological concept of disaster resilience. Mainly, for sociologists, disasters are the result of the intersection of environmental hazards and social systems that impact the ability of communities to anticipate, resist, and recover from these events (Pais & Elliott, 2008). Communities mean large groups of individuals in their own habitats and familiar surroundings (McFarlane & Norris, 2006) and not simply an aggregate of individual characteristics and microlevel interactions.

A large body of sociological research has concentrated on the identification of social attributes such as gender (Morrow & Phillips, 1999; Tierney, 2006), race (Elliott & Pais, 2006), socio-economic status (Bolin, 2006; Fordham, 1999), among others that render some communities more vulnerable to disaster situations. These social attributes result in social inequalities, not because they point to the existence of social differences, but rather because they are a result of the differential social value, power, and treatments that are structurally linked with representatives of specific classes, genders, and races/ethnicities.

While the vulnerabilities are well known (see Chapter 4 by Kusenbach and Christmann), sociological studies tend to be modest as they relate to disaster resilience. On the other hand, geographers are advancing frameworks of disaster resilience with core measures of community and social resilience. The work of Cutter and her colleagues (Cutter et al., 2003; Cutter et al., 2010) is noteworthy in considering the impact of social vulnerabilities on disaster resilience. Specifically, they argue that "social vulnerability is a multidimensional concept that helps to identify those characteristics and experiences of communities (and individuals) that enable them to respond to and recover from environmental hazards" (Cutter et al., 2003: 257). Additionally, Cutter, Burton, and Emrich (2010) find that communities in hazard-prone areas with high levels of social and economic well-being may in fact be more resilient than communities with relatively few environmental hazards but suffer from higher degrees of social and economic problems.

Considering the work of Cutter and her colleagues (2003, 2010), we argue that it is very important to move forward the sociological concept and study of community disaster resilience. In the following section we review the sociological literature and answer the following questions: What is community resilience? How has it been defined in the sociological literature? What are some of the barriers/facilitators of resilience found in the literature? What have we found? Where do we need to go? What do we need to consider?

What Is Resilience? How Has It Been Defined in the Sociological Literature?

A search of academic databases of peer-reviewed literature primarily in sociology was used to gather the research articles included in the review. Key words used in the search were "resiliency or resilient" in combination with "disasters." Studies were identified using the following inclusion criteria: (1) analyses of a "natural disaster"; (2) inclusion or discussion of resilience; (3) study conducted in the United States; (4) articles in English. Search results yielded a total of fifty-six articles. After a careful review the total number of articles that met the inclusion criteria was eighteen articles.

Most of the articles reviewed did not clearly define resilience; often the authors referred to the standard idea of "bouncing back" after a stressful situation. The only exception to this trend was the study by Green, Kleiner, and

Montgomery (2007) in which the authors stated: "On the other end of the continuum, people and their communities are considered resilient when they have access to the resources needed to prevent or respond effectively to a crisis situation" (30). We argue that there is a need to develop a sociological definition of resilience which contrasts the traditional psychological and engineering definitions of flexibility, bending, and ability to bounce back. Green et al.'s (2007) definition is a good start, but as sociologists we must adopt a working definition that identifies the pre-disaster social and economic inequalities that render a community unable to develop and sustain resiliency. Some post-disaster studies have allowed researchers to understand and identify some of the barriers and facilitators of community resilience and post-disaster recovery. Below we summarize the barriers and facilitators identified in these studies.

What Are Some of the Barriers of Resilience Found in the Literature?

In a review of the association between vulnerabilities and health outcomes in disaster situations, Ursano, Fullerton, and Tarhakopian (2008) identified low socio-economic position and mistrust of governmental agencies as potential barriers to community resilience. Additionally, Ojerio et al. (2011) found that one's poverty status significantly affected one's participation in federal wildfire risk mitigation programs in Arizona. Furthermore, in their study of the aftermath of Hurricane Katrina, Green and colleagues (2007) observed that one major barrier to resilience was the lack of access to assets that provide the ability to respond to a disaster or disaster threat; particularly, the lack of access to environmental/natural, physical/built, human, social, and financial assets.

Racial and gender inequality also have significant effects on disaster resilience. For instance, Elliott and Pais (2006) found that blacks were less likely to take hurricane warnings seriously and were almost seven times more likely to lose their jobs than whites at the aftermath of the storm. Additionally, the gendered nature of household, care, and emotion work cause women to experience and respond to disasters much differently than men. For example, women may take on additional responsibilities such as caring for children and the elderly before and after disasters (Morrow & Phillips, 1999). As always, social class also mediates women's experiences of disaster with women of lower socio-economic status facing greater barriers to disaster resilience (Fordham 1999).

What Are Some of the Facilitators of Resilience Found in the Literature?

The literature identifies several facilitators to disaster resilience. For example, Ursano et al. (2008) suggested the promotion of collective efficacy and social capital as facilitators of resilience. Quick response and replacement of community resources and knowledge of pre-disaster community vulnerabilities also help build resilience. Others suggest the development and maintenance of community networks in every aspect of disaster planning (Green et al., 2007; Kim & Kang, 2010). Specifically, Kim and Kang (2010) found that integration into community-level communication resources including community organizations and interpersonal networks impact the likelihood of engaging in hurricane preparedness activities.

Additionally, in their study of East Biloxi, Alabama after Hurricane Katrina Green, Gill, and Kleiner (2006) recommended that vulnerability must be addressed in planning for disaster preparedness, response, recovery, and redevelopment. They also put forward the need to provide fast and dignified disaster relief in order to build trust with the community. Fordham (1999) also supports providing dignified disaster relief, noting that the relationships developed during post-disaster relief efforts may build future disaster resilience. Ultimately, in order for communities to be resilient during and after a disaster, steps must be taken to mitigate their vulnerabilities.

Another recurrent theme was the recommendation to include local civic and faith-based organizations in the planning and response to disasters (Berke & Campanella, 2006; Brennan & Flint, 2007; Elliott & Pais, 2006; Green et al., 2006). The inclusion of local groups can ameliorate issues of mistrust and can accelerate and promote disaster resiliency.

What Have We Found? Where Do We Need to Go? What Do We Need to Consider?

There are several considerations in the literature. (1) The need to understand long-term recovery efforts and how they serve to mitigate or amplify pre-existing vulnerabilities (Pais & Elliott, 2008; Vale & Campanella, 2005). This is particularly important as there is evidence that suggests that as communities recover from natural disasters, pre-existing residential segregation is expanded and solidified (Pais & Elliott, 2008). (2) The identification of pre-disaster community assets, needs, and vulnerabilities (Carroll et al., 2006; Green et al., 2007). (3) The inclusion of local community residents, service providers, and diverse organizations in advocating for the needs of the vulnerable, and planning for future crisis events in order to facilitate the process of resiliency (Green et al., 2007). As suggested by Wilkinson (1991), "The policy goal should be to assist local actors in developing the capacity to solve local problems" (107). (4) Finally, utilizing community-based research models to conduct needs assessments, asset-mapping, and evaluations of response and recovery effectiveness (Green et al., 2006).

SOCIOLOGICAL DISASTER COMMUNITY RESILIENCY FRAMEWORK

Based on the review of the literature we propose the following model as a complement to other models of disaster resiliency (see Chapter 10 by Deyle and Butler). Specifically our model emphasizes sociological determinants of resiliency.

As the model suggests, pre-disaster barriers and facilitators must be identified alongside their interaction with resiliency facilitators. We argue

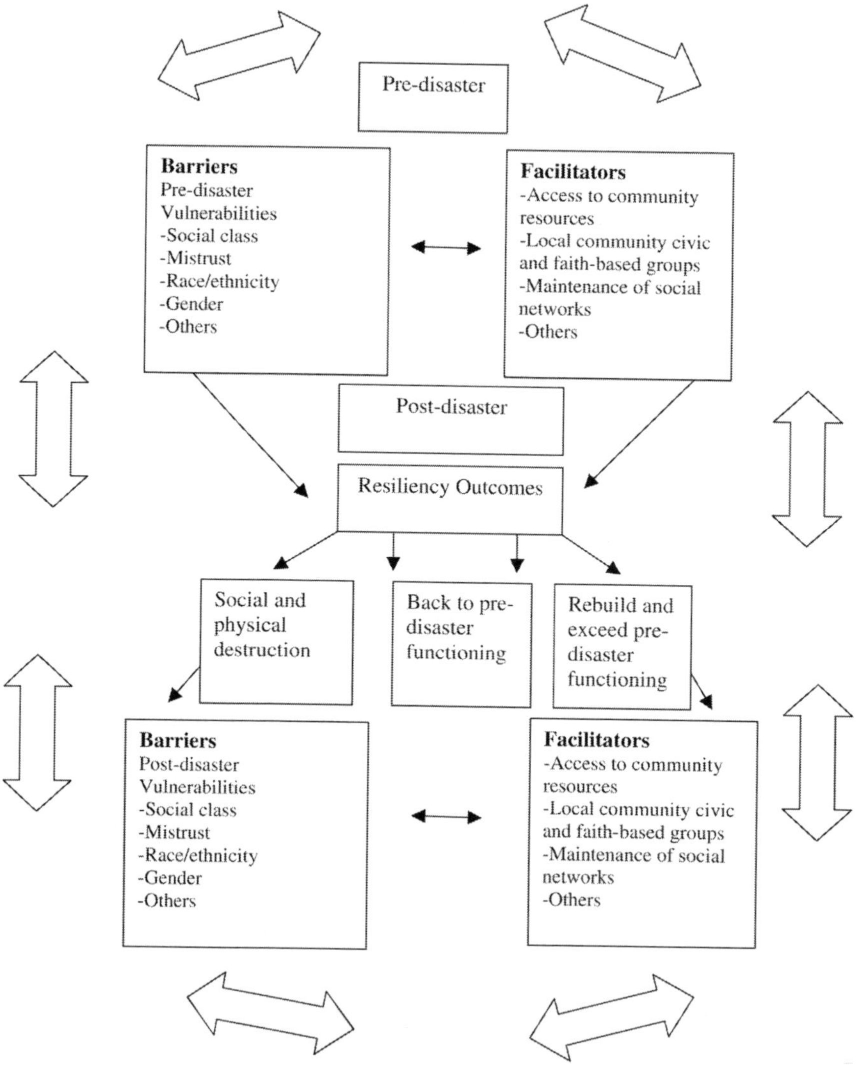

Figure 3.1 Sociological disaster community resiliency framework.

that the process is circular, thus if there is no change in pre-disaster vulnerabilities they are likely to increase post-disaster vulnerabilities or vice versa. Furthermore, if pre-disaster facilitators are limited or non-existent there is a higher probability that they will not be available or effective during or after disasters. We have identified vulnerabilities such as mistrust to government agencies, social class, race, ethnicity, and gender, as barriers to disaster resilience. By no means are these a complete set of vulnerabilities. Therefore we have also included "others" in our model. In the next section, we test some of the barriers and facilitators from our model to elucidate their potential relationship to disaster resiliency.

In our framework we identified certain pre-/post-disaster barriers (social class, mistrust, race/ethnicity, gender, and others) and facilitators (access to community resources, local community civic and faith-based groups, maintenance of social networks, and others) that closely resemble the operationalization of social capital. Social capital measures how connected people are to friends, family, and their communities. Overall, social capital is a measure of the health of a community's social fabric (Wright & Jasinski, 2005). Communities with strong social health may be more resilient in times of strain. Particularly, communities with high social capital may be the most resilient in the aftermath of disasters such as hurricanes or tornados. As stated in our review, women, racial and ethnic minorities, and individuals with low socio-economic status are the most vulnerable to such disasters (Bolin, 2006; Elliot & Pais, 2006; Ursano et al., 2008). In the following analyses we test the influence of such characteristics on some social capital components such as social trust, political distrust, and informal social interactions.

METHOD

We utilized data from the Central Florida Social Capital Survey (CFSCS) to analyze the association of demographic and social indicators on social capital. The CFSCS was a computer-assisted telephone survey conducted by the Survey Research Laboratory in the Institute for Social and Behavioral Sciences at the University of Central Florida. In total, 1,467 residents from seven Central Florida counties completed the survey. The data were weighted to compensate for sampling disproportion. For the purposes of our analyses, those cases which had missing values for any of the dependent or independent variables were removed from the sample using list-wise deletion for a total of 1,031 respondents. All analyses were conducted using Stata 9 SE, a statistical software package.

In our analyses we test the influence of living in a rural community on social trust, political distrust, and informal social interaction controlling for age, gender, race, ethnicity, level of education, and household income. For the purposes of our analyses, Orange County has been defined as an urban county, whereas Brevard, Lake, Osceola, Polk, Seminole, and Volusia counties have been defined as rural counties. We do acknowledge that it may be problematic to code entire counties as "rural" or "urban." We have considered possible alternatives to this coding scheme including zip-code and census tract-level analyses. Unfortunately, such analyses would also be limited. While zip codes and census tracts are recognized as official place designations, they do not necessarily represent contiguous communities. Furthermore, in the state of Florida counties are responsible for emergency management and planning. Because of this, we felt that county-level analyses would be more appropriate for our current study. That being said, we acknowledge that future research should focus more on local areas rather than counties at large.

When interviewed, respondents were asked to report one of the following racial categories: white, black, Asian, multi-racial, Native American, or other. Because there were few Native American respondents in the sample, this category was combined with the category of other. In the Ordinary Least Square regression analyses presented later in this chapter, black respondents are compared to all other racial categories. In addition to reporting a racial category, respondents were asked whether or not they are Hispanic. Thus respondents may be both white and Hispanic or any other combination thereof.

Social Trust

Respondents were asked a series of seven questions about how much they trust members of their community. First, respondents were asked, "Would you say most people can be trusted?" Responses were coded as follows: (1) people can be trusted, (2) you can't be too careful, and (3) depends. The second item asked respondents how much they trust people in their neighborhoods. Responses were coded as follows: (1) trust a lot, (2) trust some, (3) trust a little, and (4) not at all. This coding scheme was also used for survey questions about trust in co-workers, members of their congregation, people working in local stores, the local news media, and local law enforcement. Wright and Jasinski (2005) created a social trust index using z-scores of responses to the aforementioned survey items. Higher scores on the social trust index reflect higher social trust.

Political Trust

In addition to social trust items, respondents were also asked questions about how much they trust national, state, and local government officials. Specifically, respondents were asked, "How much of the time can you trust the National/State/Local government to do what is right?" Responses were coded as follows: (1) just about always, (2) most of the time, (3) some of the time, and (4) hardly ever. As with the social trust measures Wright and Jasinski (2005) created a z-score index of political trust. In this case it is an index of political distrust, with high scores reflecting higher levels of distrust.

Informal Social Interactions

We measured the number of informal social interactions reported by respondents using an index variable created by Wright and Jasinski (2005) in which higher scores reflect more informal social interactions. Types of interactions included in the index are socializing with neighbors, co-workers, and friends outside of formal organizations.

FINDINGS

Table 3.1 presents the weighted sample characteristics. The mean age for the sample is 50.89 years old. Seventy percent of respondents live in rural communities. The sample is nearly equally divided between men and women. Overall the sample is largely white, with just 6% of respondents being black. Almost 9% of the sample is Hispanic. Overall the sample is highly educated. The majority of respondents have attended at least some college with about 32% holding a Bachelor's degree or higher. The majority

Table 3.1 Weighted Sample Characteristics

Rural Residents	69.93
Age	50.89
Female	49.83
Black	6.00
Hispanic	8.69
Level of Education	
Less than High School	3.91
High School Diploma	23.47
Some College	39.14
Bachelors Degree	19.97
Some Grad School	2.89
Grad./Prof. Degree	10.62
HH Income in Dollars	
Less than 20,000	14.16
20–40,000	21.69
40–60,000	22.32
60–80,000	16.64
80–100,000	9.26
100–150,000	10.50
More than 150,000	5.43

Note: All values reported are percentages except for the mean age.

of respondents make between $20,000 and $80,000 per year, with 14% making less than $20,000 per year, and about 5% making more than $150,000 per year.

Table 3.2 presents the results of linear regression with OLS estimators for each of the dependent variables. In terms of political distrust, the model was not statistically significant, thus we will not discuss the model here; the results however are included in the table.

Table 3.2 OLS Regression Analyses for Social Trust, Political Distrust, and Informal Social Interaction

	Social Trust		*Political Distrust*		*Informal Interactions*	
	B	*SE*	*B*	*SE*	*B*	*SE*
Rural Residents	0.03	0.05	0.04	0.07	0.05	0.04
Age	0.01	0.00 ***	0.00	0.00	-0.01	0.00 ***
Female	0.16	0.04 ***	0.02	0.06	0.04	0.03
Black	-0.37	0.10 ***	0.14	0.13	-0.18	0.08*
Hispanic	-0.23	0.09 **	-0.02	0.12	0.02	0.07
Level of Education						
Less than High School	-0.06	0.13	-0.10	0.15	-0.17	0.10
High School Diploma (Ref.)						
Some College	0.09	0.06	-0.07	0.08	0.06	0.05
Bachelors Degree	0.10	0.07	-0.11	0.09	0.07	0.06
Some Grad School	0.06	0.10	-0.15	0.17	0.10	0.10
Grad./Prof. Degree	0.12	0.08	0.11	0.11	-0.09	0.06
HH Income in Dollars						
Less than 20,000	-0.29	0.12 **	0.09	0.15	-0.27	0.09 **
20–40,000	-0.14	0.10	0.08	0.14	-0.05	0.08
40–60,000	-0.13	0.10	0.08	0.13	-0.09	0.08
60–80,000	-0.14	0.10	0.04	0.14	-0.11	0.08
80–100,000	-0.07	0.10	-0.11	0.15	0.02	0.09
100–150,000	-0.02	0.11	-0.16	0.14	0.00	0.09
More than 150,000 (Ref.)						
Constant	-0.50	0.12 ***	0.00	0.17	0.53	0.10 ***
R^2	0.139***		0.015		0.153***	

The first model measures the influence of the independent variables on the social trust index variable. Overall the model explains roughly 14% of the variance in social trust (R^2=.139, p<.001). Older respondents have greater social trust than younger respondents, with each additional year of age accounting for about a .01 increase in social trust (B=.01, p<.001). Additionally, women appear to have greater social trust than men (B=.16, p<.001). Black respondents have lower social trust than white, Asian, multi-racial, and other races. Ultimately blacks score about .38 points lower than other respondents on the social trust measure (B=-.38, p<.001). Similarly, Hispanic respondents have lower social trust than non-Hispanics, scoring about .23 points lower in terms of social trust (B=-.23, p<.01). In terms of income, respondents at the lowest end of the income distribution (household incomes less than $20,000 per year) have lower social trust than those at the highest end of the income distribution (more than $150,000 per year) (B=-.28, p<.01). Respondents living in rural counties were not statistically different from those living in urban areas in terms of social trust.

The third model measures the influence of the independent variables on the informal social interaction index variable. The model explains roughly 15% of the variation in social interactions (R^2=.153, p<.001). Older respondents have fewer informal social interactions, with each one-year change in age resulting in .01 decrease in social interactions (B=-.01, p<.001). Similarly, black respondents have fewer informal social interactions (B=-.18, p<.05). Finally, respondents with household incomes of less than $20,000 per year have fewer informal social interactions than those with household incomes greater than $150,000 per year. Ultimately, respondents with incomes lower than $20,000 per year score .27 points lower on the social interaction measure than respondents making more than $150,000 per year. Rural county residents are not statistically different from urban residents in this model. Similarly, gender, Hispanic ethnicity, and level of education are not significant predictors of informal social interactions.

CONCLUSIONS

In this study we reviewed the sociological disaster literature and identified a number of barriers and facilitators of community resilience. From our review we constructed a sociological framework of community disaster resiliency and tested some of the components of the framework, particularly social capital. We found some differences by race, gender, age, and income levels, but overall most counties (both rural and urban) reported similar high levels of social capital.

Our sociological literature review and findings from the social capital survey suggests the importance of identifying pre-disaster and post- disaster vulnerabilities as they pertain to community resiliency. Awareness of these vulnerabilities is vital if we are to truly create resilient communities.

For communities overburdened by social inequality, the identification of these vulnerabilities can open the door to social justice practices that are likely to increase the level of disaster mitigation and subsequent community resiliency. It is not unexpected that failure to do so can render some communities unable to fully "bounce back" from an emergency situation.

Our review identifies a standard definition of resilience, based primarily on the idea of communities "bouncing back" after a stressful situation. This begs the question on whether we need to define resilience in another fashion or simply emphasize resiliency outcomes. Our framework suggests a continuum of post-disaster outcomes ranging from the destruction of the community (both physically and socially) to the ability of a community to continue to exist after a disaster ("bouncing back"), and finally to the maintaince and rebuilding of the community to exceed pre-disaster functioning. Essentially, communities of the latter type more than simply "bounce back" from a disaster, they leap forward. Within this continuum, communities can utilize the disaster experience to address pre-disaster vulnerabilities and work to change them, or communities can simply work together to reach the same level of pre-disaster functionality without proposing changes to existing social inequalities.

We also recognize that there are different pathways and mechanisms to reach community resiliency. In other words, the path to resiliency is not linear, as communities with high levels of social capital sometimes are unable to bounce back, while others with low levels might be able to be resilient. Preliminary evidence from focus groups in Central Florida supports these phenomena. For example, in the Villages (the largest retirement community in the U.S.) located in Sumter County, emergency managers told us about the high levels of social capital (income, access to resources, and community involvement). Additionally, they commented about how the community has organized several Community Emergency Response Teams (CERTs), yet when the summer hurricane season nears most residents return to the home states, leaving emergency managers with an unprepared and less resilient community. On the other hand, in Volusia County we heard from a group of farmworkers, who during the 2004 hurricane season were virtually ignored by aid workers. Even though these workers traditionally have low levels of social capital (mistrust of governmental officials, limited access to resources, and small social networks) in recent years they were able to form a CERT and have since built an emergency shelter in their community.

Finally we must discuss the political and social implications of our findings. Our model addresses the importance of identifying pre- and post-disaster vulnerabilities as a way to increase community resiliency. The model also implies that steps must be taken to mitigate such vulnerabilities, but it leaves the possibility that for some communities where vulnerabilities are not addressed or changed, or funding is not available, resilience might be impossible to achieve. On the other hand, if communities are able to be resilient without addressing or changing social inequalities, it might open the false argument to justify reduced funding for emergency operations.

We must note that the data was collected during a non-emergency context and thus we are unable to assess if these levels of social capital will remain the same during an emergency event. Nonetheless, we are hopeful that our review and model will impact the way community disaster resiliency is defined in a sociological context, and that addressing pre-disaster social inequalities are likely to increase the resilience of communities after any traumatic event, including natural disasters.

ACKNOWLEDGEMENTS

This study was supported by the U.S. Department of Agriculture under Award No. 2010–67023–21698. Any opinions, findings, and conclusions or recommendations contained in this document are those of the authors and do not necessarily reflect the views of the agencies that provided support for the project.

REFERENCES

Berke, P.R., & Campanella, T.J. (2006). Planning for post-disaster resiliency. *Annals of the American Academy of Political and Social Science*, 604, 192–207.

Bolin, B. (2006). Race, class, ethnicity, and disaster vulnerability. In H. Rodriguez, E.L. Quarantelli, & R. Dynes (Eds.), *Handbook of disaster research*. New York: Springer.

Brennan, M.A., & Flint, C.G. (2007) Uncovering the hidden dimensions of rural disaster mitigation: Capacity building through community emergency response teams. *Southern Rural Sociology*, 22, 111–126.

Carroll, M.S., Higgins, L.L., Cohn, P.J., & Burchfield, J. (2006). Community wildfire events as a source of social conflict. *Rural Sociology*, 71(2), 261–280.

Cutter, S.L., Boruff, B.J., & Shirley, W.L. (2003). Social vulnerability to environmental hazards. *Social Science Research*, 84, 242–261.

Cutter, S.L., Burton, C.G., & Emrich, C.T. (2010). Disaster resilience indicators for benchmarking baseline conditions. *Journal of Homeland Security and Emergency Management*, 7(1), 1–22.

Donoghue, E. (2007). Social science constructs in ecosystem assessments: Revisiting community capacity and community resiliency. *Society & Natural Resources*, 20, 899–912.

Dynes, R. (2003). Finding order in disorder: Continuities in the 9–11 response. *International Journal of Mass Emergencies and Disasters*, 21, 9–23.

Elliott, J., & Pais, J. (2006). Race, class, and Hurricane Katrina: Social differences in human responses to disaster. *Social Science Research*, 35, 295–321.

Fordham, M. (1999). The intersection of gender and social class in disaster: Balancing resilience and vulnerability. *International Journal of Mass Emergencies and Disasters*, 17, 15–37.

Green, J.J., Gill, D.A., & Kleiner, A.M. (2006). From vulnerability to resiliency: Assessing impacts and responses to disaster. *Southern Rural Sociology*, 21, 89–99.

Green, J.J., Kleiner, A.M., & Montgomery, J.P. (2007). The texture of local disaster response: Service provider's views following Hurricane Katrina. *Southern Rural Sociology*, 22, 28–44.

Kim, Y.C., & Kang, J. (2010). Communication, neighborhood belonging and household hurricane preparedness. *Disasters*, 34, 470–488.

McFarlane, A.C., & Norris, F.H. (2006). Definitions and concepts in disaster research. In F.H.

Norris, S. Galea, M.J. Friedman, & P.J. Watson (Eds.), *Methods for disaster mental health research* (pp. 3–19). New York: Guilford Press.

McManus, S.S., Seville, E., Vargo, J., & Brunsdon, D. (2008). Facilitated process for improving organizational resilience. *Natural Hazards Review*, 9, 81–90.

Morrow, B.H., & Phillips, B. (1999). What's gender got to do with it? *International Journal of Mass Emergencies and Disasters*, 17, 5–11.

Norris, F. (2009). Looking for resilience: Understanding the longitudinal trajectories of responses to stress. *Social Science & Medicine*, 68, 2190–2198.

NRC (National Research Council). (2010). *Private-public sector collaboration to enhance community disaster resilience: A workshop report*. Washington, DC: The National Academies Press.

O'Brien, G., O'Keefe, P., Rose, J., & Wisner, B. (2006). Climate change and disaster management. *Disasters*, 30, 64–80.

Ojerio, R., Moseley, C., Lynn, K., & Bania, N. (2011). Limited involvement of socially vulnerable populations in federal programs to mitigate wildfire risk in Arizona. *Natural Hazards Review*, 12, 11–20.

Pais, J., & Elliott, J. (2008). Places as recovery machines: Vulnerability and neighborhood change after major hurricanes. *Social Forces*, 86, 1415–1453.

Tierney, K. (2006). *Facing hazards and disasters understanding human dimensions*. Washington, DC: The National Academies Press.

Ursano, R., Fullerton, C.S., & Terhakopian, A. (2008). Disasters and health: Distress, disorders, and disaster behaviors in communities, neighborhoods, and nations. *Social Research*, 75, 1015–1028.

Vale, L.J., & Campanella, T.J. (2005). *The resilient city: How modern cities recover from disaster*. New York: Oxford University Press.

Wilkinson, K.P. (1991). *The community in rural America*. New Work: Greenwood Press.

Wright, J.D., & Jasinski, J.L. (2005). The ties that bind: The Central Florida social capital survey. Institute for Social and Behavioral Science, Department of Sociology, University of Central Florida.

4 Understanding Hurricane Vulnerability

Lessons from Mobile Home Communities

Margarethe Kusenbach and Gabriela Christmann

INTRODUCTION

Over the last decades, a social science framing of disasters has emerged. Disasters are now understood as intersections of geophysical events with vulnerable populations, as the result of natural and social processes that continually unfold over time (Hoffman & Oliver-Smith, 2002). Research on disaster vulnerability, therefore, has evolved into a comprehensive discipline that investigates not only geophysical processes but also social, economic, and political forces (Tobin & Montz, 1997; Wisner, Blaikie, & Cannon, 2004). This shift toward a more complex *social* conception of hazards and disasters in general, and of vulnerability in particular, is significant and important. However, as we argue and demonstrate in this chapter, it has not gone far enough in current research.

Our goal in the following section is to propose an innovative understanding of the concept of vulnerability that takes human agency and other complexities of the social world more clearly into account. Next, we briefly describe the methods and data of the research to which our conceptual ideas will be applied, a study of mobile home communities in Florida. The main section demonstrates the limits of previous understandings and illustrates the utility of our conceptual critiques for grasping constructions of vulnerability in real-world contexts by showing brief analyses of these data. In particular, we look at how subjective perceptions of risk and social inequalities deviate from the views of experts such as disaster managers and social scientists. Overall, the chapter highlights the importance of human agency and social complexity for the advancement of scholarly research in disaster resiliency and disaster policies.

REDEFINING VULNERABILITY

This section largely follows arguments and ideas developed in publications by Christmann and her associates (Christmann & Ibert, forthcoming;

Christmann et al., 2011). We begin by recalling an important difference between the social and the natural world. Humans are clearly objects and parts of the material environments in which they live, yet at the same time they are also subjects who construct and change these environments through individual and collective action. In order to understand disasters we need to take seriously the importance of human decision making which builds on social processes of perception and interpretation of selves, others, and environments. In relation to the concept of vulnerability, this means that people and communities are not fixed in place but can actively change their level of exposure to potentially harmful events, and that these changes depend on people's assessments of themselves, other people, and environmental conditions at hand. In other words, the human ability to interpret and react to one's own and others' vulnerability must be a component in the overall definition of vulnerability.

However, this important human element is largely missing from customary definitions of vulnerability, or even "social vulnerability" as many social scientists prefer to call it. For instance, Morrow (2008) states that "social vulnerability is a catchall phrase that has become part of the discussion related to how social and cultural conditions place some at higher risk to environmental impacts such as climate change or natural hazards. Simply stated, social vulnerability occurs when unequal exposure to risk is coupled with unequal access to resources" (4).

In this definition, social vulnerability is determined by the presence of physical risk on the one side and the absence of, or limited access to, resources on the other; both conditions together create and increase vulnerability. Many social scientists rely on a similar understanding and define vulnerable groups as people who lack financial, material, personal, human, social, and/or other capital (Morrow, 1999)—in short people who are exposed to social inequality and who are concentrated in areas with a high risk of exposure to geophysical hazards. The latter component is often assessed rather simply by the general location of people's homes. Vulnerability thus becomes a measurable characteristic of certain individuals or groups, a property that they own, largely independent of surrounding social contexts.

There is no question that these and similar conceptions of vulnerability are valid, as they are broadly supported by personal experience and by research. Some (groups of) people really do suffer much more and longer than others from exposure to geophysical hazards when they occur, and we need to understand how and why this happens. Current conceptions of vulnerability appear well suited to provide a compact answer: inequalities in resources and exposure to risk are indeed major material factors in capturing this condition. Nevertheless, we believe that a critical element is missing from these understandings. In our view, the *socially constructed* nature of vulnerability cannot be expressed entirely and sufficiently through quantifiable indicators of structural inequality plus geophysical risk.

Such views neglect that, to a considerable extent, vulnerability is rooted in the knowledge and interpretations (and thus the potential for change) that people have regarding their own and others' placement in the natural and social world. For instance, it is an act of perception and interpretation to define, individually or collectively, certain weather and climate patterns as a potentially catastrophic hurricane or a dangerous heat wave. Definitions of conditions, not the conditions themselves, determine the actions that people construct and implement to alleviate problematic situations. Saying that certain problems "exist" and that they, in turn, mandate certain solutions is a common way of speaking—however, it seems more accurate to say that conditions must first be perceived and framed as problems before they can be attended to in meaningful, problem-solving ways (Stallings, 1995). Many conditions, even physical ones, will simply not become problems without human agency. Similarly, we can find perceptions of serious problems, with very real consequences, which are not supported by factual conditions.

It is important to understand that perceptions and interpretations of vulnerability are social processes which are, by and large, different from the social processes that create and perpetuate inequalities. It is therefore possible that, despite clear indicators of vulnerability, some affected individuals and groups will not share this view regarding themselves, and vice versa. We do not have to assume an absence of consciousness, or even the presence of false consciousness, to validate the existence of vulnerability as a matter of perception because it arises from other social processes and conditions, some of which may encourage us to discount or completely overlook "clear" indicators of vulnerability. This point also helps one understand how diverse and competing perceptions of vulnerability might arise between and within affected groups, and how social conflicts might follow from such disagreements.

Despite the ongoing efforts of social scientists toward emphasizing the social dimensions of disaster vulnerability, several shortcomings of the customary views can be diagnosed.

- Insufficient tribute is paid to the *social construction* of these ideas, i.e., the fact that the interpretation of one's condition does not have to coincide with objective indicators (Janssen & Ostrom, 2006). Human perceptions and interpretations often originate in cultural contexts that exhibit only spurious links to measurable structural inequalities.
- The existing *normativity* of the concept is rather limiting. Vulnerability is always viewed as negative, as a lack of something important (various forms of capital, or resources). Resources are always viewed as positive. Neither view accounts for unintended side effects, multiple outcomes, ironies, and reversals that occur in the real world.

- There is a lack of a consideration of the complexity of *space*, beyond the general idea of exposure to risk via a simple presence in dangerous places. Space and place are viewed as containers which somehow contain people. In reality however, risk is all but simple and depends on the local interplay of physical, institutional, and interactional structures and dynamics that can hardly be captured via general location.
- There is a lack of consideration regarding the complexities of *time*. What vulnerability means can and does change at any point in time before, during, and after the impact of a particular hazard. We therefore suggest an understanding of vulnerability as a *social process* (Folke, 2006), meaning a series of social action and interaction, rather than a static set of conditions.

Building on these critiques, we suggest that vulnerability should *refer to a concept that denotes a social practice in which a certain unit (a subject, a group, or any kind of system) is placed at the center of a complex analysis of injury.* Put differently, vulnerability should be used as shorthand for a certain kind of calculation. The unit in question is placed in relation to other, interdependent units, and then analytically compared. There are three ingredients to this definition: the basic unit, the unit's relationships and interactions with other units in the environment, and the particular dimensions via which the unit in question is compared to other units.

How does this definition of vulnerability respond to the identified weaknesses of prior understandings? First, none of the definition's ingredients are pre-selected in any way, they must be actively decided upon and placed into a meaningful relationship to each other. Because social agents can focus on different units, contextual relationships, and points of comparison, vulnerability outcomes can greatly vary, even for similar situations. Second, the definition allows for both material objects, such as specific physical conditions and resources, and immaterial or cultural aspects, such as ideologies and policy regulations, to be selected for the calculation; one focus does not have to exclude or dominate the other. Third, the proposed reconceptualization of vulnerability is firmly rooted in the dynamics of real place and time. The specific calculations (and responsive actions, which might be a first step toward redefining "resilience") can change at any moment and are influenced by dynamic spatial conditions.

Regarding time, vulnerability calculations are conceptualized as "real time" events, meaning they take place under actual conditions. There is a chance of failure or success that cannot be anticipated with certainty, and each calculation counts, meaning it cannot be reversed. For example, once a household has decided not to evacuate in the available time frame due to a negative assessment of personal risk, it will not get another chance. For each new action, new vulnerability conditions occur and must be calculated accordingly. Lastly, regarding the limiting normativity mentioned above, the proposed definition does not affix a value to calculations of vulnerability or their outcomes. A calculation that seems positive at a time can be re-evaluated as negative at a later stage, and something that is initially viewed as catastrophic can turn out to have been a blessing in disguise.

METHODS

Data in this chapter stem from an interdisciplinary NSF-funded study titled "Community Resources and Disaster Resilience in Florida Mobile Home Parks," conducted between 2008 and 2010 in the Tampa Bay area on the Gulf coast of Florida. Research included the collection of qualitative and quantitative data, as well as the analysis of U.S. census data and other community-level information. The goal of the study was to investigate community and disaster-related issues in social contexts that include working- and middle-class Florida families living in manufactured housing. Original research focused on four mobile home communities and their local environments. Mobile home parks in the region significantly cluster in certain areas (Figure 4.1).

Our four research sites are located within the two circled clusters in North Hillsborough County (clusters in South Hillsborough largely consist of senior communities). Additionally, research sites were selected for their diversity in location, community size and amenities, and residential profiles (ethnicity, household types) (Table 4.1). Easy access for student researchers and the cooperation of community owners and managers were important selection factors as well. By choice, we focused on so-called family communities which accommodate residents of all ages, as opposed to so-called senior communities which typically restrict occupancy to people age fifty-five and above. Furthermore, in order to target middle- and working-class families, we concentrated on land lease communities in which all or most residents (technically) own their home but rent a lot from the park owner.

Overall, the research team conducted in-depth interviews with 103 mobile home households in the four research sites (between twenty and thirty each), as well as seven background interviews. All interviews were recorded and transcribed verbatim; nine interviews were conducted in Spanish and later translated into English. Interview participants in the four settings were recruited through a variety of leads, including responses to flyers, personal contacts, chance encounters, and third-person referrals.

In addition, we conducted forty-four follow-up interviews with previous participating households in the two largest research sites. These mobile "go-along" interviews (Kusenbach, 2003) were not tape-recorded but described in detailed ethnographic field notes. Two graduate research assistants lived separately in two of the four research sites for a time period of six months. They were thus able to gain insider access and knowledge as residents, and they recorded their observations and experiences in a large body of ethnographic field notes. Furthermore, we collected quantitative

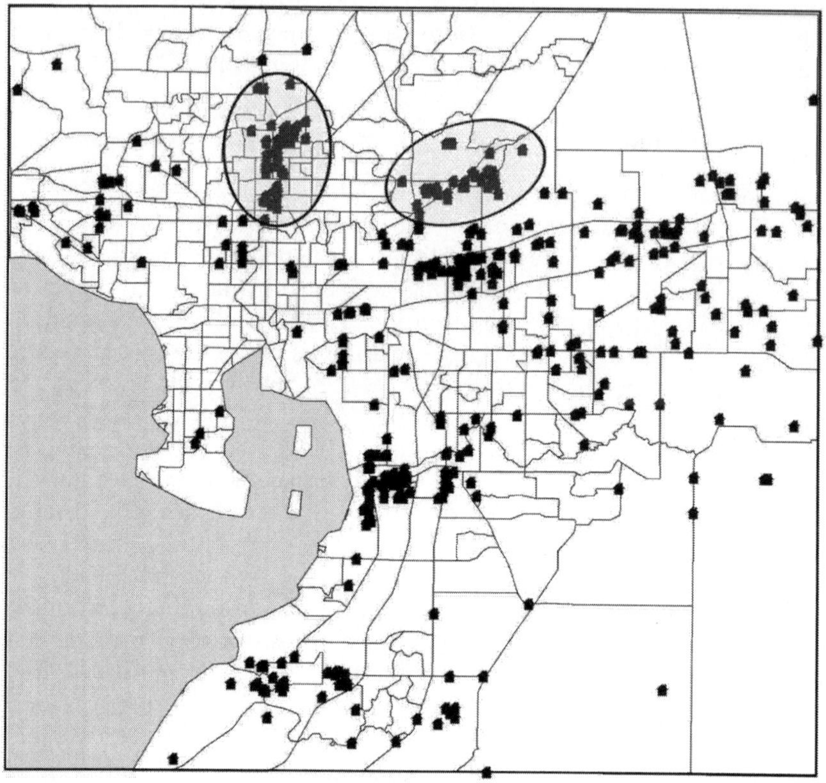

Figure 4.1 Hillsborough County mobile home parks.

data from all 103 participating households. Background and secondary research on all four research sites was conducted as well, on topics such as community history, infrastructure, and public life. Despite the non-random character of our household sample, we are confident that the diverse data we obtained allows for characterizations and conclusions that apply to a substantial portion of our target population, working- and middle-class Florida families living in manufactured housing.

In this chapter, we mainly focus on one particular household and interview in order to illustrate the real-life complexities of vulnerability in necessary detail. The goal of this case-based approach is to highlight the kinds of patterns and problems that we observed in the data at large, even though specific contexts and conditions may vary individually and overall. To compensate for the limitations of this approach, we strive to go beyond the selected case whenever possible by adding general characterizations as well as additional examples.

Table 4.1 Selected Community Characteristics

Research Site Characteristics	*"Countryside Village"*	*"Happy Place"*	*"Siesta Club"*	*"Shady Pines"*
Neighborhood and Place	Rural, Small Town	Suburban, Unincorporated County	Urban, City of Tampa	Urban, City of Tampa
Size of Park, Number of Spaces	Medium, <200	Large, 400+	Large, 300+	Small, <100
Largest Population	White Families	White Families	Latino Families	White (male) Singles
Amenities	Few (sm. clubhouse, playground, laundry, fence)	Many (clubhouse, pool, playground, laundry, riverfront park, fence, gates)	Many (clubhouse, pool, playground, laundry, gym, pond, park, fence, gates)	Few (sm. laundry, fence)
Census Tract Characteristics			(located in same tract)	
Total Population	2,969	5,024	8,347	
White (percent)	90.4	61.4	42.5	
African American (percent)	2.4	24.6	36.1	
Hispanic or Latino (percent of total population)	14.1	19.4	41.8	
Median Age	41.1	31.0	29.3	
Total Occupied Housing Units	1,255	1,687	3,343	
Owner Occupied (percent)	65.0	56.4	20.7	
Renter Occupied (percent)	35.0	43.6	87.6	
Mobile Homes (percent)	77.9	18.1	12.3	
Median Household Income in 2005–2009	$34,864	$54,514	$19,750	
Individuals in Poverty in 2005–2009 (percent)	23.9	19.8	44.1	

Source of Tract Information: U.S. Census 2010 and ACS 2005–2009 (most recent available).

UNDERSTANDING HURRICANE VULNERABILITY

The following analytic section is divided into two major parts. The first part first illustrates the limitations of the customary framing of risk, here understood as a combination of physical factors and location, and uses a series of excerpts from one household interview to illustrate the importance of understanding risk perception as a form of human agency. The second part is devoted to the other major aspect of social vulnerability, unequal access to resources or, in other words, social structural inequalities. Likewise, we begin with a discussion of the limits of traditional understanding of resources as indicated by our data, and then move toward including issues of perception and other social complexities for an understanding of vulnerability that is more firmly grounded in real-life contexts.

Understanding Risk

According to the latest American Community Survey estimates (U.S. Census Bureau, 2010), of the 8.5 million mobile homes in the United States, roughly 10% (840,000) are located in Florida, housing 1.6 million residents. Approximately 9% of all Florida homes are mobile homes. Within Florida, the highest concentration of mobile homes can be found in the central Gulf region, the area in which the research used in this chapter was conducted. According to the United States Landfalling Hurricane Probability Project (2012), the chance of a hurricane hitting the central Gulf region is about 6% per year, and it is close to 17% for a tropical (named) storm.

Previous experience and research leave no doubt that all mobile homes and their inhabitants experience a considerable risk of suffering adverse effects during high-wind events such as hurricanes or tornadoes due to the physical weaknesses of such buildings. There is some research evidence for the claim that newer mobile or manufactured homes, built after stricter safety codes were implemented in 1994 in the aftermath of Hurricane Andrew, tend to suffer less damage during high-wind events than earlier models (Montz & Tobin, 2005; Simmons & Sutter, 2008). While this may be good news for owners of newer homes and for mobile home manufacturers, who regularly put forth misleading information on the "outstanding" safety of their products, this result needs to be evaluated very carefully due to the many local and contextual factors, such as the home's attachments to the ground (tie-downs) and the immediately surrounding structures and natural features, that determine the actual *in-place* risk of these buildings.

Regardless of any differences in risk between homes and local contexts, Florida emergency managers routinely issue evacuation orders to the residents of all mobile homes in an area during anticipated high-wind events, typically along with, and sometimes even in advance of, evacuation orders issued to residents living in the first coastal storm surge (flood) zone. In other words, evacuation orders are issued solely based on an assessment of structural and geographical risk, regardless of localized conditions, and also regardless of any indicators of social inequality.

There are very sound policy reasons for a “one size fits all” approach to dealing with mobile homes and their residents during a hurricane—clarity being one of them—and we do not advocate changing this approach in any way. A mandatory evacuation issued for all mobile home residents in an area usually does not strike experts (policy makers, first responders, or academics) as being complicated or difficult to understand. It is viewed as a simple message, communicating that if you live in a certain type of home, in a particular area, you must leave now.

However, the reality is much more complicated. First, there are always those who refuse to heed orders by public officials out of principle. In our study, the percentage of people who told us they are planning to disobey any evacuation order was 9%, with another 3% appearing very reluctant to leave. This total of 12% may well be on the low side, given the results of previous studies on similar populations (Baker, 1991; Kusenbach, Simms, & Tobin, 2010).

Next, there were about 30% of respondents who added certain qualifications to their willingness to evacuate after a mandatory order was issued. Most qualifications involved storm strength, storm direction (some say they would only leave when a direct hit was imminent), or the type of evacuation ordered (some said they would only leave if the evacuation was strictly mandatory, or if they were personally ordered to leave). Other evacuation qualifications involved finding a place that would be pet-friendly or have generator power. In addition, there is a considerable, but difficult to determine, percentage of people who appear willing to evacuate and have made some preparations, however the plans they had made do not seem realistic or sufficiently safe.

All of the above complications are, more or less, “compliance” issues but they already point to the complex set of issues discussed next, namely the problem of not understanding an evacuation order, and the even more basic problem of not sharing outsider’s assessments of exposure to risk. We summarily refer to these latter kinds of problems as issues of *risk perception*. We use “perception” as an abbreviation for the complex, place, and time-sensitive patterns of searching for, learning, interpreting, and transforming into action information and messages regarding one’s personal exposure to hurricanes—in other words, calculations of vulnerability.

Consider the case of Bill and Bev, a white married couple in their late forties/early fifties living in a rural mobile home community together with a teenage daughter. Bill and Bev moved to Florida only a couple of years before they were interviewed in June 2010, because of Bill’s job at that time and because Bev’s elderly mother needed assistance. They first lived in an RV in the mother’s mobile home park in a nearby coastal town. After about a year, they moved to the three-bedroom mobile home they are currently occupying. Bev grew up in Florida and remembers evacuating for

Hurricane Elena in 1985 but Bill has never personally experienced a hurricane. Both, however, have ample experience with other hazards such as tornadoes, earthquakes, and wildfires.

In the following interview segment, Bill communicates that he is generally quite concerned about hurricanes and worries about neighbors who do not take them seriously.

Bill: I've talked to a lot of people about hurricanes in this area, especially when the wind is blowing and, you know, the conversation just goes there. Most people I've talked to have got a markedly cavalier attitude toward it, towards hurricanes, you know? And so, it kinda concerns me that they don't care, and I think it is something that they should care about.

Bill and Bev do not take the threat of hurricanes lightly. Early in the interview, we learn that they are generally willing to evacuate and that they have made numerous preparations. However, when the two student interviewers inquire about Bill and Bev's exact plans for evacuation, the conversation turns to a mandatory evacuation order that had been issued to all mobile home residents during the previous hurricane season (2009), and it turns out that Bill and Bev did not leave their home at that time.

Bev: I think our closest place to go is Town High School. I think that's what they said, if there was an evacuation. But outside of that, you never know. You know, when they gave the evacuation order last year, it's like, do we? Is that where we really have to go? All mobile homes . . . well, we live in a mobile home, does that mean we have to go? The wind is not even blowing, it's not even cloudy, does that mean we have to go? We didn't go at that time but . . . I think knowing exactly when is the right time is kind of difficult around here 'cause you don't always get that information. The weather men are always [saying]: "The sky is falling, the sky is falling," and one day, the sky really is going to fall and no one's gonna listen 'cause they are always wrong.

Interviewer: Um, how do you guys get that information, if you have to be evacuated, from the news or does the park offer it?

Bill: Well, we check a number of different sources that collaborate [sic] the information anyway. So, we listen to the radio, we watch it on TV, and get [it] on the Internet.

Bev: And our phones have weather and stuff, so we can get weather bulletins and stuff. That's more reliable than anything. Usually, we will look at the weather radar and make the assessment ourselves, pretty much. The one time they did something here for the last year was just, um, we had to put everything in that was outside, put it either in the house or in the shed and make sure the house was secure. But I mean, I don't think we even left. [Bill interjects: No, we didn't.] It didn't even rain. It did get windy, but, I mean, it didn't even rain!

This long passage provides a number of interesting clues to the process of risk perception. The first is that Bev was confused about the target group of the evacuation order, mobile home residents. Even though she knows that she lives in a mobile home, she is somehow not sure if this means that she needs to leave. Her confusion resurfaces again at a later point in the interview.

Bev: We hadn't even lived here four months and here comes, all of a sudden, they're like *all* mobile homes! They kinda said it where it didn't sound like it was "all" mobile homes, it's just, if you qualify this or this, so I called [and asked]: "Does this mean us?" "Well, you're in a mobile home, aren't you?" But, but, but, but . . . you know? So, we didn't really go anywhere, we just kept wondering, should we? Should we be going? So you know, it's just one of those not knowing [situations]. I don't know what it would take to be clear and say, "Hey, we should really get outta here."

Again, Bev conveys the confusion she felt upon receiving the evacuation order. Even after receiving additional confirmation, she remains unsure whether this order *really* applies to her or not.

Bev is not alone in her confusion. Another respondent in the study explained to us that these orders do not apply to her because she lives in a "manufactured home." Other people said that they do not live in a "flood zone" and therefore do not have to evacuate, reflecting the fact that evacuation orders are most frequently mentioned in combination with various storm surge zones. The gleaned insight here is that not all residents of mobile homes receive and process the evacuation order as applying to them even though "all mobile homes" were mentioned very clearly.

The second insight from the long interview segment relates to timing; Bev talks about the unclear timing of the evacuation order. She says it is difficult to know exactly when to leave, implying that she did not understand that mandatory evacuation orders are always immediate. Bev instead bemoans that the "right time" was not clearly communicated, and furthermore, it appeared counterintuitive to her to start evacuating while the sun was still shining. This confusion may point to a possible gap in the communication of the evacuation order, but it may also be a reflection of Bev's misunderstanding of the nature of hurricanes, or of the complexity of an evacuation that potentially involves tens of thousands of people.

The third intriguing concern that surfaced in the earlier quote is the lack of trust in the validity of the information she received. She points out that, in her view, media messages are always exaggerated ("The sky is falling") and that you can therefore safely discount some, if not most, of them. At a later point in the interview, Bill delivers a cynical monologue about the terrible inaccuracy of even regular weather forecasts. He claims, somewhat seriously, that he can do a better job of predicting the weather by simply looking at the sky than meteorologists can with all their technology and formal education.

Returning to the above quote, Bev describes how she triangulates information from a variety of sources and how she considers some of them to be more trustworthy than others. Next, she stresses that, based on assessing multiple inputs, she and Bill rather "make the assessment ourselves, pretty much." In other words, Bev and Bill actively seek out information, process that information and then decide on a course of action that they consider is best for them, largely independent of any official messages they receive and understand.

The couple's reliance on personal decision making, as opposed to official communication, is validated in hindsight by their experiences surrounding the last evacuation order. The retrospective observation that "it did not even rain" lends support to Bev's conviction of having made the right decision, which was, in this particular case, "doing nothing." Bev and Bill did follow the instructions of the park manager and put away loose items around their house which might become projectiles, yet they evaluate this instance of "doing something" and following orders as a waste of time.

This last insight supports research results by Arlikatti, Lindell, and Prater (2006) and Siegel, Shoaf, and Afifi (2003) which indicate that risk perception and successive protective actions are strongly shaped by previous disaster experiences—and maybe, as in this case, also by non-experiences of disasters. Somewhat ironically, it seems that being first put on alert but then not suffering any harm erodes trust in the source of the warning and may contribute to a false sense of safety in the future. Moreover, it trains people to rely on their own assessments rather than official communication.

Another passage from the same interview clearly indicates that vulnerable populations may not share the external assessment of being exposed to high risk in the first place.

Bill: One of the things you had asked about the park, one of the reasons we decided to move to this particular park is, if you look around, right? If you look around the park, you're going to notice something about this park. It's not unusual, but it's just one of those things you'll have to look at, and that's the density of the old royal oaks that are in this park. Alright, they're not here for a long time for no reason at all . . . The big thing is that the weather hasn't destroyed them because for some reason this area seems to . . . it avoids this area. Now eventually, you know, the odds are that something's gonna bust through it and come through, but, uh, this particular area has over a hundred years of history of not having any tornadoes coming anywhere close to here. So, that's

> just one of those things you have to think about, one of those intangibles. You go to a park where there's no trees, or there's a lot of broken trees, well, chances are, 'cause they have had weather going through there and weather likes to follow itself. I mean, if there's a particular charade, that may, you know, channel certain weather characteristics, this one here seems to be one of those areas it seems to avoid. So, that's one of the things that we looked at.

As the conversation continues, it seems that Bill operates under the impression that he does not really experience high risk, despite living in a mobile home, due to the placement of his community in what he views as a "hurricane-averse" location. Bill does not clearly understand the reasons for why his neighborhood has, apparently, been spared a hit during the previous hundred years (unlike less lucky neighborhoods who are disproportionally targeted by bad weather), yet he firmly believes in the predictive power and accuracy of this pattern. It is quite possible that Bill and Bev will heavily draw on this particular "folk belief" when making evacuation-related decisions in the future.

Somewhat contradictory to the earlier impression of their overall concern regarding hurricanes, Bev and Bill at times seem to resign to both the inevitability and the random character of hurricanes. Elsewhere, they talk about how preparations can only go "so far," and that they do not try to think too much about what they cannot control in life. And in another moment, Bev adds that "we don't want to live our whole life in fear."

Interviewer: How about this upcoming hurricane season, are you guys concerned about a hurricane hitting this year?

Bill: No.

Bev: No, not really . . . I think, at some point, Tampa's gonna get it. I mean, it's inevitable, okay? But, there isn't a whole lot you are going to do about it if it's gonna come, and you really can't know until the last minute 'cause it can change any second. All we can do pretty much is just watch.

The dose of passivity and fatalism expressed in this quote, and the previously mentioned unwillingness to be preoccupied with negative thinking, are other cultural beliefs that encourage Bev and Bill's tendency to discount their risk of exposure despite their general concern.

In our research, we came across numerous similar beliefs which caused people to downplay their exposure to risk at a very basic level. For instance, while scouting for potential research sites, we came across the owner of a waterfront mobile home community who did not see the need to participate in our research because his park was located on a patch of land considered sacred by Native Americans in Florida and would therefore be protected from hurricanes. One older male study participant pointed to a large oak tree near his mobile home and firmly claimed that this tree will protect his home during a hurricane, and a middle-aged mother told us that she believes God will protect her and her family during adverse events and it is therefore unnecessary to take any protective action or make plans for evacuation. In sum, it is safe to assume that these and similar magical beliefs reduce perceptions of risk. They buffer the felt urgency for self-protection and therefore delay, if not inhibit, remedial action.

Understanding Structures and Resources

Most definitions of social vulnerability emphasize social inequality, which in this context means the unequal presence of, and access to, certain resources that divide groups and other populations segments. The fast-growing literature on social vulnerability has identified numerous characteristics at the individual and household level that have been associated with increased vulnerability to natural hazards, such as non-white race or ethnicity, female gender, non-adult and advanced age, low income and wealth, low level of education, presence of disabilities, pet ownership, as well as several others (e.g., Cutter & Finch, 2008; Enarson & Scanlon, 1999).

Summary descriptions of the households participating in our study indicate elevated levels of many of these customary indicators of social vulnerability.

Race/ethnicity, immigration, and general mobility: Overall, the racial/ethnic make-up of the sample includes approximately 70% white participants and about a fifth Latino (one in five). The remaining share were identified as American Indian, African American, or mixed race. Fourteen percent of primary informants (usually heads of households) were foreign born, mostly in Central or South America, followed by Western Europe. The vast majority of Latino and immigrant participants resided in the large urban mobile home park. Only 15% of all primary participants were Floridians, and only 10% were from the Tampa region, indicating a high level of mobility by place of origin among all participating households. Nearly half of all participating households had lived at their current address for less than three years, also supporting the impression of high residential mobility among mobile home dwellers.

Gender: Just under 40% of interview participants were male, just over 60% were female. However, the gender ratio varied greatly among research sites, ranging from roughly 70% male participation in the small urban community in which many military veterans lived, to just over 20% in the suburban setting. Our sample is consistent with the overall pattern of mobile home residents more likely to be female than male.

Age and families: The age of all participants ranged from eighteen to eighty-nine years. Roughly one-third of primary informants were over the age of sixty, just under half were between forty and fifty-nine years, and the rest (roughly one-fourth) were younger than forty years. The average age of our sample was fifty-two years, higher than the average in all surrounding communities, despite many households with children. Over 40% of all households included children under the age of eighteen. Very few children lived in the small urban community, meaning the other three research sites had higher percentages of households with children, above 50% on average.

Work: Over 90% of participating households technically owned their homes, however, the vast majority carried mortgages or private loans. Less than a third of primary informants said that they were currently employed (including part-time positions). If we deduct the third of the sample that had reached early retirement age (sixty years), 50% of participants between the ages of twenty and sixty remain who were neither working nor in school. Employment clustered around blue-collar jobs (e.g., maintenance man or cashier) as well as some service or office positions (call center job, medical technician). Only two informants held jobs that require college degrees (teacher, nurse), and two other informants held middle-management positions (bank, insurance company).

Education: Regarding education, over 80% of informants reported completing high school or obtaining a GED, and nearly half said they had started college or received other professional training. If educational self-reports can be trusted, it must be noted that having post–high school education did not lead to employment in positions that require higher education credentials.

Employment: Self-reported household incomes in our sample ranged from zero to $6,700 monthly. Average incomes varied greatly between the four research sites but overall remained below the county and state averages. Reported median household incomes in the suburban community were similar to the median household income in the same census tract. Average incomes of participants in the two urban locations exceeded the average incomes reported for the surrounding census tract. However, in the rural park, the average income was clearly below the census tract average. Overall, incomes in the urban and rural areas were lower than incomes in the suburbs.

Health, pets, transportation: Over half of all participating households had at least one member with either a disability or a serious health problem. About 15% had two or more members with such problems. While the number of participants and household members suffering from incurable illnesses (emphysema, liver cancer) or other significant problems (severe heart disease, fibromyalgia) was small, we came across many sufferers of disabilities or chronic health problems that severely limit daily functioning, such as obesity, chronic diabetes, or severe back pain. A sizable percentage of people reported serious mental health conditions within their households, ranging from ADHD to alcoholism, from PTSD to anxieties, from depression to phobias.

Roughly sixty-six of participating households owned at least one pet, slightly above the national average of 62%. By far the most common pets were cats and dogs. About one-third of all households had one pet, yet another third of all households, or half of all pet owners, had more than one pet. About 10% of households in the sample had three or more pets, up to over a dozen.

Regarding transportation, one in five of all participating households did not own a car. This presents a major obstacle to evacuation especially in non-urban areas that do not offer public transportation. In addition, by far not everyone who had access to a car was capable of driving in an emergency situation, and some cars could not be relied upon even for local trips. In sum, the number of households who lack reliable transportation is considerably higher than the number of carless households, even though the exact rate is difficult to determine.

All in all, when looking at commonly assessed dimensions of vulnerability within the examined group, an image of high overall vulnerability emerges. While we expected to find low employment, income, and wealth, partially due to the serious impact of the housing and economic crises on the region, health and transportation problems loomed much larger than anticipated.

Yet again, there is a curious disconnect between a social science (and maybe even common sense) framing of this population as "vulnerable" and the self-perceptions of strengths and weaknesses among research participants which greatly affected their hurricane-related planning and preparations. Again, we use "perception" as an abbreviation for the complex, place, and time-sensitive patterns of searching for, learning, interpreting, and transforming into action information and messages regarding one's strengths and weaknesses, in other words, calculations of vulnerability.

The case of Bill and Bev allows for a deeper understanding of the human elements and other complexities which tend to be missing from customary descriptions of social vulnerability patterns. Overall, when considering the above indicators, Bill and Bev's resource situation was not good. In addition to their high residential mobility, at the time of the interview, Bill had been unemployed for over a year, Bev even longer. Their approximate monthly household income of $1,100 consisted of a military pension (Bill is a disabled veteran) and was supplemented by (dwindling) savings. Each month, Bill and Bev pay $425 in lot rent and make a $300 payment to the owner of their mobile home in a rent-to-own type of arrangement. Together, they suffer from a combination of allergies, asthma, and type 2 diabetes. Both require daily medication and both also are clearly overweight. The family has several pets, including a dog and a cat. Bev and Bill own a 1992 van which is converted for camping, they also own an older RV which is stored about an hour away, and they both have cell phones. Bill has some college education and Bev has a high school degree. In sum, when considering traditional indicators of social vulnerability, Bill and Bev exhibit a number of weaknesses, including low income and net worth, health problems, and multiple pets, but they also possess certain strengths, including (but not limited to) transportation and some education, in addition to plenty of life experience.

But how do Bill and Bev view their own strengths and weaknesses, and how do these assessments affect their preparations for hurricanes? We already mentioned that Bill and Bev generally appear to be concerned about hurricanes, and that they generally seem to be willing to evacuate. When asked about the possibility of evacuating to a public shelter, Bev replies in the following way:

Bev: We checked on that because of his asthma, and he has to have a certain medication every day and we both have CPAC machines we use when we sleep, so those are one of those things we would let them know we need help with, any of those kind of things, if he needed a breathing treatment or anything. We like to know about those things and we like to know if we have to go to a shelter, but pretty much as long as we can run our van, we have converters, we can run anything that we need to from the van, for a short period of time, so we probably don't have to worry about it too much. We'd rather be self-sufficient.

The quote reveals an interesting paradox. While Bill and Bev believe that they may need additional help during a hurricane, they also have a clear preference for coping on their own: "We'd rather be self-sufficient." Because they do worry about hurricanes on some level, Bill and Bev have made a number of plans for evacuation. Most importantly, they rely on their travel-ready van and a number of other supplies. Like many study participants, Bill and Bev describe some advance planning and preparation which is generally positive, of course, and indicates an understanding of their own weaknesses.

The real problem is that, like many others, Bill and Bev feel that the limited preparations and plans they have put in place are sufficient in neutralizing their weaknesses, and may even somehow turn weaknesses into strengths. Because of the van and the converter, Bev feels well prepared and (carefully) claims that "we probably don't have to worry about it" anymore. The same sense of independence and desire for controlling their own decisions that Bill and Bev exhibited regarding their personal risk comes into play when considering, and alleviating, their own weaknesses. Having made some preparations gives Bev and Bill a sense of agency and control regarding hurricanes, and it greatly diminishes their perceived vulnerability.

At one point during the interview Bev admits that they have actually gathered none of the items that are officially recommended for hurricane preparation, aside from the usual supplies that they keep in the van (which include radios, flashlights, batteries, and water containers). Right after stressing that "you have to keep watch[ing], be prepared, and know what to do," Bev mentions that "the only thing that we don't have is, like, all that stuff, that whole long list of stuff they say you need to have in case of, you know [a hurricane]."

Following this admission, both Bill and Bev quickly revert back to emphasizing how much they feel in control of a potential crisis and how much they prefer to rely on their own decisions and preparations.

Again, this pattern appears to be common among mobile home participants. While some people miscalculate their social vulnerabilities—like the older gentleman who assured us that his health was "excellent," especially considering that he had survived several heart surgeries (while proudly showing us one of his former pacemakers in a plastic bag)—many respondents quickly move from acknowledging their limited resources and strengths to claiming a sense of agency and control based on some preparations they have made, even though many of these precautions appear minimally effective at best. Despite their many differences and despite the fact that the majority relies on some form of public support, mobile home residents are a group of people who greatly emphasize independence and self-sufficiency, as well as a desire to help others in need. Very few admit that they will be in great need of assistance in a crisis situation; less than 5% even consider the option of evacuation to a public shelter.

One example of the absence of valuable resources concerns the issue of home insurance. Less than half of all homeowners in our sample carried any kind of insurance on their home or possessions. Only one-fourth of all participants was certain that their policy included wind damage (hurricanes), and only one in eight was certain that they were covered against flood damage. Bill and Bev were among the majority who did not have home or renter's insurance, and they offer a ready explanation of why this is not really such a big problem.

Bev: We have chosen to not [have insurance], mostly because of the cost at this time; we're trying to keep our head above water. But the owner, the person who really owns the home, because we are owner-financing, so, he has said, don't waste your money on insurance, I'd rather you pay your bill, basically and, you know . . . he's the one who would really ultimately lose the house and lose income if something happened to the house. So, he hasn't seen the threat [to be big] enough to want the payments for something they probably wouldn't pay out on anyway.

Bill: The fact of the matter is, if the worst thing should happen and the house gets destroyed, he would put us in another house.

Bev: Right, and we would just end up paying for that one.

Here, Bev and Bill almost manage to turn the fact that they do not have any insurance on their future home or on their belongings into a strength. They firmly believe that the "fantastic" and "great guy" who still owns their home will honor their rent-to-buy agreement in case a hurricane destroys their place and simply transfer it to another mobile home. It is a considerable perceptual feat to feel equally or even better off without any insurance of one's belongings than with at least some sort of coverage. This view is not uncommon; and it may be an indication of the power of positive thinking that is typical for this group.

A last interesting point can be made regarding the particular resource of social networks, something that is almost always viewed as a strength in the vulnerability puzzle. Bill and Bev have made many friends in the park since moving in, they are very outgoing and helpful people. By objective measures, and by their own account, Bill and Bev have strong family and neighborhood ties and therefore appear to be less vulnerable than others who lack these supports.

However, a quick check on the presence of networks can easily overlook the fact that, in the case of mobile home residents during a hurricane, as maybe in other groups that are struggling economically and facing a crisis, strong networks can be a liability as much as they can be helpful. One of the biggest problems is that, typically, most of the network members of mobile home residents also live in mobile homes and therefore cannot offer shelter.

Bill: We have friends in the park, if any of them needed, you know, assistance getting out, we would definitely take care of that.

Bev: Like our friend Nancy, they only have one car, and they could shove everybody in the car if they needed to, but if they wanted to evacuate with us and we were gonna take the travel trailer, then we would make sure we were in contact with them and all the kids, and we would just all go somewhere together if we had to.

In addition to worrying about friends and neighbors, Bill and Bev express concerns for Bev's mother and her husband who live in a mobile home park in a nearby coastal town yet are not planning to ever evacuate. Throughout the interview, Bill and Bev repeatedly discuss their willingness and abilities to help other people in case of an emergency. However, they think little about how other people might be able to help them in a situation of need. The only time they mention how they may benefit from other people's resources is when they talk about a second home that is owned by Bev's parents in the larger Atlanta area, about ten driving-hours away, which might serve as a destination for an evacuation.

In short, pooling resources with others can be a very valuable strategy for people who have little, but sharing with people who have even less might considerably *increase*, instead of decrease, one's already strained supplies and energies. Of course, there are emotional benefits to intact social networks. We know, for instance, that families who stayed together during Hurricane Katrina experienced less emotional strain than the ones who separated, regardless of whether they evacuated or not (Haney, Elliott, & Fussell, 2007). However, it is also possible that in a crisis the social network of members beyond the immediate family bring about obligations and worries that outweigh the benefits these ties might otherwise provide. Generally speaking, the multiple and complex impacts of social networks on disaster vulnerability need to be discussed with much more nuance than is usually done (Ersing & Kost, 2012; Kusenbach & Taylor, 2012). During the study we came across complex perceptions of one's own social vulnerabilities that deviate from outside assessments, mostly erring on the side of greater safety; and we also found that not all resources or forms are capital have positive effects in a crisis.

CONCLUSION

This chapter pursued two related goals: first, to develop an innovative definition of disaster vulnerability that overcomes the identified weaknesses of past definitions, and second, to demonstrate the applicability and utility of the leveled critiques in analyses of primarily qualitative data from a study of mobile home communities. Throughout the chapter, we emphasized the gaps and contradictions between the customary academic definition of vulnerability as "risk plus unequal resources" and the subjective perceptions of risk and resources at the ground level. The case of Bill and Bev provided the details and nuances which are necessary to illustrate the many complexities of vulnerability as a matter of perception, yet it also served to discuss the kinds of patterns and problems that we observed within the larger dataset and in the larger population. Overall, issues of human agency emerged and were expressed in patterns of seeking and interpreting information, distancing from official messages, striving for autonomy, and mitigating one's own weaknesses. Other insights include the important role of local contexts for, and the time sensitivity of, vulnerability calculations, as well as the uncertain value attached to these operations.

Not explored in this chapter are aspects and dimensions of vulnerability that go beyond individual- and household-level characteristics. In the case of mobile home parks, we must distinguish between microcommunities at the block level, the park community, and the surrounding social area. In yet another step, larger social and spatial contexts (such as urban, suburban, or rural) or entire regions (for instance, coastal versus inland, or south versus north) deserve to be investigated more closely regarding their particular vulnerability patterns in the future.

Besides the considerations of Christmann and her colleagues (2011) which parallel the arguments and ideas leveled in this chapter, a recently suggested framing of "resilience as learning" and the concept of "adaptive resilience" appear to be productive innovations that go beyond the customary framing of resilience as "bouncing back" (see Chapters 10 and

12). Any new conceptual ideas will need to be applied to, and validated in, the real world. As with vulnerability, the human aspects as well as the temporal, spatial, and normative complexities of resilience deserve our serious attention.

We conclude with some tentative policy recommendations that build on the particular insights developed in this chapter. Emergency managers need to acknowledge the existence of a tension between advocating individual emergency planning and preparedness on the one hand and requests for compliance with official orders on the other. People who are trained to take responsibility for their own decisions will not follow evacuation orders unless they agree with the implied assessments of their personal risk and vulnerabilities. Because mobile home residents, like most people, prefer to rely on their own assessments, evacuation orders should be accompanied by additional information that carefully explains the reasons and concerns officials have for making evacuation decisions. Hearing these considerations might convince at least some listeners that the resulting orders are reasonable.

Official communications of evacuation orders must include more details on the recommended timing of the evacuation. Further, they should refer to "mobile homes," "manufactures homes," and "trailers" in order to avoid any misunderstandings. Mobile home residents respond negatively to sensational media messages and tend to eventually discount most of them. Therefore, publicly communicated disaster information should be as balanced, nuanced, and honest as possible. Overall, people must be reminded that they are preparing for worst-case scenarios and not certain impacts. This is not likely to downplay the urgency of such messages yet it will boost their overall credibility.

Mobile home residents need to evacuate for high-wind events and not shelter in place. Gathering the long list of supplies recommended in most brochures is not only financially difficult for people who struggle to meet daily needs, it is also the wrong strategy for mobile home dwellers. Instead, mobile home residents must be instructed to assemble mobile supply kits and then asked to evacuate to safe locations in the area; they do not need to leave the region. Generally, hurricane pamphlets must be revised to better reflect the life circumstances of population groups that deviate from the "average" upper-middle-class suburban homeowner who owns a boat, a pool, and a double-car garage. Generally, the lack of trust in public communication is a difficult obstacle to overcome, especially among people who tend to be distrustful of government at all levels. Channeling messages through more relevant and respected authorities and institutions might make a big difference and lead to greater compliance with orders. Mobile home residents, like most people, consider pets to be family members. Some will risk the well-being of human family members over concerns for their pets. In order to save lives, emergency managers must develop more realistic shelter options for the majority of American households that own pets; there is simply no alternative.

REFERENCES

Arlikatti, S., Lindell, M.K., & Prater, C.S. (2006). Risk area accuracy and hurricane evacuation expectations of coastal residents. *Environment and Behavior*, 38, 226–247.

Baker, E.J. (1991). Hurricane evacuation behavior. *International Journal of Mass Emergencies and Disasters*, 9, 287–310.

Christmann, G., & Ibert, O. (forthcoming). Vulnerability and resilience in a socio-spatial perspective: A social-scientific approach. In Vulnerability and resilience in a socio-spatial perspective, special issue, *Raumordnung und Raumforschung*.

Christmann, G., Ibert, O., Kilper, H., & Moss, T. (2011). Vulnerabilitaet und Resilienz in Sozio-Raeumlicher Perspektive: Begriffliche Klaerungen und Theoretischer Rahmen. IRS working paper no. 44, Institute for Regional Development and Structural Planning (IRS), Erkner, Germany. Retrieved April 26, 2012 from http://www.irs-net.de/publikationen/veroeffentlichungen/working-papers/index.php.

Cutter, S.L., & Finch, C. (2008). Temporal and spatial changes in social vulnerability to natural hazards. *Proceedings of the National Academy of Sciences USA*, 105, 2301–2306.

Enarson, E., & Scanlon, J. (1999). Gender patterns in flood evacuation: A case study in Canada's Red River Valley. *Applied Behavioral Science Review*, 7, 103–124.

Ersing, R. L. & Kost, K. A. (2012). Approaching practice: Social networks in the context of disasters. In Ersing, R. L. & Kost, K. A. (Eds.) *Surviving disaster: The role of social networks*. (p. 1–14). Chicago, IL: Lyceum Books, Inc.

Folke, C. (2006). Resilience: The emergence of a perspective for social-ecological system analyses. *Global Environmental Change*, 16, 253–267.

Haney, T.J., Elliot, J.R., & Fussell, E. (2007). Families and hurricane response: Evacuation, separation, and the emotional toll of Hurricane Katrina. In D.B. Brunsma, D. Overfelt, & J.S. Picou (Eds.), *The sociology of Katrina: Perspectives on a modern catastrophe* (pp. 71–90). New York: Rowman and Littlefield.

Hoffman, S.M., & Oliver-Smith, A. (Eds.) (2002). *Catastrophe and culture: The anthropology of disaster.* Santa Fe, NM: School of American Research Press.

Janssen, M.A., & Ostrom, O. (2006). Resilience, vulnerability, and adaptation: A cross-cutting theme of the international human dimensions program on global enviromental change. *Global Environmental Change*, 16, 237–239.

Kusenbach, M. (2003). Street phenomenology: The go-along as ethnographic research tool. *Ethnography*, 4, 449–479.

Kusenbach, M., Simms, J.L., & Tobin, G.A. (2010). Disaster vulnerability and evacuation readiness: Coastal mobile home residents in Florida. *Natural Hazards*, 52, 79–95.

Kusenbach, M., & Taylor, C.K. (2012). Hurricane evacuation among mobile home residents in Florida: The complex role of social networks. In R. Ersing & K.A. Kost (Eds.), *Surviving disaster: The role of social networks* (pp. 63–83). Chicago, IL: Lyceum.

Montz, B.E., & Tobin, G.A. (2005). Socio-economic patterns of vulnerability: Senior living developments and Hurricane Charley. *Papers of the Applied Geography Conferences*, 28, 216–225. Retrieved April 26, 2012 from http://applied.geog.kent.edu/AGCPapers/2005/AGC2005.htm.

Morrow, B.H. (1999). Identifying and mapping community vulnerability. *Disasters*, 23, 1–18.

Morrow, B.H. (2008). Community resilience: A social justice perspective. Community and Regional Resilience Institute (CARRI) research report no. 4. Retrieved April 26, 2012 from http://www.resilientus.org/publications/reports.html.

Siegel, J.M., Shoaf, K.I., & Afifi, A.A. (2003). Surviving two disasters: Does reaction to the first predict response to the second? *Environment and Behavior*, 35, 637–654.

Simmons, K.M., & Sutter, D. (2008). Manufactured home building regulations and the February 2, 2007 Florida tornadoes. *Natural Hazards*, 46(3), 415–425.

Stallings, R.A. (1995). *Promoting risk: Constructing the earthquake threat*. New York: Aldine DeGruyter.

Tobin, G.A., & Montz, B.E. (1997). *Natural hazards: Explanation and integration*. New York: Guildford Press.

U.S. Census Bureau. (2010). Physical Housing Characteristics For Occupied Housing Units: 2010 American Community Survey 1-Year Estimates. Retrieved July 25, 2012 http://factfinder2.census.gov/faces/tableservices/jsf/pages/productview.xhtml?pid=ACS_10_1YR_S2504&prodType=table

United States Landfalling Hurricane Probability Project. (2012). Retrieved April 26, 2012 from http://www.e-transit.org/hurricane/welcome.html.

Wisner, B., Blaikie, P., & Cannon, T. (2004). Vulnerability of coastal communities to sea level rise: A case study of Cape May County, New Jersey. *Climate Research*, 22, 255–270.

5 Building Collaborative Capacity for Disaster Resiliency

Beth Gazley

INTRODUCTION

In the wake of unprecedented recent disasters, it has become evident to policy makers that a wide range of private nonprofit agencies play unanticipated but valued roles in disaster response and recovery (Boris & Steuerle, 2006; DeVita, 2007; GAO, 2008; Homeland Security Institute, 2006; Kapucu, Augustin, & Garayev, 2009; Pipa, 2006). These community organizations differ from Voluntary Organizations Active in Disaster (VOADs) (see Appendix for additional definitions). Nonetheless, their community-oriented missions of social and human service provision, economic development, volunteerism, or other local services make them inevitable participants in some level of disaster planning, response, and recovery. Some of these organizations serve vulnerable populations (such as agencies serving individuals with disabilities; child care and after-school organizations; area agencies for the elderly; shelters for the homeless; and animal welfare agencies for domestic pets). Some participate in community disaster response and recovery on a temporary, ad hoc, or spontaneous basis. Some are called on because they bring particular expertise or resources, such as historic preservationists, stewards of natural resources, or community foundations.

Although the emergency planning documents for a wide range of federal and state agencies now recognize the importance of this broad group of private sector actors, strategies have only recently been implemented to define who they are and to formally involve them in community emergency planning (Levy, 2011). Local planning groups organized as Community Organizations Active in Disaster (COADs) offer a key mechanism by which local government officials and community organizations can accomplish these goals of coordinated planning and response. Although COADs have existed for more than twenty years, they are still forming in many communities and the timing is ripe for greater practitioner and researcher attention to their collaborative capacity and contributions to community resilience.

This chapter begins by describing the role of COADs within the public policy framework, then moves to defining their role at the local level.

From a practical perspective, COADs offer a potentially important role in FEMA's recently developed "whole community" concept (see Chapter 2 by Edwards). From a conceptual perspective, collaborative theory has been developing rapidly and there are still many opportunities to test its propositions in the emergency management context. The chapter uses examples of collaborative activity from recent disasters in Vermont and elsewhere to apply theoretical concepts to real events and offer practical lessons about effective future COAD formation and performance measurement. This chapter incorporates selections from more than a dozen interviews with local, state, and federal representatives of COAD and VOAD.

THE ROLE OF COADS IN EMERGENCY MANAGEMENT

In the past decade, the emergency management field has been characterized by a greater emphasis on proactive planning and prevention, as well as a recognition by public officials that private sector actors should be involved in all stages of emergency management planning, including preparedness and mitigation as well as response and recovery (BENS, 2007; Brudney & Gazley, 2009; McEntire, 2004; McEntire et al., 2002; Waugh, 2000). According to FEMA (2011b),

> [A] government-centric approach to emergency management is not enough to meet the challenges posed by a catastrophic incident. Whole Community is an approach to emergency management that reinforces the fact that FEMA is only one part of our nation's emergency management team; that we must leverage all of the resources of our collective team in preparing for, protecting against, responding to, recovering from and mitigating against all hazards; and that collectively we must meet the needs of the entire community in each of these areas. This larger collective emergency management team includes, not only FEMA and its partners at the federal level, but also local, tribal, state and territorial partners; non-governmental organizations like faith-based and nonprofit groups and private sector industry; to individuals, families and communities, who continue to be the nation's most important assets as first responders during a disaster.

This perspective is shaped only in part by the public sector's recognition that responses to major disasters are beyond the capacity of both governmental agencies and principal first-responder charities such as the American Red Cross (ARC). It also reflects a more sophisticated understanding of community dynamics, specifically a recognition that top-down planning and response will be ineffective at identifying each community's most salient needs or resources, and can even impede the empowerment of citizens to act on their own behalf. Thus, the "whole community" concept also reflects the desire to build resilient, self-sufficient communities which are more able to support one another to meet disaster recovery needs. From a practical point of view, community organizations cannot assist in whole community recovery if they are occupied with their own recovery.

A key policy document guiding the implementation of this approach is the National Disaster Recovery Framework (NDRF), last updated in September 2011. The NDRF is a conceptual guide for public and private agencies that applies the principles of individual empowerment, partnership, and inclusiveness to locally led recovery organizations and processes. The NDRF addresses all levels of participation (individual, family, private sector, and various levels of government). With respect to nonprofit involvement, the NDRF (FEMA, 2011: 21) suggests the potential for a very broad level of participation, including social, human, cultural, health, and infrastructure services such as "community recovery planning, case management services, volunteer coordination, . . . psychological and emotional support, technical and financial support, housing repair and construction to meet accessibility . . . standards, and project implementation."

The NDRF was developed with substantial stakeholder participation, but the document still contains some untested and overly general assumptions about nonprofit participation. For example, the NDRF describes the nonprofit role as supplementary to government efforts: to "fill gaps where government authority and resources cannot be applied," including "grantors and investors" who will "inject needed financial resources to meet recovery needs . . . that otherwise are not funded by a government program" (FEMA, 2011: 21). In reality, nonprofit missions, priorities, and capacities vary enormously. Only some private sector agencies will be comfortable with this "supplementary" philosophy.

As they implement the NDRF, each community will have to determine a planning approach that ensures public officials understand what is possible and what is not from the private sector. This process becomes imperative when essential disaster response services are involved. For example, in the several days in August 2011 following Hurricane Irene's arrival, the statewide 2–1-1 information system operated by Vermont United Way quickly exceeded its capacity despite sophisticated emergency backup and continuity plans that had been updated after severe spring flooding. The problem was attributed to new state officials, who were not familiar with 2–1-1's mission or its role in the state's operating emergency plan and who assumed 2–1-1 could be redirected to supplement an overburdened 9–1-1 system. When citizens were told to use the 2–1-1 system for non-essential purposes such as volunteering, the misunderstanding nearly collapsed this essential social service referral system (personal communication with Mary Ellen Mendl, March 16, 2012). The story suggests that public officials must also learn to navigate the whole community concept and to familiarize themselves not only with a new taxonomy of private sector actors, but also with how to work together. Those who have been trained in the rigid hierarchical protocols of the National Incident Management System (NIMS) will require more time and experience to engage most effectively with its more inclusive and amorphous elements, and to learn the language of negotiation (rather than command and control) that is required to work within a network of autonomous actors (McGuire et al., 2010).

THE ROLE OF COADS WITHIN THE WHOLE COMMUNITY

COADs offer a means by which the community-level planning recommended by FEMA can be accomplished. COADs operate at a local level in coordination with local emergency managers to engage in pre-disaster planning. They are different than other emergency planning efforts such as Local Emergency Planning Committees (LEPCs, which work to prevent and plan responses to industrial and chemical accidents) or first-responder and volunteer activities organized through CERT teams and Citizen Corps Councils, or long-term recovery committees organized after past disasters. A principal distinction is their organizational (rather than citizen) level of membership, but they also are characterized by their relatively greater autonomy and mission to plan proactively, helping to identify response and recovery resources in advance of emergency events, and addressing all four emergency management functions. COADs can include both faith-based and secular service providers and community organizations.

Many COADs have been recreated from Long Term Planning or Recovery Committees (LTPCs, LTRCs) operating in communities after disasters. In some communities, COAD creation is not pursued where Long Term Recovery Committees already exist. In Vermont, for example, despite two major disaster declarations after floods in 2011, the state still depends on a loose structure of LTRCs that supplement the planning committees which operated prior to 2011. In many other states, COADs are already active or are being created out of whole cloth by state VOAD directors and state offices of faith-based and community initiatives.

Regardless of their origin, there is a clear distinction between the broader mission of a COAD and the more focused long-term recovery efforts that follow a disaster. For example, in the case of Missouri's organizing activities, COADs are considered the successors to the Unmet Needs Committees from the 1993 Midwest floods in the sense that they also represent many human services providers and a concern for the relief and recovery of victims of disasters. However, Missouri's emergency planning document goes on to stress that "the similarity . . . ends there. COADs have a much broader mission—to strengthen area-wide disaster coordination by sharing programs, policies, information, joint planning, and training" (Missouri Disaster Recovery Partnership, 2002: 3). A related challenge for COAD formation, therefore, is how those that have evolved from long-term recovery committees effectively broaden their mission while still retaining the experiential knowledge of the original group.

Some of the key actors in COADs, both from an experiential point of view and also by virtue of their mission, are VOADs. For the purposes of this discussion, VOADs are defined as single charitable or religious organizations with a mission of disaster response and recovery. Local VOADs such as a Red Cross chapter may play central roles in local emergency management activity, so that coordinating their efforts through COADs or Long Term Recovery Committees offers advantages. However, disaster researchers have also observed the challenge of mixing highly experienced disaster relief organizations with less knowledgeable or committed actors into one planning group (Robinson & Gerber, 2007).

Although COADs are expected to coordinate with the highly centralized state emergency management system, they represent a distinctly bottom-up, community-based approach to emergency planning and response, different than the hierarchical Citizen Corps. Their grassroots nature is reflected in their diverse membership, which is selected by each COAD and could range from animal welfare organizations to youth-serving agencies to developmental disabilities providers. National VOAD officials estimate that many COADs are led by Red Cross or Salvation Army personnel (agencies with disaster response missions), but COADs have also been staffed by county emergency management personnel, an Area Agency on Aging, a Volunteer Resource Center, or a United Way (personal communication with James McGowan, National VOAD, February 15, 2012). They have been funded by state emergency planning grants, county budgets, funds left over from past disaster recovery, private foundations, fund-raising efforts, or have no budget at all.[1] This diversity in structure and membership is considered one of their strengths, reflecting the expectation that COADs will draw on the best resources available to each community (personal communication with Steve Cain, Indiana VOAD, March 1, 2012).

Perhaps due to the youth of many statewide COAD organizational activities, but also their decentralized and unregulated nature, very little is known about COADs, including how many groups are operational, their geographic distribution, scope of responsibilities, membership, and activities. Also yet unmeasured are their accomplishments, such as their ability to leverage additional private sector voluntary resources or to empower their own constituents to participate more effectively in emergency planning. In one such example of how this potential empowerment occurs, individuals with disabilities in a program in Hawai'i are taught not only individual preparedness skills but also how to become peer trainers (personal communication with Neil McDevitt, FEMA Office of Disability Integration). If handled correctly, these opportunities for citizen empowerment can result in greater community resilience and less reliance on emergency support functions during disasters. Also still an open question is the nature of the relationship between COADs and national VOAD structures, and also between COADs and public officials, including the level of communication, information sharing, autonomy, and coordinated planning that can be expected. For example, COADs are not yet explicitly named as participants in the NDRF although their potential role is described indirectly in that policy document.

APPLYING THEORIES OF COLLABORATIVE CAPACITY TO EMERGENCY PLANNING AND MANAGEMENT

From a conceptual perspective, COADs can be viewed as collaborative networks, with the same challenges and opportunities which other networked forms of public service delivery are subject to. Collaboration or "co-labor" describes the process of operating in multi-organizational arrangements to solve problems that require joint effort (Agranoff & McGuire, 2001; Thomson & Perry, 2006). Scholars have produced an enormous literature on cross-sector and interorganizational collaboration over the past two decades, driven by the increase in privatized forms of public service delivery. Collaboration in emergency management is still an emerging field of scholarship, although McGuire, Brudney, and Gazley (2010) suggest that emergency management offers many opportunities for understanding the complexities of collaborative management.

Especially, scholars share an interest in the "cohesion factors" (Agranoff & McGuire, 2001: 311) that bind partners together and in how collaborative capacity is built (i.e., the human and material resources that build effective joint action). Given its reliance on organizational and human attributes, collaborative activity is best studied through a multi-theoretical and multi-dimensional lenses, borrowing from social network, resource dependence, ecological, agency, group, and life cycle theories of organizational behavior, but also from political, cognitive, psychological, and gender theories that address the human attributes of joining and cooperative behavior (Gazley, 2008, 2010).

COADs, like any coalition of organizations, are driven to collaborate for statutory, financial, or mission-related reasons, but also because the leaders of these various organizations have developed personal relationships, experiences, and shared norms that build trust in partners. The "cohesion factors" central to network formation can be understood not just as network or organizational-level qualities but also as qualities an entire community shares. For example, each member organization brings staffing, a mission, and resources that contribute to or limit its participation. Each leader of each organization brings knowledge, biases, and historical baggage that accomplish the same. And collaborative capacity also depends on citizen resources, a community's historical experience with disasters, civic engagement, and infrastructure. Finally, it is useful to note that the concepts of "collaborative capacity" and "community resilience" are rarely linked, but depend on similar qualities which deserve more exploration in research (Goldstein, 2011).

Five of these most salient "cohesion factors" or qualities expected in collaborative emergency management are discussed further below. For the practitioner, these are expressed as prescriptions to support COAD development. For the researcher, these statements can be reworded and explored as research questions or hypotheses.

1. *The quality of the interorganizational relationships within COADs should be fostered in a way that supports an engaged membership, healthy trust levels, and other measures of collaborative capacity.* Many collaborative management studies observe that trust between partners is an important element in informal cooperative systems, including those that characterize local emergency management (Gazley, 2008, 2010; Kapucu, 2006; McGuire, 2006; Moynihan, 2009; Waugh, 2000). In post–September 11 interviews, disaster responders identified lack of communication and trust as barriers to a more effective response (Kapucu & Van Wart, 2006). In general collaboration research, public managers cite past cooperative experiences as a key factor in building trust across the sectors (Gazley, 2008, 2010). Using an expression borrowed from the management literature, "Repeated exchange encourages effective exchange" (Poppo & Zenger, 2002: 710). Healthy COAD development, therefore, depends in part on its origins, longevity, and past performance.

Unrelated partnerships that exist prior to disaster can offer unexpected but effective pathways to collaboration when disasters strike. Public and nonprofit officials describe incidents in upstate New York prior to Hurricane Irene in 2011, and in Tuscaloosa, Alabama prior to the 2011 tornadoes where earlier, unrelated community collaboration supported a more timely community response to these disasters (personal communications with James McGowan, National VOAD, and David Myers, DHS Center for Faith-Based & Neighborhood Partnerships, February 15, 2012). Many observers also note the importance of COAD planning in advance of disasters, and the relatively greater difficulty of putting response and recovery systems in place while addressing immediate emergency needs. In a report commissioned by NVOAD to evaluate the voluntary sector's response to floods along Alaska's Yukon and Kuskokwim Rivers in 2009, Eller and Gerber (2010: 8) wrote:

> It is not at all uncommon for government officials in the areas of emergency management . . . to lack well-developed working relationships with voluntary nonprofit organizations. The same is true from the nonprofit sector side—while many voluntary organizations engage in disaster recovery and relief service provision when a disaster strikes a given community, many do so on an ad hoc basis and do not develop routine relationships with key government agencies.

Eller and Gerber (2010) related this situation, retrospectively, to the difficulty in estimating the value of the voluntary sector's response. But the lesson can be applied prospectively to COAD formation to suggest that broad relationship building in advance of disasters offers multiple benefits, including not only a more realistic assessment of material resources, but also the development of cohesion, trust, and norms of behavior within the emerging network. William Elwell, Vermont's VOAD director, noted the lack of prior local emergency planning in many parts of the state, leading to the absence of local organizations at the table when Irene struck and causing coordination problems with state response efforts (personal communication, April 10, 2012).

2. *Shared goals and risks should be identified and built upon.* Another element of collaborative capacity, both within and outside the emergency management field, is a sense of shared risk built on a common set of goals and expectations (Agranoff & McGuire, 2003; Comfort, 1999; Gazley, 2008; Kapucu, 2006b; Thomson & Perry, 2006; Waugh & Streib, 2006). We cannot assume this quality exists even within a community that shares the same risk of a tornado, earthquake, or flood because cognitive understandings of risk can vary so much from individual to individual, and thus from organization to organization (Tierney, Lindell, & Perry, 2001). Accurate perceptions of risk are built on information, training, and actual experience (Drabek, 2003). Observers note that even active and engaged disaster coalitions are still at risk of falling apart if their collaborative capacity has not been adequately nurtured: "If it's not strong before, it'll get weaker, but if strong before it'll get even stronger" (personal communication with David Myers, DHS Center for Faith-Based & Neighborhood Partnerships, February 15, 2012).

3. *Coalitions should be both stable and adaptable.* As a network forms, emergency management systems are not alone in their reliance on both ordered structure and flexibility, but this dual quality has particular salience in the unpredictable arena of disaster response (Comfort et al., 2004). Emergency management systems thrive on stable staffing and low turnover, stable funding, and routinized protocols. Stable staffing is tied to performance because frequent turnover impedes the development of interpersonal relationships and may also cause changes in missions or the scope of services among partners. In Vermont, long-term recovery activity post-Irene has been slowed by incompatible software among VOAD partners and a decentralized system of emergency planning. And yet, there were numerous stories of successful disaster support functions that were created ad hoc. A statewide community foundation became an unexpected but essential actor in response and recovery when it volunteered as the fiscal agent for several disaster relief funds.

While interparty agreements such as MOUs have a role in this process because they can support an understanding of what can change and what should not, Eller and Gerber (2010) also note that the flexibility and efficiency achieved in Alaska's successful collaborative response was due in part to FEMA's decision to bypass formal procurement contracts when they involved voluntary organizations. Because there are many forms of mutual agreement, research on COADs must assess what level of formal agreement achieves the best balance of structure and flexibility. Past research suggests that formal agreements may improve the perceptions of collaborative effectiveness more than actual performance (Gazley, 2010).

Stable but flexible emergency management also depends on effective socio-technical systems for sharing information, which allow communities to build on a stable platform of communication but reallocate resources quickly as circumstances change (Comfort et al., 2004). COADs may serve these needs because the capacity to communicate is built not only through technical means but also by creating regular opportunities for human and organizational interaction. In a survey conducted through the National Association of Counties, Brudney and Gazley (2009) found that a county emergency manager's perception of preparedness increased based on the frequency of interactions with voluntary organizations. Eller and Gerber (2010) suggest that regular communication between partners and sectors also improves trust levels.

4. *Breadth of participation must be balanced against depth and focus, and power imbalances must be managed.* One of the more complex dilemmas facing COADs as they develop is how broad and inclusive they should be. Federal documents clearly favor broader representation from the private sector (see, for example, reports from the GAO, 2008, and the National Commission on Children and Disasters, May 11, 2010). Such breadth of participation is also consistent with concepts of resiliency because they offer redundancy, a term used most often in the context of physical infrastructure but equally applicable to social systems.

Brudney and Gazley (2009) categorize the potential participants in disaster planning and response as "primary," "secondary," and "tertiary" organizations. Primary or first-responder organizations have an express mission of emergency planning and response. Secondary responders include organizations providing philanthropic, health, and social and human services to displaced persons or others affected by disasters, such as United Way agencies, shelters, churches, youth agencies, mental health counseling agencies, foundations, and civic associations and their staff and volunteers. Although these actors would be involved on a non-routine basis, their contributions might still be essential after large-scale disasters and thus earn them a seat at the planning table. Tertiary responders include agencies serving large numbers of constituents who might be adversely affected by emergencies. They have a rationale to engage in emergency planning at least internally and perhaps also jointly with public officials. This group includes a broad array of private sector actors, including museums, schools and colleges, libraries, neighborhood groups, membership associations, charities that serve dependent and frail populations, along with businesses such as banks, food distributors, and manufacturers.

Local officials are likely to be divided on whether COADs should involve only organizations with an explicit disaster focus, or whether organizations should nominate themselves to participate regardless of mission differentiation. As the story of the Vermont Community Foundation illustrates, it was precisely the general and non-specific, community-focused nature of this foundation's mission that allowed it to participate so easily as the fiscal agent for response and recovery activities. This story not only suggests that potential participants in disaster response and recovery cannot be identified through mission alone, but also that many more community resources are likely to exist than are documented through emergency planning. Therefore, an inclusive COAD model may offer an improved approach to community resilience when compared to a closed or VOAD-dependent model because it offers a way of considering—and accessing—a broader array of community resources.

Inclusivity raises a second dilemma: how to involve and what to expect from community organizations that do not have a disaster response mission compared to those that do. It is possible that a planning approach incorporating general, longstanding community resources can help to stabilize community emergency planning activities, buffering communities against the "ebb and flow" of VOAD resources and capacity. For this reason, COADs would appear to be natural homes not only for CERT or Red Cross representatives but also for United Way agencies, area agencies on aging, and other permanent pieces of the community fabric. As these coalitions evolve, it is worth observing how COADs that are attempting a more inclusive approach to membership avoid creating two classes of membership, VOADs versus non-VOADs?

Robinson and Gerber (2007) catalog a number of potential obstacles to the integration of non-VOAD organizations, including structural, professional, and cultural differences. Therefore, inclusivity places new expectations on non-VOAD organizations, as well. Some observers note that resiliency is not yet a part of the common language of the social services sector (personal communication with David Myers, DHS Center for Faith-Based & Neighborhood Partnerships, February 15, 2012). Some community organizations may have a stronger incentive to develop institutional-level emergency plans than others (perhaps due to regulatory or fiscal requirements). We can also expect that some service providers operating in the same subsectors will find it challenging to suspend any industry competition they are engaged in to participate in collaborative emergency planning.

Finally, COAD organizers should note whether physical, political, or cultural barriers make it difficult for certain kinds of agencies to participate. As Lai (2012) observes about medical responses to pandemics, the network of participating health agencies can exclude important actors as readily as it includes others. Although a number of reports have been produced that document the value of integrative planning to address the needs of minority populations, the participation within COADs of organizations that can represent these minorities (broadly defined to include individuals with disabilities, non-English-speaking populations, undocumented residents, ethnic minorities, etc.) may still be quite limited in many communities (National Council of La Raza, 2011; White, Fox, & Rooney, 2007). Therefore, COADs may find it useful to assess the extent to which preparedness activities across physical and cultural barriers are fair and equitable.

5. *Dominant logics must be created.* The stories I relay from recent disaster experiences describe multiple instances where perceptions about the relative roles of an emergency management partner were misaligned. A tremendous amount of effort has been expended on defining roles and relationships within the alphabet soup of public sector emergency agencies, but that work remains unfinished when it comes to public-private role definition. COADs will be in the middle of this challenge and, even at the local level, the power imbalances that exist at other jurisdictional levels may still appear and make it difficult to build effective coalitions.

Scholars have also noted the difficulty in creating coalitions where competing institutional logics operate (Friedland & Alford, 1991; Reay & Hinings, 2009). Institutional logics are built on past institutional practice, create institutional cultures, and help to explain how various actors interpret actions or establish rules of behavior. When dominant logics compete, Bryson, Crosby, and Stone (2006) suggest, collaboration is more difficult. Competing logics may be particularly troublesome across the public and private sectors.

Theories of isomorphic behavior suggest that coalitions can work out these differences over time and eventually establish dominant logics of their own. Koliba, Zia, and Mills (2011) describe the need for a "hybridized" approach to disaster response that balances the command-and-control aspects of emergency management against its collaborative nature (Waugh, 2009). But in an open system such as a COAD, where all partners remain committed to multiple sectors and logics, the process for achieving this level of agreement must be purposeful and persistent. The public sector requires training to understand the difference between public and nonprofit accountability, mission constraints, and autonomy, which can be overlooked in a drive for speed and efficiency when a disaster strikes. The language that creeps into some public reports and emergency management guidelines must be carefully worded to characterize the voluntary, independent nature of the collaborative relationship with private sector actors. Discussions must be held within local jurisdictions to decide who leads and how much direction to provide to COADs.

CONCLUSION

A welcoming public policy environment exists for greater voluntary sector involvement in all four stages of emergency management. This chapter suggests that there is a potential for COADs to help organize local community needs and resources related to disaster response, especially in communities where no similar networks yet exist. Where COADs, Long Term Recovery Committees, and similar networks have already formed, these organizing activities deserve greater attention in research and practice to learn from model efforts. Where they are still under development, attention should be paid to creating intersector partnerships with a strong collaborative capacity.

In all instances, either of ongoing or emergent activity, organizers and scholarly observers of COADs will find this chapter useful in understanding how collaborative performance can be measured. Identifying areas where communication is poor, relationships are not well developed, and goals are unaligned allows emergency planners to more efficiently forge the common blueprint for community response described in federal documents. This proactive strategy is consistent with the mission of emergency management. Collaborative emergency planning can resolve potential disagreements in advance of an emergency, strengthen mutual goals, build trust in partners, and identify and address gaps in resource needs.

This chapter also observes that the scholarship on emergency management would benefit from more research on collaborative performance. There is great opportunity to test general theories of collaborative capacity and performance in the unique context of emergency management. The normative position that local planning efforts should be commensurate with the potential severity of a community's hazards could be tested. Logic models could be developed to link collaborative capacity to performance (Brudney & Gazley, 2009; Nolte & Boenigk, 2011). The strength of collaborative relationships could be analyzed employing social network analysis.

Such endeavors could employ multi-theoretical and multi-level approaches. Studies could make use of individual behavioral and attitudinal data, organizational capacity and behavioral data, and community-level data such as demographic and geophysical information. The dimension of time is a particularly difficult quality to incorporate into performance studies, but time is a key ingredient in the process of building coordinated action. In the emergency management context, time could be included through panel studies that observe how and whether communities effectively use collaborative activity generated before disasters occur to respond to those events later on.

If the challenges enumerated in this chapter can be addressed as COADs are created, communities have the opportunity to achieve a higher general level of resilience. Resilience, described in greater detail in some of the accompanying chapters in this volume, allows communities not only to bounce back or recover from disasters, but possibly to "bounce forward" by strengthening relationships, increasing community engagement, and achieving greater levels of social capital on which future community action might depend (Manyena et al., 2011).

NOTES

1. A similar diversity is seen among Local Emergency Planning Committees where, according to a 2008 survey, no pattern is observed with respect to membership, meeting frequency, how integrated their plans are with other local or state emergency plans, who is responsible for implementing a plan, or how closely these committees coordinate with other emergency planning and response bodies (USEPA, 2008).

REFERENCES

Agranoff, R., & McGuire, M. (2001). Big questions in public network management research. *Journal of Public Administration Research and Theory*, 11(3), 295–396.

Agranoff, R., & McGuire, M. (2003). *Collaborative public management: New strategies for local governments*. Washington, DC: Georgetown University Press.

BENS. (2007). *Getting down to business: An action plan for public-private disaster response coordination; The report of the business response task force*. Washington, DC: Business Executives for National Security.

Boris, E.T., & Steuerle, C.E. (Eds.) (2006). *After Katrina: Public expectation and charities' response*. Washington, DC: Urban Institute.

Brudney, J.L., & Gazley, B. (2009). Planning to be prepared: An empirical examination of the role of voluntary organizations in county government emergency planning. *Public Performance and Management Review*, 32(3), 372–399.

Bryson, J.M., Crosby, B.C., & Stone, M.M. (2006). The design and implementation of cross-sector collaborations: Propositions from the literature. Special issue, *Public Administration Review*, 66, 44–55.

Comfort, L.K. (1999). *Shared risk: Complex systems in seismic response*. Amsterdam: Pergamon.

Comfort, L.K., Dunn, M., Johnson, D., Skertich, R., & Zagorecki, A. (2004). Coordination in complex systems: Increasing efficiency in disaster mitigation and response. *International Journal of Emergency Management*, 2(1–2), 62–80.

DeVita, C.J. (2007). Preparing for the next disaster. In C.J. DeVita (Ed.), *After Katrina: Shared challenges for rebuilding communities* (pp. 27–31). Washington, DC: Urban Institute.

Drabek, T.E. (2003). *Strategies for coordinating disaster responses*. Boulder, CO: Institute of Behavioral Science, University of Colorado.

Eller, W.S., & Gerber, B.J. (2010). Voluntary nonprofit organizations and disaster recovery: Assessing the value of the nonprofit contribution to the 2009 Alaskan rivers flood recovery effort. Unpublished report prepared for National Voluntary Organizations Active in Disaster.

FEMA (Federal Emergency Management Agency). (2011a). National Disaster Recovery Framework. September. Retrieved from http://www.fema.gov/recoveryframework/.

FEMA (Federal Emergency Management Agency). (2011b). Whole community. December. Retrieved from http://www.fema.gov/about/wholecommunity.shtm.

Friedland, R., & Alford, R.R. (1991). Bringing society back in: Symbols, practices and institutional contradictions. In W.W. Powell & P.J. DiMaggio (Eds.), *The new institutionalism in organizational analysis* (pp. 232–263). Chicago: University of Chicago Press.

GAO (Government Accountability Office). (2008). FEMA should take action to improve capacity and coordination between government and voluntary sectors. Report no. GAO-08–369. February 27. Retrieved from http://www.gao.gov/products/GAO-08–369.

Gazley, B. (2008). Inter-sectoral collaboration and the motivtaion to collaborate: Toward an integrated theory. In L.B. Bingham & R. O'Leary (Eds.), *Big ideas in collaborative public management* (pp. 36–54). Armonk, NY: M.E. Sharpe.

Gazley, B. (2010). Linking collaborative capacity to performance measurement in government-nonprofit partnerships. *Nonprofit and Voluntary Sector Quarterly*, 39(4), 653–673.

Goldstein, B.E. (2011). *Collaborative resilience: Moving through crisis to opportunity*. Cambridge, MA: MIT Press.

Homeland Security Institute. (2006). *Heralding unheard voices: The role of faith-based organizations and nongovernmental organizations during disasters*. Arlington, VA: Department of Homeland Security Science and Technology Directorate.

Kapucu, N. (2006). Public–nonprofit partnerships for collective action in dynamic contexts of emergencies. *Public Administration*, 84(1), 205–220.

Kapucu, N., Augustin, M., & Garayev, V. (2009). Interstate partnerships in emergency management: Emergency management assistance compact (EMAC) in response to catastrophic disasters. *Public Administration Review*, 69(2), 297–313.

Kapucu, N., & Van Wart, M. (2006). The evolving role of the public sector in managing catastrophic disasters: Lessons learned. *Administration and Society*, 38(3), 279–308.

Koliba, C.J., Zia, A., & Mills, R.M. (2011). Accountability in governance networks: An assessment of public, private, and nonprofit emergency management practices following Hurricane Katrina. *Public Administration Review*, 71(2), 210–220.

Lai, A. (2012). Organizational collaborative capacity in fighting pandemic crises: A literature review from the public management perspective. *Asia-Pacific Journal of Public Health*, 24(1), 7–20.

Levy, D.A. (2011). It takes a village: Community approach to managing large scale incidents. Presentation to the U.S. HHS Integrated Training Summit, May 3. PowerPoint slides. Retrieved from http://www.integratedtrainingsummit.org/presentations/2011/main_training_summit/17_-_it_takes_a_village_community_approach_to_managing_large_scale_incidents.pdf.

Manyena, B., O'Brien, G., O'Keefe, P., & Rose, J. (2011). Disaster resilience: A bounce back or bounce forward ability? *Local Environment: The International Journal of Justice and Sustainability*, 16(5), 417–424.

McEntire, D.A. (2004). The status of emergency management theory: Issues, barriers and recommendations for improved scholarship. Paper presented at the FEMA Higher Education Conference, Emmitsburg, MD, June 8.

McEntire, D.A., Fuller, C., Johnston, C.W., & Weber, R. (2002). A comparison of disaster paradigms: The search for a holistic policy guide. *Public Administration Review*, 62(3), 267–281.

McGuire, M. (2006). Collaborative public management: Assessing what we know and how we know it. Special issue, *Public Administration Review*, 66, 33–43.

McGuire, M.C., Brudney, J.L., & Gazley, B. (2010). The "New Emergency Management": Applying the lessons of collaborative governance to 21st century emergency planning. In R. O'Leary, D. Van Slyke, & S.H. Kim (Eds.), *The future of public administration, public management and public service around the world: The Minnowbrook perspective* (pp. 117–128). Washington, DC: Georgetown University Press.

Missouri Disaster Recovery Partnership. (2002). Missouri COAD guidance manual. Revised January 1. Retrieved from http://sema.dps.mo.gov/docs/programs/Executive/Disaster%20Recovery%20Partnership/COAD.pdf.
Moynihan, D.P. (2009). The network governance of crisis response: Case studies of incident command systems. *Journal of Public Administration Research and Theory*, 19(4), 895–915.
National Commission on Children and Disasters. (2010). Progress report on children and disasters. May 11. Retrieved from http://www.acf.hhs.gov/nccd/20100511_NCCD_Progress_Report_FINAL.pdf.
National Council of La Raza. (2011). Emergency manager's toolkit: Meeting the needs of Latino communities. Retrieved from http://www.nclr.org/index.php/publications/emergency_managers_tool_kit_meeting_the_needs_of_latino_communities/.
Nolte, I.M., & Boenigk, S. (2011). Public-nonprofit partnership performance in a disaster context: The case of Haiti. *Public Administration*, 89(4), 1385–1402.
Pipa, T. (2006). *Weathering the storm: The role of local nonprofits in the Hurricane Katrina relief effort*. Washington, DC: Aspen Institute.
Poppo, L. & Zenger, T. (2002). Do formal contracts and relational governance function as substitutes or complements? *Strategic Management Journal,* 23, 707–25.
Reay, T., & Hinings, C.R. (2009). Managing the rivalry of competing institutional logics. *Organization Studies*, 30(6), 629–652.
Robinson, S.E., & Gerber, B. (2007). A seat at the table for nondisaster organizations. *The Public Manager*, 36(3), 4–7. Retrieved from http://www.feiaa.org/files/File/TPM_v36_n3.pdf#page=6.
Thomson, A.M., & Perry, J.L. (2006). Collaboration processes: Inside the black box. *Public Administration Review*, 66(6), 20–32.
Tierney, K.J., Lindell, M.K., & Perry, R.W. (2001). *Facing the unexpected: Disaster preparedness and response in the U.S*. Washington, DC: Joseph Henry Press.
USEPA (United States Environmental Protection Agency). (2008). 2008 nationwide survey of local emergency planning committees (LEPCs). Retrieved from http://www.epa.gov/osweroe1/docs/chem/2008_lepcsurv.pdf.
Waugh, W.L. (2000). *Living with hazards, dealing with disasters: An introduction to emergency management*. Armonk, NY: M.E. Sharpe.
Waugh, W.L. (2009). Mechanisms for collaboration in emergency management: ICS, NIMS, and the problem with command and control. In R. O'Leary & L.B. Bingham (Eds.), *The collaborative public manager: New ideas for the 21st century* (pp. 157–176). Washington, DC: Georgetown University Press.
Waugh, W.L., & Streib, G. (2006). Collaboration and leadership for effective emergency management. *Public Administration Review*, 66(6), 131–140.
White, G.W., Fox, M.H., & Rooney, C. (2007). Nobody left behind: Report on exemplary and best practices in disaster preparedness and emergency response for people with disabilities. March 15. Retrieved from http://www.nobodyleftbehind2.org/powerpoint/index.shtml.

6 Resilient Rural Communities

Housing Recovery Patterns Following Hurricane Katrina

Nazife Emel Ganapati, Shaoming Cheng, and Sukumar Ganapati

INTRODUCTION

Despite an emerging literature on post-disaster housing recovery, our understanding of rural recovery remains limited. Existing literature mainly focuses on urban areas, with little attention paid to rural housing recovery. The urban bias is especially evident in post-Katrina literature, most of which is devoted to the city of New Orleans (e.g., Bergal et. al., 2007; Campanella, 2006; Richardson, Gordon, & Moore, 2008; Olshansky & Johnson, 2010). We address this gap in the literature by identifying the housing recovery patterns and determinants influencing such recovery in rural counties affected by Hurricane Katrina.

Housing recovery is a critical component of a community's resilience and recognized as one of the six core Recovery Support Functions in FEMA's NDRF (FEMA, 2011). Loss of housing due to a disaster disrupts the daily life of a community's residents, limiting their ability to carry out their normal activities and routines. Post-disaster housing recovery is generally a precursor for the community's recovery along other dimensions (Peacock, Dash, & Zhang, 2007). It also aids in psychological recovery (Kilmer et al., 2010) and helps revive the social life of the community (Comerio, 1998).

In the following section, we review the literature on post-disaster housing recovery, specifically focusing on what enables or hinders such recovery. Next, we outline our methodology and research design. We then present the results of our analysis on housing recovery and interpret our findings. Finally, we lay out an agenda for future research on post-disaster housing recovery in rural areas and discuss the policy implications of our research.

POST-DISASTER HOUSING RECOVERY

Post-disaster housing is significant in the often long, overall recovery process of the affected communities (Bolin, 1986; Campanella, 2006; Comerio, 1998). According to Quarantelli's (1982) seminal study, there are four phases of post-disaster housing recovery: emergency sheltering (immediately following a disaster); temporary sheltering (short-term living arrangements, e.g., in tents); temporary housing (medium-range living arrangements, e.g., in trailers); and permanent housing. However, as Quarentelli (1982, 1995) noted, the distinctions between the phases are not clear as to when the temporary sheltering should transition into temporary housing or permanent housing. Despite its shortcomings, Quarentelli's typology has been often used by researchers to study post-disaster housing recovery (e.g., Ganapati & Ganapati, 2009; Peacock et al., 2007). Our study relates to the permanent housing phase in this typology.

The literature that exists on post-disaster housing recovery (NRC, 2006) has two limitations that we would like to emphasize in this chapter. The first is that recovery is often measured as post-disaster indicators returning to pre-disaster levels (Lu et al., 2007; Vigdor, 2008; Zhang & Peacock, 2010). This is a problematic approach because returning to pre-disaster levels does not mean building back for the better. More importantly, such an approach does not account for changes in the broader economic context (e.g., housing market meltdown, economic recession). To fill in this methodological gap in the literature, we use a quasi-experimental design, with a set of matched counties which had similar characteristics and hence statistical equivalence or comparability with that of Katrina-affected counties prior to the hurricane event. The counterfactual scenarios suggest what would have occurred to the affected counties in the absence of the hurricane. The benefit of this method is that the comparison with the counterfactuals allows us to identify the recovery process in a prospective manner (i.e., looking forward), rather than in a retrospective way (i.e., looking backward in terms of what was before). Such a prospective outlook is better suited for examining a community's resilience in the broader context of economic changes during the recovery process.

The second limitation of the post-disaster housing recovery literature is that it is broadly based on urban areas. Rural areas have little or no visibility in this literature. This bias is also evident in the broader disaster literature in which the terms "Resilient Cities" and "The Resilient City" are frequently used (Godschalk, 2003; Vale & Campanella, 2005) as well as in Katrina literature which mainly focuses on the city of New Orleans (Bergal et. al., 2007; Campanella, 2006; Richardson et al., 2008; Olshansky & Johnson, 2010). In the literature, studies that have an exclusive focus on rural communities and that highlight the distinctiveness of rural areas during recovery are very few: Régnier, Neri, Scuteri, and Miniati's (2008) study which deals with recovery in rural Aceh (Indonesia) and Tamil Nadu (India) after the Indian Ocean tsunami; Dyer and McGoodwin's (1999) study which focuses on rural fishing communities in South Florida and Louisiana after Hurricane Andrew; and Saenz and Peacock's (2006) study which examines the hidden costs of Hurricane Katrina for rural people and places. To fill in this gap in the literature, we focus on the recovery of rural areas affected by Hurricane Katrina.

The literature on post-disaster housing recovery highlights six broad factors that affect post-disaster housing recovery, albeit in the context of urban areas given disaster scholars' lack of attention to rural areas: population characteristics, housing characteristics, employment and income, human and social capital, rural characteristics, and disaster damages. These factors and their significance for rural areas are briefly enumerated and discussed below.

Population characteristics. Demographic characteristics affect housing recovery in different ways. Ethnic and racial minorities, women, and elderly face more barriers and are more disadvantaged during post-disaster recovery. Studies conducted after Hurricane Andrew and Katrina show that minority groups (African Americans and Hispanics) could face discrimination during recovery (Girard & Peacock, 1997; Weil, 2009) and that predominantly minority communities are slow in recovering from disasters (Dash, Peacock, & Morrow, 1997; Zhang & Peacock, 2010). In their review, Fothergill, Maestas, and Darlington (1999) find that racial and ethnic communities are vulnerable to natural disasters, due to factors such as language, housing patterns, community isolation, and cultural insensitivities. Women often play an active role during disaster relief and recovery (Enarson, 1999). Yet, their recovery could take longer than men because women and women-headed households could have limited access to formal relief and recovery mechanisms after disasters (Morrow & Enarson, 1996). As for the elderly, they also face barriers in recovery due to limited physical and economic resources (Morrow, 1999).

Employment and income. Researchers draw attention to the vulnerability of low-income groups to disasters (Fothergill & Peek, 2004) and show that low-income neighborhoods recover more slowly than high-income neighborhoods (Bolin, 1994; Dash, Peacock, & Morrow, 1997; Kamel & Loukaitou-Sideris, 2004; Levine, Esnard, & Sapat, 2007; Peacock & Girard, 1997). This is because the traditional mechanisms used to finance post-disaster housing recovery in the U.S. private sector insurance and the federal government programs fail to meet the needs of low-income groups (Comerio, 1998; Mueller et al., 2011). However, Zhang and Peacock's (2010) study on Hurricane Andrew finds that income does not influence the pace of the recovery process.

Housing characteristics. Research on post-disaster recovery with respect to housing characteristics shows that there are differences in the scale and pace of recovery with respect to single-family versus multi-family housing units. Multi-family housing takes longer to recover than single-family housing (Comerio, 1998; Lu et al., 2007; Wu et al., 2007). This is explained in part due to the bias toward single-family, owner-occupied housing in post-disaster assistance programs (Comerio, 1998; Peacock, Zhang & Dash, 2005; Wu & Lindell, 2004). Although multi-family housing in rural areas may not be as prevalent as urban areas, rural areas could still be vulnerable to disasters due to the mobile homes they accommodate.

Human and social capital. The disaster research has increasingly recognized the significance of human and social capital (see Chapter 3) in the disaster recovery process, highlighting their role in the resilience of communities (Wisner, 2003). Chamlee-Wright and Storr (2009) show that church provision of club goods fostered social cooperation and community redevelopment in a New Orleans neighborhood following Hurricane Katrina. Brouwer and Nhassengo (2006) highlight the importance of social capital for recovery of poorer households after the 2000 floods in the Mabalane District, Mozambique. In their studies of 1995 and 2001 earthquakes in Kobe (Japan) and Gujarat (India), respectively, Nakagawa and Shaw (2004) suggest that communities with social capital and leadership can make a successful and speedy recovery after disasters. It should be noted, however, that the form of social capital in rural areas, based on kinship and strong ties, could be distinctive from that in urban areas based mainly on weak ties (Hofferth & Iceland, 1998; Lannoo et al., 2012).

Rural characteristics. There are no systematic studies which examine the impact of the degree of ruralness on housing recovery. However, the literature often depicts rural areas as more vulnerable to disasters because of their unique characteristics, such as reliance on a single-sector economic base (e.g., agriculture, fishing) with no alternative employment sources (Cutter & Emrich, 2006; Saenz & Peacock, 2006). Even measures of vulnerability reflect such thinking about rural areas. Cutter, Boruff, and Shirley's (2003) widely used Social Vulnerability Index (SoVI), for instance, posits a positive correlation between the rural characteristics of a place and its vulnerability.

Disaster damages. The extent of disaster damages influences the long-term recovery of communities. Extant literature highlights the influence of property damage on the ability of businesses and households to recover from disasters (Bolin, 1994; Kroll et al., 1991). Zhang and Peacock (2010) argue that the property damage from Hurricane Andrew had long-term effects. The volume of home sales increased dramatically in the first year after the hurricane. Property abandonment occurred to a significantly high degree in "heavily damaged, lower income, rental, and minority, especially Black, neighborhoods" (Zhang & Peacock, 2010: 21). Disaster damages have indirect consequences (e.g., on utilities) that could impact the recovery process (Tierney, Nigg, & Dahlhamer, 1996).

In sum, while the body of knowledge on housing recovery continues to grow, it mainly deals with urban areas, emphasizing the importance of such population characteristics as race, ethnicity, age, and gender in addition to that of single- versus multi-family housing, income, social capital, and disaster damages in recovery. It also depicts rural areas as inherently vulnerable. This study examines housing recovery in rural areas affected by Hurricane Katrina by using a quasi-experimental research design.

METHODS

Although much research has focused on the city of New Orleans, Hurricane Katrina was a catastrophic disaster that impacted a large-scale area (Comfort et al., 2010). According to the Spatial Hazard Events and Losses Database for the U.S. (SHELDUS) (Hazards and Vulnerability Research Institute, 2010), Katrina affected a total of 288 counties. Of these, 197 were rural (non-metropolitan) counties, all of which are included in our study. These counties are distributed across the states that were most heavily impacted by Hurricane Katrina, such as Alabama, Louisiana, and Mississippi, as well as other states that had some impact, such as Arkansas, Florida, Georgia, Indiana, Kentucky, Michigan, Ohio, and Tennessee. The unit of analysis is the county because this is the lowest geographical administrative unit for which most data are available on an annual basis.

The dependent variable in this study is housing recovery, which we conceptualized in terms of the housing stock. Admittedly, there are several ways of measuring housing recovery, such as housing prices and quality (Zhang & Peacock, 2010). We chose the housing stock (i.e., number of housing units) because this is a key measure of the extent to which the housing has been rebuilt following damage from Hurricane Katrina. We used the following indicators from the U.S. Census Bureau to measure the extent of housing stock recovery: (1) net change of housing units, because this gives the incremental addition to the housing stock; and, (2) total housing units, because this gives the total housing stock at any time. We examined these two variables for three years (2006, 2007, and 2008) following Hurricane Katrina.

Our study focuses on the aforementioned six groups of factors which affect housing recovery in affected communities: population characteristics, housing characteristics, employment and income, human and social capital, rural characteristics, and disaster damages (a total of twenty-one independent variables). The variables are listed in Table 6.1, along with the hypothesized effect (positive or negative) on housing recovery based on the literature.

The *population characteristics variables* (total of seven) encompass migration (net domestic and international), race (% of black population) and ethnicity (% of Hispanic population), gender (% of female population and % of female-headed households), and elderly (% of population over sixty-five). *Employment and income variables* (total of four) include unemployment (%), income (income per capita, median household income), and poverty (poverty rate). *Housing characteristics variables* (total of four) are the nature of housing structures (% of multi-unit structures, % of mobile homes), housing value (median house value), and vacancy (% of vacant homes). *Human and social capital variables* (total of two) are education (high school diploma and above) and civic associations (# of nonprofit organizations), respectively. *Rural characteristics variables* (total of two) consist

Table 6.1 Variable Names and Definitions

I. Dependent Variable

Variable names	*Definitions*	*Data sources*
Net change of housing units	Housing unit estimates—net change, April 1, 2000 (base), 2006–2008	U.S. Census Bureau, USA Counties data files
Total housing units	Total housing unit estimates, 2006–2008	

II. Independent Variables

Variable names	*Definitions*	*Hypothesized effect*	*Data sources*
Population characteristics			
Domestic migration	Net domestic migration, 2006–2008	+	U.S. Census Bureau, USA Counties data files
International migration	Net international migration, 2006–2008	+	
% of population over 65	% of population over the age of 65 years, 2006–2008	-	
% of black population	% of black population, 2006–2008	-	
% of Hispanic population	% of Hispanic population, 2006–2008	-	
% of female population	% of female population, 2000	-	
% of female-headed households	% of female-headed households, 2000	-	
Employment and income characteristics			
Unemployment rate	Civilian unemployment rate, 2006–2008	-	U.S. Census Bureau, USA Counties data files
Income per capita	Income per capita, 2006–2008	+	
Median household income	Median household income, 2000	+	
Poverty rate	Poverty rate, all age	-	
Housing characteristics			
Housing structure, multiple units	% of housing units by units in structure—multi-dwelling structure, 2000	-	U.S. Census Bureau, USA Counties data files

(*continued*)

Housing structure, mobile homes	% of housing units by units in structure—mobile home, 2000	-	
Median house value	Median house value, 2000	+	
% of vacant homes	% of vacant homes, 2000	-	
Human and social capital			
High school diploma and above	% of adult pop (>25) with high school diploma and above, 2000	+	U.S. Census Bureau, USA Counties data files
# of associations	# of nonprofit organizations, 2000	+	National Center for Charitable Statistics (NCCS)
Rural characteristics			
Agriculture earnings	Earnings in agriculture, forestry, fishing, and hunting (NAICS 11), 2006–2008	-	U.S. Census Bureau, USA Counties data files
Rural-urban continuum	1 to 9 scale; 1 = central counties of metro areas of one million population or more; 9 = completely rural counties, 1993 definition	-	Economic Research Services, USDA
Disaster damages			
Property damage	$ value (log), 2005	-	SHELDUS†
Crop damage ($, log)	$ value (log), 2005	-	

†Spatial Hazard Events and Losses Database for the U.S. (SHELDUS) (Hazards and Vulnerability Research Institute 2011, http://webra.cas.sc.edu/hvri/products/sheldus.aspx).

of rural earnings (earnings in agriculture, forestry, fishing, and hunting) as well as the degree of ruralness based on the 1993 Rural-Urban Continuum Codes.[1] *Disaster damage variables* (total of two) are property and crop damage in dollar value.

We used data from multiple sources. Most of our data were from the U.S. Census Bureau, as indicated in Table 6.1 for each variable. In addition, we relied on the National Center for Charitable Statistics (NCCS) for social capital data and on the Economic Research Services (USDA) data for the degree of ruralness. Lastly, we used the SHELDUS database for damage data from Hurricane Katrina. This is a county-level dataset for the U.S. compiled by the Hazards and Vulnerability and Research Institute at the University of South Carolina. It is important to note that the SHELDUS data on Hurricane Katrina is as of December 2005 as reported by the National Climate Data Center. It may over- or underestimate the true losses from Katrina. Therefore, the data should be interpreted with some caution. However, SHELDUS remains the best available data source on hazard-related losses, which is available to the public free of charge.

We analyzed the data in three phases. Our *first phase* involved using a quasi-experimental design, in which we identified a matched county for each affected rural county (pair-wise matching) and established statistically equivalent counterfactual scenarios for measuring housing recovery progress. This methodological approach has been applied to hazard mitigation before (Deyle, Chapin, & Baker, 2008) but is relatively new for disaster recovery, with a notable exception of Xiao (2011). It complements earlier approaches which measure recovery as post-disaster economic indicators returning to pre-disaster levels, pre-disaster growth rates, or prior growth patterns (Vigdor, 2008; Zhang & Peacock, 2010).

The application of a quasi-experimental design in our study involved the identification of a matched county for each rural county that was affected by Hurricane Katrina. A matched county should ideally resemble the affected county on several characteristics of interest. The matched counties are geographic places that closely resembled affected counties prior to the disaster on a number of pre-selected characteristics, including income and earnings, rural characteristics, poverty, migration and age, employment, crime, and county size (Table 6.2).

The advantage of the quasi-experimental design is that it allowed us to measure the unobserved counterfactual—what would have happened in the affected counties if Hurricane Katrina had not taken place. The counterfactuals served as controls for macroeconomic and industrial factors that may

Table 6.2 Criteria Used for Identifying Matched Control Counties

Income and earnings	*Employment*
Income per capita, 2004 Earnings in agriculture, forestry, fishing,	% of employment in agriculture, forestry, fishing, and hunting (NAICS 11), 2004
and hunting (NAICS 11), 2004	Civilian labor force unemployment rate, 2004
Rural characteristics	Migration and age
Metro/rural dummy variable	Net domestic migration, 2000
Average value of land and buildings per acre, 2002	Resident population: Median age, 2000
Farms by size—under ten acres, 2002	Crime
Number of violent crimes known to police,	
Poverty Poverty rate, all age, 2004	2002
Size	
	Land area, 2000

have influenced disaster-affected counties from reaching their pre-disaster economic, social, and environmental situation. This is particularly important for housing recovery under current housing market conditions and economic recession.

Ideally the matches (counterfactuals) should have all the same features as the affected counties prior to Hurricane Katrina. Finding a perfect match in the real world, however, is almost impossible. The alternative is to find an acceptable or optimal match under specific criteria. Of course, a similar growth pattern in the pre-treatment period does not necessarily imply that the control and treatment counties will continue to experience the same growth in the post-treatment period in the absence of the treatment.

We selected the optimal matches based on minimal Mahalanobis distance between the pair of case and control counties. The Mahalanobis distance is given by:

$$d(X_T, X_M) = (X_T - X_M)' - R^{-1} (X_T - X_M)$$

where X_T and X_M are vectors of pre-selected matching criteria of affected (treatment) county *T* and matching county *M*, and *R* is the variance-covariance matrix related with variables *X*. The smaller the Mahalanobis distance, the more alike is a pair of counties. We used the optimal matching algorithm provided by Isserman and Rephann (1995) to select the best case-control pairs that minimize the sum of the Mahalanobis distances. Table 6.2 presents the matching criteria used in the optimal matching algorithm and Figure 6.1 shows the locations of Hurricane Katrina-affected rural counties and their matched rural counties.

Our *second phase* in analyzing data involved using the group of matched control counties as a benchmark to evaluate whether the group of affected counties as a whole has reached the levels of their counterfactuals. Analysis

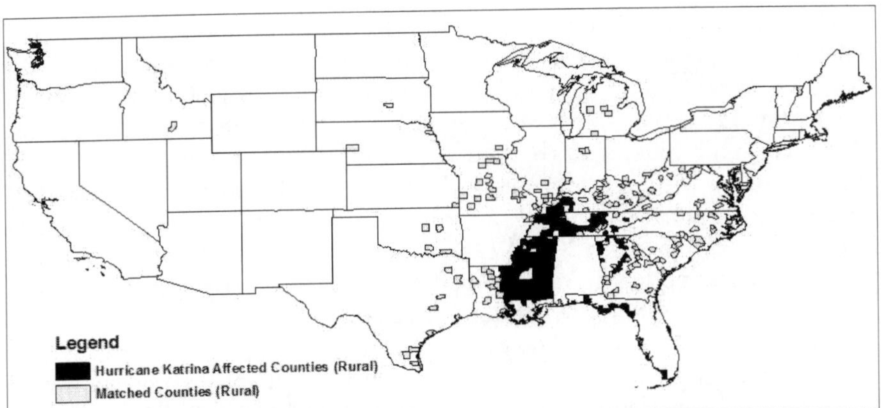

Figure 6.1 Hurricane Katrina-affected counties and their matched counties.

of individual affected and matched counties is not appropriate for direct comparison because individual counties' idiosyncratic observed and unobserved characteristics which are not controlled in the matching procedure may affect the recovery process and final outcome. Intergroup comparisons, to a great extent, are able to cancel out the potentially unaccounted idiosyncratic county characteristics especially in relatively large samples. Yet, bias may still exist and may not be fully detected or eliminated, particularly when some of the uncontrolled idiosyncratic county characteristics are not directly observable.

To assess the counterfactual recovery of the Hurricane Katrina-affected counties, we conducted *t*-tests for the years prior to (2004) and after the disaster event (2006, 2007, and 2008) for all affected rural counties (197) as well as for rural counties in four states which house a majority of the rural counties affected by the disaster, namely Alabama, Louisiana, and Mississippi. The purpose of the *t*-tests was to detect whether the means of the groups of affected and matched rural counties are statistically equivalent or different. Statistical indifference of group means in year 2004 would suggest equivalence between the affected and matched rural counties a year before Hurricane Katrina occurred and would indicate the appropriateness of the matching criteria and procedure used. Statistical difference, if any, in years 2006, 2007, and 2008, would indicate annual differences in growth patterns between the affected and control groups, and hence, would suggest whether or not the affected counties have recovered to the counterfactual levels.

Our *third and last phase* of data analysis involved conducting multivariate panel analysis of the housing stock recovery for the period 2006 to 2008 to identify factors that affect housing recovery. The Housing Recovery (HR) in a county (i) over a period of time (t) is given by:

$$HR_{it} = x_{it}\beta_k + z_i\delta_p + \mu_i + \varepsilon_{it}$$

in which x_{it} is a 1xk vector of variables that vary over individual counties, and time β is the kx1 vector of coefficients on x, z_i is a 1xp vector of time-invariant variables that vary only over counties, δ is a px1 vector of coefficients on z, μ_i is the county-specific effect, and ε_{it} is the disturbance term. We tested and evaluated both fixed- and random-effect panel model specifications using the Hausman test (Hausman, 1978). Results suggest that random-effect panel models are appropriate for both dependent variables. In comparison to the fixed- and random-effect panel model specifications, the fixed-effect models are able to control potential correlation between μ_i and x_{it} and/or z_i, while the random-effect models assume no correlation. The random-effect models, however, are able to estimate coefficients, δ, for time-invariant variables, z_i, when the correlation between μ_i and x_{it} and/or z_i is minimal or moderate, while fixed-effect models are not. The two variables used in the second phase, *t*-tests, were used as dependent variables in the panel analyses for identifying determinants influencing housing recovery in rural counties after Hurricane Katrina.

RESULTS

We group our results under two categories: on differences in terms of housing stock between affected and matched rural counties (based on the one-tailed *t*-tests) and on drivers of post-Katrina housing recovery in affected rural counties (based on the multivariate panel analysis). Our main findings are the following: (1) There was more net change of housing units in Katrina-affected rural communities during the recovery period than in their counterfactuals, which suggests their resilience. (2) Housing growth in these rural communities was mainly driven by domestic migration, percentage of mobile homes, and social capital during the recovery period.

Differences between the Affected and Matched Rural Counties in Housing Stock: More Net Change in Housing in Affected Counties

The results of the *t*-tests of the group means of the affected and matched rural counties are shown in Table 6.3. The null hypothesis is that there is no statistically significant difference between the two group means, i.e., $H_0: \mu_{matched} = \mu_{affected}$, which implies that Hurricane Katrina did not have any effect on the affected counties. The alternative hypothesis is that the group means of the matched rural counties is smaller than that of the affected counties for a given recovery indicator, i.e., $H_1: \mu_{matched} < \mu_{affected}$. We tested the other one-tailed alternative hypothesis, i.e., $H_1: \mu_{matched} > \mu_{affected}$, but we found no statistically significant difference for this H_1.

For both the indicators of housing stock recovery (dependent variables), there is no statistically significant difference between the two groups' means in 2004, thus suggesting statistical equivalence and comparability between the groups of affected and matched rural counties prior to Hurricane Katrina. The Hurricane Katrina-affected counties and their matched counties were thus statistically similar in terms of the dependent variables, which is a positive confirmation of the validity of the selection of the matched counties.

The *t*-test results show that the differences in net change of housing units are positive and significant for all three years. Katrina-affected rural counties exhibit recovery above counterfactual levels. This suggests a sustained net change of housing units from 2006 to 2008. The *t*-tests show no statistical difference at the state level. With respect to the total housing units, however, the *t*-test results show no statistical difference between the affected and the matched counties. We are thus unable to reject the null hypothesis for the total housing units. That is, although the net change is significantly high for affected counties as compared to matched counties,

Table 6.3 Statistical Significance of the One-Tailed *t*-test for the Katrina-Affected and Matched Counties

	Net change of housing units	*Total housing units*
All rural counties (197), H_0: $\mu_{matched} = \mu_{affected}$; H_1: $\mu_{matched} < \mu_{affected}$		
2004	0.167	0.657
2006	0.072	0.613
2007	0.047	0.587
2008	0.031	0.566
Alabama rural counties (21), H_0: $\mu_{matched} = \mu_{affected}$; H_1: $\mu_{matched} < \mu_{affected}$		
2004	0.286	0.471
2006	0.348	0.471
2007	0.343	0.465
2008	0.326	0.459
Louisiana rural counties (18), H_0: $\mu_{matched} = \mu_{affected}$; H_1: $\mu_{matched} < \mu_{affected}$		
2004	0.617	0.929
2006	0.302	0.913
2007	0.227	0.902
2008	0.199	0.896
Mississippi rural counties (72), H_0: $\mu_{matched} = \mu_{affected}$; H_1: $\mu_{matched} < \mu_{affected}$		
2004	0.902	0.352
2006	0.745	0.351
2007	0.656	0.345
2008	0.518	0.332

Note: Alternative H_1: $\mu_{matched} > \mu_{affected}$ has also been tested, but none of the tests showed a statistically significant difference between the two group means to support the H_1.

the total housing stock is similar. The *t*-tests show no statistical difference at the state level.

Most Important Drivers of Housing Recovery: Domestic Migration, Mobile Homes, and Social Capital

To determine the drivers of recovery, we used a multivariate panel analysis, testing the effect of the independent variables identified through a literature review on the dependent variables (housing recovery indicators). We evaluated both fixed- as well as random-effect model specifications for each dependent variable. Because the Hausman (1978) test results support the random-effect specifications for all models tested, we have presented only these results in Table 6.4. Overall the regression results are quite robust. The R-square value

Table 6.4 Panel Analysis Results

	Net change of housing units	*Total housing units*
Domestic migration	0.28***	0.09***
International migration	1.62*	2.62**
% of population over 65	-101.78***	-96.09*
% of black population	-5.80	-48.74***
% of Hispanic population	79.34**	430.14***
% of female population	28.27	73.16*
% of female-headed households	0.00	-0.02
Unemployment rate	-25.07	-3.00
Income per capita	-0.02	0.03*
Median household income	0.06***	0.01
Poverty rate	13.91	9.85
Housing structure, multiple units	32.58*	353.67***
Housing structure, mobile homes	0.24***	2.64***
Median house value	-0.01*	-0.03*
% of vacant homes	0.17**	0.34*
High school diploma and above	7.40	80.85*
# of associations	4.75***	26.46***
Agriculture earnings	-0.01	0.01
Rural-urban continuum	216.90***	-700.24***
Property damage ($, log)	0.98	33.57
Crop damage ($, log)	-4.38	35.55
Constant	-3750.83**	-2552.16
Summary statistics		
# of observations	591	591
# of affected counties	197	197
Overall R-square	0.64	0.92

Note: A ***, **, or * indicates that the coefficient is significantly different from zero at 1%, 5%, or 10% levels respectively.

is high: .64 for net change of housing units and .92 for the total housing units. Although many of the independent variables included in our analysis are strongly associated with our dependent variables as explained below, the coefficient is significantly different from zero at the 1% level for domestic migration, percentage of mobile homes, and social capital.

Population characteristics. We hypothesized that migration would have a positive effect while all other indicators of population characteristics would have a negative effect on housing recovery. We found that, along with both domestic and international migration, the percentage of Hispanic population has a significant positive effect on both housing recovery indicators. Senior population over sixty-five years of age is negatively associated with housing recovery, as hypothesized. Race and gender variables show less consistent results. They are significantly associated with the total housing units but not with the net change. The sign of association is negative for the percentage of black population and positive for the percentage of female population. The percentage of female-headed households is not significant.

Employment and income characteristics. Our study exhibits mixed effects of employment and income. Unemployment rate and poverty can be reasonably hypothesized to have a negative effect on housing recovery indicators. However, neither of these variables is significantly related to housing recovery. The remaining indicators in this category are significantly associated either with the net change in housing units or the total housing units. Income per capita has a significant positive effect on the total housing units. Median household income has a positive effect on the net change in housing units.

Housing characteristics. All housing indicators affect the recovery process in a significant way. We hypothesized that housing structures with multiple units, mobile homes, and vacant homes would be negatively associated with housing recovery, whereas median house values would be positively associated. However, the results show otherwise. The percentage of housing structures with multiple units, mobile homes, and vacant homes each have a statistically significant positive effect on both housing recovery indicators. Interestingly, the median house value has a negative effect on both net change in housing units and total housing units.

Human and social capital characteristics. Literature on disasters indicates that human and social capital is generally conducive to the recovery process. Our results confirm this hypothesis to the large extent that these have a positive effect on the recovery. Education has a significant positive effect on total housing units. The number of nonprofit associations has a positive effect on both the indicators.

Rural characteristics. Literature on housing recovery surmises that the more rural an area, the less likely they are to recover. The regression results show that the agricultural earnings do not have a significant effect on the housing recovery indicators. Ruralness has a significant positive effect on the net change of housing units, but a negative effect on total housing units. The results thus show ambiguous effect of rural characteristics on the housing recovery indicators.

Disaster damages. We hypothesized that the higher the degree of disaster damages, the less likely the areas will be to have housing recovery. We used two indicators—property damage and crop damage—to indicate the disaster damages. Neither property nor crop damage has a significant effect on the housing recovery indicators.

DISCUSSION

Our one-tailed *t*-tests of the group means of the affected and matched rural counties show that Katrina-affected rural counties experienced more growth than the counterfactual levels in the net change of housing units. Such growth started in 2006, suggesting an immediate recovery or "bounce-back" right after Katrina, and continued through 2008. Especially the net change of housing units was significantly high for affected counties as compared to their counterparts, possibly due to higher demand for housing in rural areas following the disaster (e.g., demand from urban areas). With respect to the total housing units, however, affected counties recovered to the counterfactual levels, possibly due to the damages to their housing stock.

At the broad level, our findings from the *t*-tests indicate that the rural counties are indeed resilient. While further research into the underlying patterns is required, such recovery is counter to the presumption in the literature that rural areas are not resilient. This presumption is perhaps most explicit in the well-respected Social Vulnerability Index (SoVI) (Cutter et al., 2003). SoVI assumes a positive correlation between the rural characteristics of a place and its vulnerability and misses the possibility of rural resilience in the face of disasters.

Our multivariate panel analyses show the significance of migration, elderly population, and Hispanic in post-disaster housing recovery.Both international and domestic migration has a positive effect while elderly population has a negative effect on post-disaster housing recovery in rural areas affected by Katrina. The positive sign is to be expected for migration and confirms the hypothesis because migration increases the demand for housing in the disaster-affected areas, so that there is pressure for new housing units to be built in the areas. The negative sign is also to be expected for the elderly population due to their unique vulnerabilities and confirms the findings of earlier studies (Morrow, 1999). The strong positive association between the Hispanic population and housing recovery indicators was not expected due to the findings of earlier studies which suggest that Hispanics face disadvantages during recovery (Weil, 2009). However, Hispanic populations may be higher in rural areas that experience growth in their housing stock following disasters because they are disproportionately employed in the construction sector.

With respect to race and gender, the results of our study are mixed. Black population has a negative effect on the total housing units, confirming the literature which suggests that blacks were disadvantaged in the post-Katrina New Orleans context (Hartman & Squires, 2006). As for gender, the percentage of female population positively affects the total housing units. This finding again contrasts with the literature, which argues that women take a longer time to recover from disasters (Morrow & Enarson, 1996).

Part of our findings on employment and income contradicts the existing literature on disasters. Despite studies that highlight the disadvantages of low-income neighborhoods during recovery (Kamel & Loukaitou-Sideris, 2004; Levine et al., 2007), poverty and unemployment rate does not have an effect on housing recovery in rural counties included in this study. This may suggest the difference between housing recovery and other recovery perspectives. Housing recovery does not only depend on community employment and income conditions, but also is subject to external forces, for example, influx of redevelopment and/or speculations. So in a short(er) run, the disadvantageous internal housing recovery dynamics may be partially offset by strong and continued external forces. This could be in part due to the urban bias in the literature as explained earlier. The positive significant association of income per capita with the total housing units and of median household income with the net change in housing units also contradicts with Zhang and Peacock's (2010) study which suggests that income does not influence the pace of the recovery process.

The strong positive association between the percentage of housing structures with multiple units, mobile homes, and vacant homes may suggest that communities that house such units may be susceptible to housing redevelopment. Homeowners or renters of these units may be more footloose than the single-family homeowners and may be more likely to be displaced, thereby allowing for development of new housing units in the area. Post-Katrina development trends in urban areas offer interesting insights on what may have happened in rural areas. In New Orleans, thousands of public housing units were singled out for demolition following Katrina to make room for newer housing units (Eaton, 2007; Falk, Hunt, & Hunt, 2006; Quigley, 2007; Saulny, 2006). It is not our intention to discuss possible motivations behind this demolition (e.g., to reduce the poor and make room for developers). However, former Louisiana Congressional Representative Richard H. Baker stated the following immediately after the hurricane: "We finally cleaned up public housing in New Orleans. We couldn't do it but God did" (Saulny, 2006). It is possible that similar demolition projects have taken place in rural areas affected by Katrina, paving the way for newer housing development. As for the median house value, its negative effect on housing recovery indicators suggests that rural communities with lower median house values exhibit higher levels of growth in their housing stock.

Among the remaining indicators, our findings on human and social capital confirm the findings of earlier studies that emphasize the benefits of such capital for recovery (Chamlee-Wright & Storr, 2009). The ambiguous effects of the rural characteristics of a place and no significant effects of disaster damage warrant further examination.

CONCLUSION

In this chapter, we identified both housing recovery patterns and determinants influencing the recovery in all 197 rural counties affected by Hurricane Katrina. The chapter complemented earlier research that focuses on post-disaster recovery in urban centers, and filled in a research gap of recovery patterns and processes in rural areas. Furthermore, it introduced an alternative methodological approach for measuring recovery with the use of quasi-experimental design.

The key results of our study are twofold. First, Katrina-affected rural counties experienced housing growth during their recovery period above their matched counterparts in terms of the net change of housing units, starting immediately after the hurricane. This finding is in contrast with the existing literature which suggests that rural areas are vulnerable (Cutter et al., 2003). Second, domestic migration, the percentage of mobile homes, and social capital has the most significant positive effect on housing during post-disaster recovery in rural areas.

There is a need for further research on several fronts. First, this study examined post-Katrina data from 2006 to 2008, which is a relatively short time frame when it comes to recovery. Future research should consider a longer time frame in studying post-disaster housing recovery trajectories of rural areas. Second, this study focused on Katrina, which was a unique disaster event. There is a need for studies that focus on other disaster events (e.g., floods, earthquakes). Third, this study was limited to the county as the unit of analysis. Even if counties show recovery at the macrolevel, communities within the counties may not have recovered, and may be suffering from concentrated damage. Future research should focus on housing recovery at the subcounty level (e.g., community, city, and neighborhood). Fourth, studies that take into account actual disaster payments (e.g., private insurance payments or Federal Emergency Management's payments to households), as an alternative to SHELDUS data on estimated disaster damages, could provide different insights on how damage influences post-housing disaster recovery. Fifth, as mentioned earlier, much of the recovery literature focuses on urban areas. We not only need more studies on rural areas but also studies which compare the housing recovery trajectories of urban and rural areas. Sixth, Rubin, Saperstein, and Barbee (1985) observe that both speed and quality are the metrics of recovery. This study did not address the quality of recovery per se. Future studies should focus on this aspect of recovery. Lastly, this study was based on secondary data and is unable to provide a "thick description" of recovery processes in rural areas. There is a need for contextual qualitative case studies that can better capture the complex recovery processes on the ground.

This study offers several policy insights for post-disaster housing recovery in rural areas. Its most important insight is that rural areas that have higher proportions of elderly require more support in terms of housing recovery (e.g., in the form of housing recovery programs that focus exclusively on this disadvantaged population). The study also highlights the need to better track migration patterns in and out of rural areas following disasters, especially those areas that are near the urban areas. These areas may be transformed into higher-density suburban areas due to migration in the aftermath of disasters. Such transformation may have an indirect impact on the vulnerability of urban areas for future disasters (Cutter et al., 2007). Lastly, there is a need to better track households that live in mobile homes and multi-family structures in rural areas because these households may be pushed out following disasters to make room for future growth.

ACKNOWLEDGEMENTS

This chapter is based on research supported by U.S. National Science Foundation (NSF) Grant No. SES-1045624. The findings and opinions reported are those of the authors and are not necessarily endorsed by the funding organization or those who provided assistance with various aspects of the study.

NOTES

1. The 1993 Rural-urban continuum codes of the Economic Research Service, developed by the U.S. Department of Agriculture, classify U.S. counties into ten categories (0–9). Metropolitan (urban) counties are coded 1 to 3 (based on population size) and non-metropolitan (rural) counties are coded 4 to 9 (based on the scale of urban population and adjacency to metro areas). At the extremes, code 0 represents central counties of metro areas with a population of 1 million or more while code 9 represents completely rural counties. For more extensive discussion on the 1993 rural-urban continuum codes, please refer to: http://www.ers.usda.gov/Briefing/Rurality/RuralUrbCon/priordescription.htm.

REFERENCES

Bergal, J., Shipley Hiles, S., & Koughan, F. (2007). *City adrift: New Orleans before and after Katrina*. Baton Rouge: Louisiana State University Press.

Bolin, R. (1986). Disaster impact and recovery: A comparison of black and white victims. *International Journal of Mass Emergencies and Disasters*, 4(1), 35–50.

Bolin, R. (1994). *Household and community recovery after earthquakes*. Boulder, CO: University of Colorado.

Brouwer, R., & Nhassengo, J. (2006). About bridges and bonds: Community responses to the 2000 floods in Mabalane District, Mozambique. *Disasters*, 30(2), 234–255.

Campanella, T.J. (2006). Urban resilience and the recovery of New Orleans. *Journal of the American Planning Association*, 72(2), 141–146.

Chamlee-Wright, E., & Storr, V.H. (2009). Club goods and post-disaster community return. *Rationality and Society*, 21(4), 429–458.

Comerio, M.C. (1998). *Disaster hits home: New policy for urban housing recovery*. Berkeley: University of California Press.

Comfort, L.K., Birkland, T.A., Cigler, B.A., & Nance, E. (2010). Retrospectives and prospectives on Hurricane Katrina: Five years and counting. *Public Administration Review*, 70(5), 669–678.

Cutter, S.L., Boruff, B.J., & Shirley, W.L. (2003). Social vulnerability to environmental hazards. *Social Science Quarterly*, 84(2), 242–261.
Cutter, S.L., & Emrich, C.T. (2006). Moral hazard, social catastrophe: The changing face of vulnerability along the hurricane coasts. *Annals of the American Academy of Political and Social Science*, 604, 102–112.
Cutter, S.L., Johnson, L.A., Finch, C., & Berry, M. (2007). The U.S. Hurricane coasts: Increasingly vulnerable. *Environment*, 49(7), 8–20.
Dash, N., Peacock, W.G., & Morrow. B.H. (1997). And the poor get poorer: A neglected Black community. In W.G. Peacock, B. H. Morrow, & H. Gladwin (Eds.), *Hurricane Andrew: Ethnicity, gender and the sociology of disasters* (pp. 206–225). London and New York: Routledge Press.
Deyle, R.E., Chapin, T.S., & Baker, E.J. (2008). The proof of the planning is in the platting: An evaluation of Florida's hurricane exposure mitigation planning mandate. *Journal of the American Planning Association*, 74(3), 349–370.
Dyer, C.L., & McGoodwin, J.R. (1999). "Tell them we're hurting": Hurricane Andrew, the culture of response, and the fishing peoples of South Florida and Louisiana. In A. Oliver-Smith & S.M. Hoffman (Eds.), *The angry earth: Disaster in anthropological perspective* (pp. 213–231). New York: Routledge.
Eaton, L. (2007). In New Orleans, plan to raze low-income housing draws protest. *New York Times*, December 14. Retrieved January 17, 2011 from http://www.nytimes.com/2007/12/14/us/nationalspecial/14orleans.html.
Enarson, E. (1999). Women and housing issues in two U.S. disasters: Hurricane Andrew and the Red River Valley Flood. *International Journal of Mass Emergencies and Disasters*, 17(1), 39–63.
Falk, W.W., Hunt, M.O., & Hunt, L.L. (2006). Hurricane Katrina and New Orleanians' sense of place: Return and reconstitution or "Gone with the Wind"? *Du Bois Review*, 3(1), 115–128.
FEMA (Federal Emergency Management Agency). (2011). Housing recovery support function. In *National Disaster Recovery Framework*. Retrieved from http://www.fema.gov/pdf/recoveryframework/housing_rsf.pdf.
Fothergill, A., Maestas, E.G., & Darlington, J.D. (1999). Race, ethnicity and disasters in the United States: A review of the literature. *Disasters*, 23(2), 156–173.
Fothergill, A., & Peek, L.A. (2004). Poverty and disasters in the United States: A review of recent sociological findings. *Natural Hazards*, 32, 89–110.
Ganapati, N.E., & Ganapati, S. (2009). Enabling participatory planning in post-disaster contexts: A case study of World Bank's housing reconstruction in Turkey. *Journal of the American Planning Association*, 75(1), 41–59.
Girard, C., & Peacock, W.G. (1997). Ethnicity and segregation: Post-hurricane relocation. In W.G. Peacock, B.H. Morrow, & H. Gladwin (Eds.), *Hurricane Andrew: Ethnicity, gender and the sociology of disasters* (pp. 191–205). London and New York: Routledge Press.
Godschalk, D.R. (2003). Urban hazard mitigation: Creating resilient cities. *Natural Hazards Review*, 4(3), 136–143.
Hartman, C.W., & Squires, G.D. (2006). *There is no such thing as a natural disaster: Race, class, and Hurricane Katrina*. New York: Routledge.
Hausman, J.A. (1978). Specification tests in econometrics. *Econometrica*, 46(6), 1251–1271.
Hazards and Vulnerability Research Institute. (2010). *The spatial hazard events and losses database for the United States, Version 8.0* [online database]. Columbia, SC: University of South Carolina. Retrieved December 17, 2010 from http://www.sheldus.org.
Hofferth, S.L., & Iceland, J. (1998). Social capital in rural and urban communities. *Rural Sociology*, 63(4), 574–598.

Isserman, A.M., & Rephann, T. (1995). The economic effects of the Appalachian Regional Commission: An empirical assessment of 26 years of regional development planning. *Journal of the American Planning Association*, 61(3), 345–364.

Kamel, N.M.O., & Loukaitou-Sideris, A. (2004). Residential assistance and recovery following the Northridge earthquake. *Urban Studies*, 41(3), 533–562.

Kilmer, R.P., Gil-Rivas, V., Tedeschi, R.G., & Calhoun, L.G. (2010). *Helping families and communities recover from disaster: Lessons learned from Hurricane Katrina and its aftermath*. Washington, DC: American Psychological Association.

Kroll, C.A., Lands, J.D., Shen, Q., & Stryker, S. (1991). *Economic impacts of the Loma Prieta earthquake: A focus on small business*. Working paper no. 91–187. Berkeley: University of California at Berkeley, University of California Transportation Center and the Center for Real Estate and Economics.

Lannoo, S., Verhaeghe, P.-P., Vandeputte, B., & Devos, C. (2012). Differences in social capital between urban and rural environments. *Journal of Urban Affairs*. (Early view http://onlinelibrary.wiley.com/doi/10.1111/j.1467–9906.2011.00592.x/).

Levine, J., Esnard, A-M., & Sapat, A. (2007). Population displacement and housing dilemmas due to catastrophic disasters. *Journal of Planning Literature*, 22(1), 3–15.

Lu, J-C., Peacock, W.G., Yang, Z., & Dash, N. (2007). A comparative study of single family and multifamily housing recovery following 1992 Hurricane Andrew in Miami-Dade County, Florida. In *Proceedings of the Hazards and Disasters Researchers Meeting* (pp. 22–26). Boulder, CO.

Morrow, B.H. (1999). Identifying and mapping community vulnerability. *Disasters*, 23(1), 1–18.

Morrow, B.H., & Enarson, E. (1996). Hurricane Andrew through women's eyes: Issues and recommendations. *International Journal of Mass Emergencies and Disasters*, 14(1), 1–22.

Mueller, E.J., Bell, H., Chang, B.B., & Henneberger, J. (2011). Looking for home after Katrina: Postdisaster housing policy and low-income survivors. *Journal of Planning Education and Research*, 31(3), 291–307.

Nakagawa, Y., & Shaw, R. (2004). Social capital: A missing link to disaster recovery. *International Journal of Mass Emergencies and Disasters*, 22(1), 5–34.

NRC (National Research Council). (2006). *Facing hazards and disasters: Understanding human dimensions*. Washington, DC: The National Academies Press.

Olshansky, R., & Johnson, L. (2010). *Clear as mud: Planning for the rebuilding of New Orleans*. Chicago: American Planning Association.

Peacock, W.G., Dash, N., & Zhang, Y. (2007). Sheltering and housing recovery following disaster. In H. Rodríguez, E.L. Quarantelli, & R.R. Dynes (Eds.), *Handbook of disaster research* (pp. 258–274). New York: Springer.

Peacock, W.G., & Girard, C. (1997). Ethnic and racial inequalities in hurricane damage and insurance settlements. In W.G. Peacock, B.H. Morrow, & H. Gladwin (Eds.), *Hurricane Andrew: Ethnicity, gender and the sociology of disasters* (pp. 171–190). London and New York: Routledge Press.

Peacock, W.G., Zhang, Y., & Dash, N. (2005). *Single family housing recovery after Hurricane Andrew in Miami-Dade County*. College Station, TX: Hazard Reduction and Recovery Center, Texas A&M University.

Quarantelli, E.L. (1982). *Sheltering and housing after major community disasters: Case studies and general observations*. Columbus: Ohio State University.

Quarantelli, E.L. (1995). Patterns of sheltering and housing in U.S. disasters. *Disaster Prevention and Management*, 4(3), 43–53.

Quigley, W.P. (2007). Obstacle to opportunity: Housing that working and poor people can afford in New Orleans since Katrina. *Wake Forest Law Review*, 42, 393–419.

Régnier, P., Neri, B., Scuteri, S., & Miniati, S. (2008). From emergency relief to livelihood recover: Lessons learned from post-tsunami experiences in Indonesia and India. *Disaster Prevention and Management*, 17(3), 410–429.

Richardson, H.W., Gordon, P., & Moore, J. (2008). *Natural disaster analysis after Hurricane Katrina: Risk assessment, economic impacts and social implications*. Northampton: Edward Elgar.

Rubin, C.B., Saperstein, M.D., & Barbee, D.G. (1985). *Community recovery from a major natural disaster.* Monograph no. 41. Boulder, CO: University of Colorado, Program on Environment and Behavior, Institute of Behavioral Science. Retrieved November 27, 2009 from http://desastres.unanleon.edu.ni/pdf/2002/diciembre/pdf/eng/doc2510/doc2510.htm.

Saenz, R., & Peacock, W.G. (2006). Rural people, rural places: The hidden costs of Hurricane Katrina. *Rural Realities*, 1(2), 1–11.

Saulny, S. (2006). Clamoring to come home to New Orleans projects. *New York Times*, June 6. Retrieved January 17, 2011 from http://www.nytimes.com/2006/06/06/us/nationalspecial/06housing.html.

Tierney, K.J., Nigg, J.M., & Dahlhamer, J.M. (1996). The impact of the 1993 Midwest floods: Business vulnerability and disruption in Des Moines. In R.T. Sylves & W.L. Waugh (Eds.), *Disaster management in the US and Canada*, 2nd edition (pp. 214–233). Springfield, IL: Charles C. Thomas.

Vale, L., & Campanella, T.J. (Eds.) (2005). *The resilient city: How modern cities recover from disaster.* New York: Oxford University Press.

Vigdor, J. (2008). The economic aftermath of Hurricane Katrina. *The Journal of Economic Perspectives*, 22(4), 135–154.

Weil, J.H. (2009). Finding housing: Discrimination and exploitation of Latinos in the post-Katrina rental market. *Organization and Environment*, 22(4), 491–502.

Wisner, B. (2003). Disaster risk reduction in megacities: Making the most of human and social capital. In A. Kreimer, M. Arnold, & A. Carlin (Eds.), *Building safer cities: The future of disaster risk* (pp. 181–196). Washington, DC: World Bank.

Wu, J-Y., Chen, L-C., Liu, Y.C., & Tszeng, J-S. (2007). Relationship between the speed of post-disaster condominium housing reconstruction and household characteristics. In *Proceedings of the Hazards and Disasters Researchers Meeting* (pp. 35–37). Boulder, CO.

Wu, J-Y., & Lindell, M.K. (2004). Housing reconstruction after two major earthquakes: The 1994 Northridge Earthquake in the United States and the 1999 Chi-Chi Earthquake in Taiwan. *Disasters*, 28(1), 63–81.

Xiao, Y. (2011). Local economic impacts of natural disasters. *Journal of Regional Science*, 51(4), 804–820.

Zhang, Y., & Peacock, W.G. (2010). Planning for housing recovery? *Journal of the American Planning Association*, 76(1), 5–24.

Part II

Taking Action in Uncertain Environments

Connecting Resilience to Hazards

Ann-Margaret Esnard

The exploration of resilient communities through the lens of hazards and disasters exposes a set of multi-layered, interrelated, complex, and uncertain phenomena. The chapters in this section of the book deal with linkages between hazards and community resilience, and provide important research findings relevant to a multi-disciplinary audience of policy makers, practitioners, and researchers. The authors of the chapters also offer targeted recommendations and advice for planners, managers, and policy makers in rural communities.

Brody and Gunn's chapter reminds us about the importance of natural moderators of coastal hazards (e.g., soils and functional intact wetland), and that such natural assets should not be undermined given well-known land development patterns, urbanization trends, and rural-to-(sub)urban land conversions that can accelerate flooding. Coastal communities, faced with challenges of climate change and sea level rise, also have to recognize that conditions, social values, and perspectives change over longer term horizons of decision making, and that monitoring, continuous learning, and adaptation are necessary due to changing conditions and the need to re-evaluate interventions. The authors' broader views of "uncertainty in the social system" and attempts to account for the goals, priorities and perceptions of current and future stakeholders are especially apropos. It appears that these issues remain understudied in rural communities that face their own challenges and opportunities. For example, our cumulative public policy and planning interventions lead in some cases to adaptation, which itself raises a separate question about the interplay between adaptive actions and resilience, how they are different, whether one is a process and the other is an outcome, whether we are indeed becoming more resilient, or whether we are simply getting better at adapting. The "community resilience" puzzle is therefore incomplete without attention to planning, policy, and politics that affect what Passerini (2010) refers to as the "common good." She reminds us that while natural disasters are not a physical entity, they, like other commons, represent dynamic systems of human-natural interactions. She also notes that the "common good" can include safety, quality of life, building codes, and land use codes (Passerini, 2010: 316).

Well-designed policies undergirded by strong political institutions at the local and state levels can potentially increase resilience and reduce vulnerability to hazards (natural, physical, and socio-economic). We are reminded in the chapter by Hawkins that planning policy and implementation is essential for building community resilience given the implication for shaping land development patterns and reducing the risk to populations within hazardous areas. Additionally, there is a need for shared governance, as well as the need for local governments to find ways to value these local institutions, and incorporate them into governance systems (see Chapter 11 by Shaw). The private sector (broadly construed to encompass businesses, non-governmental organizations, volunteer, academic and technical institutions, faith-based organizations, and other civic-minded organizations) is part of this shared governance system. It has become increasingly clear that both the private and public sectors "have resources, capabilities and access to different parts of the community with collective efforts to identify interdependencies, needs, and resources in advance, a community can significantly improve its disaster resilience" (NRC, 2011: 5). As such, there is the need to rethink public and private sector roles particularly when formulating and implementing community resilience strategies, and particularly when addressing public critical infrastructure risks.

Strategic efforts must recognize three important facets of the "community resilience" puzzle, what is referred to here as the three Cs: communities, collaboration, and commitment. Despite the variation in contexts and meanings, facet one, *community,* is an underlying but subtle point of convergence in the chapters. It must be acknowledged however that the term "community" is a vague concept, often generalized for simplicity and traditionally defined based on geographic boundaries, culture, interests, or organizations (Passerini, 2010). However, this conceptualization continues to evolve among disaster planning and policy scholars. Community is defined as a group of people for whom the domain of interest is relevant (NRC, 2011), bonded where for our purposes the domain of interest is resilience. A more nuanced conceptualization offered by Passerini (2010: 310) is that of a loosely bonded "mosaic of communities" because "people living in the same geographic area can have vastly different values, sense of community, access to services, interests, religions, or sense of obligation to others." From a community stakeholder perspective, May refers to "communities of interest" among rural sectors in his study area; Shaw refers to "human resources/roles of communities and local institutions"; and Deyle and Butler refer to "communities of inquiry." From the geographical and regulatory perspective, Brody and Gunn refer to the Community Rating System (CRS) program in their discussions on minimizing flood loss, specifically that non-structural mitigation techniques under the FEMA CRS program can significantly reduce flood losses in urban settings, and should be a key consideration for expansion into rural jurisdictions. An inherent challenge is therefore working toward shared values and visions for the "common good" or for the goal of sustainable resilient communities.

Facet two, *collaboration*, is another essential piece of the community disaster resilience puzzle whereby collaborators need to "recognize that no person or entity has all the expertise, insight, information, influence, or resources to build community resilience" (NRC, 2011: 57). This sentiment is another underlying theme in the chapters in this section of the book, specifically with regard to broadening collaboration and stakeholder groups. May's chapter highlights the importance of the private sector, including the food and agriculture sector and trade and professional associations.

Brody and Gunn by default remind us about the importance of environmental planners, and our colleagues with training and expertise in ecological science and soil science. Deyle and Butler call directly for broader involvement of landowners, business owners, environmentalists, advocates of social equity, while Hawkins showcases the important role that a network of emergency management and disaster planning organizations play in development of comprehensive plans and policies that mitigate the effects of disasters. These networks are also instrumental for information exchange, diffusion, and dissemination.

Shaw's chapter has detailed case studies that among other things highlight the role of the education department, volunteer organizations, neighborhood associations' youth, and local media. Although focused on Japan, it provides useful lessons for rural aging communities that have or continue to experience sharp declines in population and economic stability. The "community-centered, human-resource" approaches to disaster risk reduction are particularly instructive and applicable. The Japanese case studies are highly demonstrative of potential approaches and tools that can be considered by policy makers and practitioners who wish to "focus on the roles of people and communities" and the roles of "local institutions" rather than a sole focus on "structural" measures. Additionally, Deyle and Butler remind us that it will become increasingly important that planners and managers relinquish some control, particularly if an adaptive/flexible governance approach is to work in situations that require long-term horizons of decision making.

This broad collaborative stakeholder network is described by the NRC (2011) as the "full fabric of the community." Similarly, Smith's disaster recovery assistance framework (Smith, 2011) is a broad (but fragmented) network and has relevance in the resilience context, because at its core it gets at institutions, networks, coordination, and negotiation. That network is described as comprising the following: "Public sector organizations (federal, state and local governments); quasi-governmental and non-governmental organizations (community development corporations, homeowner's associations, special service districts, regional planning organizations, professional associations, and colleges and universities); non-profit relief organizations (national organizations, community organizations, and foundations); private sector organizations (financial and lending institutions, consultants, developers, construction and engineering firms, insurance companies, businesses and corporations, and the media); international relief organizations and nations; and emergent groups and organizations" (Smith, 2011: 11–12).

Overall, there is increased awareness of the importance of broad collaborative networks. Ideal collaborative networks are also difficult to define, although there have been efforts made to provide guidelines for community-based private-public collaboration (NRC, 2011). What has become increasingly clear however is that awareness cannot be equated with action, which in turn cannot be characterized as commitment. Commitment is usually discussed in the context of mandates, and commitment at the local level (particularly in rural communities) to invest in protective actions and support mandates that will increase disaster resilience is generally weak. Some of May's earlier research on mandate design is relevant because it delved into potential actions and incentives that increase or influence local government commitment to and capacity for undertaking actions consistent with mandate goals that influence development in hazard-prone areas (May, 1994). Similarly, research by Smith, Ward, and Berke (n.d.: 24) assessed state capacity to assist local governments in hazard mitigation planning and policy making. They reported that the state hazard mitigation officers were frustrated with many local governments for failing to embrace a greater commitment to the implementation of hazard mitigation measures, including the reluctance of communities to use the results of the risk assessment found in their plans to drive the selection of hazard mitigation projects and policies. Smith et al. (n.d.) also reported a number of factors that have hindered the ability of states to increase awareness and commitment to hazard mitigation at the local level. They include an overreliance on consultants, a lack of hazard mitigation awareness among citizens and elected officials, and the inability to identify a local advocate for hazard mitigation. Ultimately, this discussion comes full circle because greater (and sustained) commitment requires fostering (and sustaining) broad partnerships and collaborations for the "common good" of our "mosaic communities."

REFERENCES

May, P.J. (1994). Analyzing mandate design: State mandates governing hazard-prone areas. *Publius*, 24(2), 1–16.

NRC (National Research Council). 2011. *Building community disaster resilience through private-public collaboration*. Washington, DC: The National Academies Press.

Passerini, E. (2010). The nature of human communities. In N.D. Phillips, D.S.K. Thomas, A. Fothergill, & L. Blinn-Pike (Eds.), *Social vulnerability to disasters*. Boca Raton, FL: CRC Press, Taylor and Francis Group.

Smith, G. (2011). *Planning for post-disaster recovery: A review of the United States disaster assistance framework*. Washington, DC: Island Press.

Smith, G., Ward, W., & Berke, P.R. (n.d.) The role of hazard mitigation planning in building local capacity and commitment: A tale of six states. Working paper. Chapel Hill: University of North Carolina. Retrieved April 2012 from http://www.ie.unc.edu/cscd/projects/pdf/Smith_Lyles_Berke_State_Case_Working_Paper.pdf

7 Public Risks and Disaster Resilience

Rethinking Public and Private Sector Roles

Peter J. May

INTRODUCTION

The specter of energy blackouts, crumbling infrastructure, and damage from earthquakes, flooding, and tornadoes is all too common. Notable instances of each type of peril in the U.S. are reminders of the need to reduce the potential for further harms. But, the ability to do so is limited given the diffuseness of the risks and the costs that are required to achieve meaningful risk reduction. These constraints are especially evident for rural areas that have more limited resources and abilities to undertake risk-reduction efforts. Lack of progress in reducing risks undermines the resilience of rural communities to damaging events. This chapter considers the design and implementation of policies for addressing risks as a means of enhancing community resilience.

Risks are typically conceptualized as the likelihood of harm that is associated with some physical phenomenon, human activity, or technology. Risks vary considerably with respect to the likelihood of harm and people's perception of both the likelihoods and the ensuring harms. Risks that occur often and for which the impacts can be averted through protective actions taken by individuals and firms (crime, car accidents, workplace injuries, household mishaps, and so on) are one end of a continuum of potential harms. The other end comprises "public risks" that are not so apparent or common and, as a consequence, are not as readily addressed. Peter Huber (1986) notes such risks "are centrally produced or mass-produced, broadly distributed, often temporally remote, and largely outside the individual risk bearer's direct understanding and control" (90). The public label underscores the largely non-excludable nature of the sources of harms and the non-rival aspects of their impacts. The likelihood, timing, and location of catastrophic occurrences for such risks are hard to pinpoint in advance. Because of this, the risks of harm to particular individuals or entities are best characterized as low probability but of potentially high consequence.

Clearly, not all public risks are the same nor do they invoke the same degree of harm. Consider three different examples. One is the potential for earthquakes in rural areas that do not have an extensive history of earthquakes. This is illustrated by the August 2011 central Virginia 5.8-magnitude earthquake and the November 2011 Sparks, Oklahoma 5.6-magnitude earthquake. Each was in an area of known seismic hazard but with limited history of notable events. The structural damage to unreinforced masonry buildings from the Virginia earthquake was substantial while that from the Oklahoma earthquake was more limited. A second scenario is the contamination of animal feed manufactured in the United States as happened in 2006 with melamine-tainted feed for livestock, fish, and shrimp. This led to an extensive recall of the feed materials, disruption of the food supply chain, and negative economic impacts to an important rural economic sector. A third scenario is the 2003 power outage that affected some forty-five million people in eight U.S. states. The blackout had cascading consequences that included disruptions to power and water supplies, transportation services, cellular communications, and manufacturing. The blackout and these disruptions affected urban as well as rural areas.

Each of these scenarios involved low-probability events for which it would have been hard to predict the likelihood of damaging impacts for any given location or entity. Each of these events affected rural areas in previously unforeseen ways. Given the diffuseness of the harms, it is difficult to avert them in the first place or to increase the resilience of places and people to such events. Making governments, private, or individual actors—the cities and towns of central Virginia or rural Oklahoma, livestock and other food producers, and electricity producers—aware of such uncertain risks and responsible for anticipating and preventing harm hardly seems sufficient for addressing the risks. Simply put, relatively few would take the requisite actions to reduce risks to a tolerable level or to substantially enhance resilience.

Why is this? The answer is that awareness of risks is in itself insufficient to motivate widespread action. Camerer and Kunreuther (1989), discussed in what follows, show there are noteworthy cognitive, behavioral, and organizational limits to the willingness of individuals and organizations to take protective actions for addressing low-probability, high-consequence risks. Furthermore, the isolated efforts of some will be diminished if not negated because of the unwillingness or inability of others to act. The ability of the federal government to induce or otherwise compel risk-reduction actions on the part of state and local governments and the private sector is limited. Much rests on the commitment and capacity of these entities. Given that these tend to be lower for rural communities, residents, and businesses, the challenges of bringing about risk reduction in rural areas is arguably more difficult than in other parts of the country.

As is evident from the chapters in this volume, improving disaster resilience entails a wide range of actions that either directly or indirectly strengthen the physical and social fabric of communities. Two key aspects of the physical fabric of any community are the buildings in which citizens live, work, and conduct business and the infrastructures that provide life support to a community. Efforts to enhance community resilience must address each of these by making buildings more resistant to disasters and by improving the resilience of key infrastructures. The first rests mainly on the adoption and enforcement by state and local governments of regulatory provisions for reducing risks and enhancing public safety. The second rests on motivating those private sector entities that manage and own various infrastructure to take appropriate actions to reduce risks and enhance infrastructure resilience.

The discussion of these approaches and the empirical findings about them draw from two studies that I have undertaken with others. One was a national study of building code enforcement conducted in the mid-1990s that included surveys of state building code agencies and of a national sample of local jurisdictions (May & Burby, 1998). Though the data for that study are now dated, they are still relevant for illustrating the factors that limit the implementation of regulatory provisions. The differences between larger and smaller jurisdictions and more urban and rural states are particularly notable. The second study is a study of critical infrastructure protection policy in the United States (May & Koski, 2012). This research highlights the challenges of fostering lasting public-private partnerships for addressing public risks with particular attention in this chapter to those sectors of greatest relevance to rural areas—agriculture and food, dams, energy, and transportation.

POLICY DESIGN AND IMPLEMENTATION CHALLENGES

Two challenges establish the broad contours for considering policies for reducing risks. One is the difficulty of stimulating appropriate actions among less motivated individuals and entities that would be better off in taking appropriate risk-reduction efforts. The second is the set of multi-level governance problems brought about by the fact that the federal government has inherently indirect influence over state, local and private sector entities that are central to risk reduction.

Unwillingness to Act

Despite the reminders of the potential for disruption of our lives and destruction of property from public risks noted at the outset of this chapter, there is not much of a constituency pushing for efforts to reduce these risks beyond experts within relevant policy communities. In commenting about the absence of broader publics in such policy making, May (1991b) notes: "Policy problems are perceived as remote; action in addressing the problem or lobbying for governmental action is relatively costly; and benefits from taking action are either relatively remote or sufficiently diffuse to prevent individualized action" (194). The lack of relevant publics is especially notable in rural areas where, perhaps except for agricultural organizations, there are not well-established advocacy groups or professional associations (e.g., engineering and planning professionals) who are likely to band together to advocate for risk-reduction efforts.

Why do people and organizations tend not to take protective actions? The answer to this question is provided by extensive research about decision making regarding low-probability, high-consequence events. In summarizing the relevant literature, Howard Kunreuther (2006) provides four sets of reasons why residents of hazard-prone regions tend not to voluntarily undertake loss prevention measures. One reason is underestimation of relevant probabilities. The evidence noted by Kunreuther suggests people do not seek out additional information or adequately process information about probabilities in making decisions about low-probability risks regardless of consequence.

A second reason for inaction noted by Kunreuther (2006) is the short time horizon of most people in making decisions. The myopia is framed in a time horizon of a couple of years. This is inadequate to justify decisions for which the costs are upfront and the benefits accrue probabilistically over a longer time horizon. Related to this is a third reason in people's "mental accounting" for expenditures. Kunreuther cites studies that suggest people, regardless of income, often think in terms of different categories of expenditures for which protective actions have relatively low fixed limits.

A fourth reason cited by Kunreuther (2006) is based on newer research in behavioral economics that relates to interdependencies in decision making for protective actions for diffuse risks. Not only do people base their decisions on what others do (friends, relatives, neighbors) but those decisions interact in ways that can diminish or defeat isolated protective actions. A homeowner in a flood-prone area might be reluctant to elevate the home for flood protection if no others nearby have done so.

These behavioral and cognitive limits to decision making about protective actions for low-probability, high-consequence risks are not limited to protective actions taken by individual homeowners or residents. The main barrier can be summed up as a commitment problem for many businesses, organizations, governmental entities, and communities that are unwilling or unable to take sufficient protective actions to prevent harms from public risks or to mitigate their impacts. This problem has been noted in research addressing a number of public risks that include mitigation of natural hazards (Burby & May, 1998), earthquake risk reduction (Berke & Beatley, 1992; May & Birkland, 1994; May & Feeley, 2000), and local commitment to climate change (Zahran et al., 2008).

Multi-level Implementation Complexities

Risk-reduction efforts for public risks require actions at multiple levels of government. For example, efforts to avert losses from natural disasters require changes in land use and development patterns by state and local governments and their establishment and enforcement of building regulations (Burby et al., 2000). Given this, any national risk-reduction policy must address overlapping authorities and responsibilities. Shared governance is the norm rather than the exception. As studied by May and Williams (1986) for disaster policies, the nature of this sharing can be quite varied depending on the level of government that takes the lead and the respective levels of expertise.

The commitment problem is an obstacle given the reluctance of subnational governments to be proactive in mitigating risks and the limited leverage that the federal government has to compel or otherwise induce action by them. A number of risk-reduction programs exist regarding such things as state planning for earthquake risks, local floodplain management under the national flood insurance program, and local climate change efforts. But, the implementation of these is uneven and their impacts are highly disjunctive (Burby, 2001; May, 1991a; Zahran et al., 2008). The shared governance issues are compounded when one considers efforts of states to induce local risk-reduction efforts, given the more limited resources of many states and their reluctance to compel local governments into acting.

Added to this mix is the necessity for extensive involvement of the private sector in addressing many public risks. Enhancing the resilience of transportation systems, energy systems, or other infrastructure requires investments in improvements on the part of numerous private entities (NRC, 2011). As with the challenges of shared governance among multiple levels of government, public and private partnerships establish differing expectations and complications for policy implementation (Linder, 2000).

Addressing These Challenges

The preceding discussion points to substantial cognitive, behavioral, and organizational limits to the willingness of individuals and organizations to take protective actions for addressing low-probability, high-consequence risks. The discussion also suggests a number of potential pitfalls in attempting to bring about concerted action in addressing such risks among multiple levels of government and the private sector. Efforts to design policies for addressing public risks are further limited by the diffuseness of the risks, the variety of and uncertainty over potential solutions, and the costs that are entailed.

Two different strategies for building commitment and capacity for disaster resilience are considered in what follows. One is the adoption and enforcement by state and local governments of regulatory provisions for reducing risks and enhancing public safety. The second is the creation of planning partnerships for defining and implementing common sets of solutions for risk reduction. Enhancing the resilience of the building stock entails a traditional regulatory approach involving governmental action.

Enhancing the resilience of infrastructure entails public and private planning partnerships aimed at inducing private action.

State and Local Protective Regulation

The central role of state and local governments in public safety is well enshrined in American government. That role includes building codes, planning and zoning requirements, and provisions for policing and emergency management. The federal government has had important roles in stimulating new code provisions and, especially after the September 11 attacks, in funding state and emergency planning and preparedness programs (Eisinger, 2006; Gerber, 2007; May 1991a). Nonetheless, states and by extension local governments decide the contours of building safety and land use requirements and are responsible for implementing them. This leads to substantial variation in each both across states and among local jurisdictions. Regulatory provisions regarding building safety are particularly pertinent to risk reduction because various code provisions promote protection from earthquakes, fire, floods, and tornadoes.

States typically adopt by reference private, consensus-based code provisions such as the International Building Code promulgated by the International Code Council and the National Fire Protection Association's standards regarding fire safety. Often states amend these codes or adopt portions of them. States also establish administrative procedures for updating state requirements to reflect changes in the codes. Some states, such as New York and Wisconsin, have written their own building provisions.

State policy makers mandate local governmental compliance with the state building code, establish an enabling set of state provisions that requires adherence with the state code if a locality adopts a building code but does not require such adoption, or they choose more limited provisions. With respect to local enforcement of building codes, states either authorize local enforcement for those localities that adopt state codes, assign primary responsibility to local governments for code enforcement, or have procedures for delegating local governmental enforcement. States also differ with respect to the authority of state officials to revoke local governmental authority to issue building permits, or to otherwise pre-empt that role, if a local government fails to adequately enforce the state building requirements. All of this adds up to much variation in the regulation of building safety by state and local governments. Of interest for this chapter is how that variation plays out as it relates in particular to rural areas and what that means for risk reduction in such areas.

The data for this study are comprised of information about state policies for building regulation and local government enforcement efforts as of the mid-1990s. State policies governing the regulation of buildings were characterized by collecting and coding relevant state legislation and administrative statutes for the thirty-three states with statewide provisions.[1] A survey of state building code agencies for these thirty-three states was undertaken to identify the extent to and ways with which the state agencies undertake oversight of code enforcement by local governments. The data about local code enforcement come from a mail-out survey of a national sample of local government building officials (or the relevant lead building authority) also undertaken in 1995 that was a basis for research on improving regulatory compliance with building code provisions (Burby, May, & Patterson, 1998; May & Burby, 1998). The sample frame included 990 cities and counties throughout the country, originally based on a complex weighting scheme to reflect population size and seismic risk. Responses were obtained from 819 governments (83% response rate) for which the analyses that follow compare results for the 138 jurisdictions that have less than 10,000 residents with 681 jurisdictions that have larger populations. This cutoff for smaller jurisdictions is based on balancing the need for sufficient sample size with a desire to include jurisdictions that are located in rural areas.[2] The measures for characterizing state and local efforts are discussed with the findings that follow and are detailed in an appendix.

Building Regulation in Rural and Other States

Given that states have discretion in deciding the extent of building regulation, one key issue is the extent to which regulatory provisions of states with relatively larger rural populations differ from those for states with more urban populations. To get at this, I classified rural states as those with 40% or more of their 1990 population designated by the U.S. Census as rural. Table 7.1 provides a comparison of the extent of building regulations for the rural states (underlined) with other states (not underlined). The four categories of states describe the extent to which states mandate and oversee local building regulation based on a cluster analysis of key attributes of state building provisions and oversight of local implementation (May, 1997).

The seventeen minimalist states are those without state codes or provisions that are limited to restricted situations (e.g., state provisions governing schools, public buildings, or state buildings). The eight enabling states are those with only enabling provisions, authorization of local governments to enforce the state code if they choose to do so, and limited or no state review of code administration by local governments. The thirteen mandatory states are those with mandatory local codes but limited state review or oversight of actions taken by local governments. The twelve energetic states are those with mandatory local codes and state oversight of enforcement of codes by local governments.

States with larger rural populations are far less likely to have mandatory or energetic state building regulation than the other states. Twenty-seven percent of the rural states fall into these categories of state regulation whereas 60% of the states with more urban populations fall into the more extensive categories of state building regulation. The differences in state

Table 7.1 Rural versus Other States' Building Regulation

Categories of States [a]			
Minimalist	*Enabling*	*Mandatory*	*Energetic*
Alabama	Arkansas	California	Alaska
Arizona	Georgia	Florida	Connecticut
Colorado	Idaho	Indiana	Kentucky
Delaware	Iowa	Massachusetts	Michigan
Hawaii	Louisiana	Maryland	Montana
Illinois	Minnesota	New Mexico	North Carolina
Kansas	Nebraska	Nevada	New Jersey
Maine	West Virginia	Rhode Island	New York
Missouri		Utah	Ohio
Mississippi		Virginia	Oregon
North Dakota		Washington	Tennessee
New Hampshire		Wisconsin	Vermont
Oklahoma		Wyoming	
Pennsylvania			
South Carolina			
South Dakota			
Texas			

[a] Categories of states complied by author from cluster analysis of state regulatory provisions (see May, 1997). States that are underlined have relatively larger rural populations (40% or greater rural makeup) than those that are not underlined.

building regulation, like much state regulation, stem from a combination of more limited interest pressures for regulation, ideological pre-dispositions of state legislatures against strong regulatory roles, and lower wealth (May, 1997).

Local Jurisdictions' Building Code Enforcement

Regardless of state building regulatory provisions, local jurisdictions differ in the extent to which they enforce building codes. Of interest is how smaller jurisdictions as a proxy for those in rural areas differ from larger jurisdictions in this regard. Table 7.2 shows a comparison of key differences for larger (greater than 10,000 population) and smaller (less than 10,000 population) jurisdictions based on the 1995 national survey of local building authorities. The top half of the table considers different measures of enforcement effort based on respondent reporting for each item. The bottom half of the table considers the enforcement environment based on each respondent's rating of the degree to which different categories of interests advocate for or against stronger code enforcement in the jurisdiction. The details of the measures are explained in Appendix 2.

Table 7.2 Local Jurisdictions' Building Code Enforcement

	Comparison of Means [a]		
	Larger Jurisdictions (pop > 10,000)	*Smaller Jurisdictions (pop < 10,000)*	*t-test p-values* [b]
Code Enforcement Measures			
Enforcement effort (1–5)	3.56 (.02)	3.34 (.07)	<.01
Code budget priority (1–5)	2.74 (.03)	2.53 (.07)	<.01
Perceived capacity (1–5)	3.66 (.03)	3.46 (.09)	.03
State technical assistance (0–2)	1.30 (.03)	1.14 (.07)	.05
Enforcement Environment			
Interest demand (0–100)	35.76 (1.03)	21.16 (2.12)	<.01
Interest opposition (0–100)	20.73 (.84)	18.37 (1.90)	.26
Ratio demand to opposition	17.59 (1.09)	10.92 (1.91)	<.01
Number of Observations	681	138	

[a] Cell entries are mean values for the designated variables and categories with standard errors shown in parentheses.
[b] Independent sample t-test with two-tailed p-values reported not assuming equal variances.

These findings show that larger jurisdictions have on average greater enforcement efforts, stronger budget priorities for code enforcement, and greater perceived staff capacities for this function. As well, they rely on average to a greater degree on technical assistance from state authorities. The mean differences are statistically distinguishable but they are not that large—comprising differences of 6% (perceived capacity) to 14% (state technical assistance). Nonetheless, the differences show that smaller jurisdictions systematically have less capacity and commitment to building code enforcement.

The findings in the lower part of Table 7.2 about the enforcement environment suggest how the interest-group environment shapes the willingness of local jurisdictions to enforce codes. The degree of demand for stronger code enforcement is shown to be a notable difference between larger and smaller jurisdictions. The former have a 69% higher degree of interest demand than the latter. Our earlier work on this topic suggests that opposition to code enforcement is relevant, but only as it is compared with the strength of demands for it (Burby et al., 1998). The difference in this ratio for the non-rural versus rural comparison also stands out with the former having a 61% higher ratio than the latter.

As with the state level, a number of factors constrain strong code enforcement at the local level, particularly in rural areas. One key consideration that follows from the findings about the state level is the lack in many instances of a strong state mandate to adopt and enforce building regulations. Without this and without much of a consistency pressing for such action, elected officials in rural areas are hard-pressed to make this a priority, a concrete manifestation of the commitment problem. Resistance to new codes or stronger enforcement is often based on concerns of homeowners and other building owners that they will have to bring their structure into compliance with the new codes when that is not realistic or financially feasible. There are provisions in national, model codes for application of reduced standards in such situations. But, it takes technical expertise to interpret and apply these provisions. Such limitations in staff capacity for building code enforcement in many rural jurisdictions further constrain the ability to make the building stock more disaster-resistant.

PLANNING PARTNERSHIPS FOR CRITICAL INFRASTRUCTURE PROTECTION

The regulation of building safety is based on traditional forms of regulation. Given the extensive role of the private sector in many aspects of risk reduction, it is important to consider non-coercive ways of motivating private entities to take risk-reduction efforts. One such example is the federal approach to critical infrastructure protection that is based on intergovernmental and private sector planning partnerships (Koski, 2011). The details of this approach to critical infrastructure protection are found in President Bush's *National Strategy for the Physical Protection of Critical Infrastructure and Key Assets* (U.S. Office of the President, 2003) and the Department of Homeland Security's *National Infrastructure Protection Plan* (2006, updated 2009). These documents identify eighteen different sectors for critical infrastructure planning partnerships.[3] The discussion that follows addresses four sectors that directly relate to the economic health of rural areas: Agriculture and Food, Dams, Energy, and Transportation (May & Koski, 2012). In total, the national plans call for actions by nine lead federal agencies as augmented by efforts of thirty-six federal agencies, by all states and most municipalities, and the concerted involvement of some 230 different professional associations and trade groups.

The production of plans and the planning process are means of inducing action for addressing risks to critical infrastructures. The ability to sustain and grow these efforts rests on the networks of actors that are fostered and their willingness and ability to attend to the risks at hand. Chisholm (1989) argues that formal coordination mechanisms typically fall short for such disconnected entities unless there is a common sense of purpose. In particular, he suggests "coordination is an activity that is understood to occur only after sufficient *community of interest* develops among the relevant parties, forming a foundation for later efforts" (33, emphasis added). Like Chisholm, O'Toole (2003) argues that success in addressing complex problems that entail multi-actor solutions comes from building communities of interest that work toward common goals.

A community of interest provides a hidden-hand coordination mechanism that is animated by shared purpose and common interests among affected entities. Though somewhat amorphous in concept, communities of interest are akin to policy communities that have more or less regularized players who address issues of common concern (Rhodes, 2008) and to epistemic communities that are linked by common knowledge of a given phenomenon (Haas, 1992). Members of a community of interest that are potentially vulnerable to a public risk, firms concerned about tainted feeds, managers of water systems concerned about distribution disruptions, city officials concerned about natural disasters, and so on, are linked by common fates and interdependencies among the respective firms or entities. Communities of interest surrounding public risks like threats to critical infrastructure are fluid. Membership is by affinity and at arms-length rather than through formalized voluntary associations or networks. A given community expands or contracts depending on the issues of the day, the perceived immediacy of the need to act, and the salience of the risk at hand.

Professional and trade associations provide the nuclei for creating communities of interest. They provide linkages among association members and a basis for mobilizing action among the members. In addition, such associations have legitimacy within their respective industries and have the potential to be important information brokers. But, relevant professional associations differ in their commitment to the goals for a given community and their willingness to be central players. This leads to consideration in what follows of the makeup of different communities of interest for rural-related infrastructure protection and the degree of engagement of relevant professional and industry associations.

The basic data for getting at this come from annual reports about each sector's coordinating council, comprised of private sector and other non-governmental entities, as reported annually by the Critical Infrastructure Partnership Advisory Council—CIPAC (DHS, 2010; Koski, 2011). The composition of each sector's coordinating councils is characterized from membership lists that were obtained from the annual CIPAC reports for 2008 through 2010. For the four sectors considered here, there are a total of 226 participating firms, trade associations, professional associations, or other entities. The characterization of the engagement of trade and professional associations is based on a more in-depth coding of the extent to which critical infrastructure issues appear as important agenda items for each association.

Makeup of Planning Partnerships

The foundation of the community of interest is its members. The variation in the membership of sector coordinating councils considered here is broad. Table 7.3 shows this for each sector's coordinating council when

Table 7.3 Participation in Rural-Related Sector Coordinating Councils

	Critical Infrastructure Protection Sector [a] *(Percent distribution for each sector)*			
Sector Composition [b]	*Food & Agriculture*	*Energy*	*Transportation*	*Dams*
Business	28	51	47	65
Industry Association	70	47	51	9
Other Association	2	0	1	18
Other Entities	0	2	1	9
N of members [c]	60	53	79	34

Notes:
[a] Each column is one of the defined rural-related critical infrastructure protection sectors.
[b] Columns show the percentage distribution of participating entities across categories of coordinating council membership. Businesses are individual firms or enterprises. Industry associations are industry or trade associations. Other associations are non-business-related professional associations. Other entities are non-business-related entities such as research entities.
[c] Number of organizational members in each CIP sector's coordinating council. Total N = 226 entities.

categorizing the participating entities as a business, industry association, or other association or entity.[4]

Three things stand out in Table 7.3. One is the relatively large number of participating entities in all but the Dams Sector. The Dams Sector is more specialized, for which participation includes a number of energy producers that rely on hydro-power and associations representing national and state regulatory authorities. The other sectors have coordinating councils with diverse membership within each category of entity. As discussed by Koski (2011), members of the sector coordinating councils are selected in a variety of ways. These include self-nomination and identification by lead federal agencies for each sector. Members are also selected for a variety of reasons such as their expertise, their representation of segments of a given sector, and the presumed greater financial impacts to them from harm to infrastructure. These varied selection practices lead to partnerships in which sectors have different starting points in terms of the diversity of their membership and prior working relationships.

A second notable observation from Table 7.3 is the more extensive involvement of business entities and trade associations than other types of organizations. More than three-quarters of the participants are business- or industry-related. This reflects the larger number of such entities for each sector and, presumably, a stronger private interest that businesses have in protecting their property than might be the case for public organizations or communities.

The third observation from Table 7.3 is the dominant role of industry and trade associations. With the exception of the Dams Sector, these constitute nearly 50% or more of the participants in the sector coordinating councils. The involvement of these organizations is important because of their potential role as kernels for fostering broader communities of interest.

Engagement of Industry and Professional Associations

The participation of industry and professional associations in coordinating councils will not catalyze broader communities of interest unless the associations embrace efforts to enhance the resilience of critical infrastructures. To be effective catalysts the associations need to be advocates for protecting critical infrastructures. In order to identify the degree of association engagement, we had research assistants search websites for 256 associations for the fourteen critical infrastructure protection sectors for which associations participated on coordinating councils (May & Koski, 2012). We gauged each association's engagement by the extent to which critical infrastructure issues appeared on an association's website under the main page listings for key issues, as an agenda item at a recent annual meeting, or as part of a recent annual report for the organization. Website entries we coded as having: (1) no mention of the issues; (2) a lesser concern with passing mention in the relevant listing; (3) an intermediate concern as one of a half-dozen to dozen issues or of many agenda items; and (4) a central concern that shows up as a priority issue.[5]

Table 7.4 shows differing levels of engagement among relevant associations for the four rural-related critical infrastructure sectors. The four sectors are ranked from highest to lowest with respect to the mean engagement scores (on the 1 to 4 scale) of all associations within a given sector. The listed associations are those that arguably represent rural interests given that these are more likely to catalyze such interests than broader based associations. Of relevance is the degree of overall sector engagement, the extent to which rural associations are part of each sector, and their degrees of engagement.

The Agriculture and Food Sector stands out as having associations that are more engaged than the other sectors considered in this chapter, a wider set of rural-related associations, and relatively strong engagement among them. This sector is ranked fourth among the fourteen sectors we rated in terms of overall association engagement. Rural-related associations account for 12% of all associations for this sector. On balance, a number of these have influential roles in helping to advocated for protecting the security and safety of agricultural, livestock, and food production systems.

The Transportation and Energy Sectors are ranked eighth and ninth respectively in terms of overall association engagement. Each has relatively few rural-related associations as participants in the sectors. These comprise 11% of all associations in the Transportation Sector and 24% of all associations in the Energy Sector. The two sectors are similar in that rural-related interests seem to be relatively minor roles in contrast to much more

Table 7.4 Rural-Related Association Engagement for Critical Infrastructure Issues

Degree of Engagement [b]	*Participating Rural Associations* [c]
	Rural-Related Critical Infrastructure Sectors [a]
	Agriculture & Food
Strong Engagement	American Feed Industry Association International Dairy Foods Association National Cattlemen's Beef Association National Grain and Feed Association National Milk Producers Federation National Pork Producers Association National Renderers Association
Weak Engagement	American Farm Bureau Federation National Corn Growers Association National Chicken Council National Cotton Council of America National Pork Board
	Transportation
Strong Engagement	National Waterways Alliance Association of American Railroads
Weak Engagement	American Waterways Operators American Short Line and Regional Railroad Association
	Energy
Strong Engagement	National Rural Electric Cooperative Association Interstate Natural Gas Association
Weak Engagement	International Association of Drilling Contractors U.S. Oil and Gas Association
	Dams
Strong Engagement	Association of State Dam Safety Officials National Water Resources Association
Weak Engagement	National Hydropower Association National Mining Association

[a] Listing of the four major rural-related sectors ordered from highest to lowest degree of overall association engagement.
[b] Degree of engagement of the designated rural-related associations based on a coding of relevant websites.
[c] Listing of rural-related industry and professional associations participating in each sector coordinating council.

significant national players in each sector. The one exception is the strong engagement of the National Rural Electric Cooperative Association.

The Dams Sector is notable because of the numerous dams among the nation's 75,000 dams that are located in rural areas. Many of these are earthen dams that were constructed for flood prevention and land conservation purposes. Of these, 8,100 are designated as major dams by the U.S. Army Corps of Engineers that serve hydro-power and flood control purposes (National Atlas, 2012). The overall level of engagement of associations in this sector is low with a ranking of thirteenth out of the fourteen sectors we consider in the larger project. Only nine associations are involved in this sector of which the four listed in Table 7.4 can be considered as representing rural interests. The findings show this sector as being a very nascent community of interest.

The emphasis here is the role of national associations that represent rural interests and that can be key players at the national level in catalyzing efforts to enhance commitment of the private sector to enhancing resilience of relevant critical infrastructures. Not addressed is the potential role of more localized private and public sector entities (local agriculture associations, engineering and planning professionals, drainage and water management districts, and so on) in fulfilling similar roles at the local level in advancing community resilience.

CONCLUSIONS

This chapter has considered the challenges of addressing public risks—widely dispersed risks that have less common occurrences but for which the impacts can be very damaging. The reluctance of many individuals, organizations, and communities to protect against such possibilities suggests raising awareness of the vulnerabilities is an insufficient means for achieving meaningful risk reduction. At the same time, the ability of the national and state governments to induce action is limited by resource constraints and the indirect control that these governments have over lower-level governments and private entities. These challenges are heightened for rural areas of the country because of more limited governmental capacities and fewer organized demands to address the vulnerabilities.

Consideration in this chapter of two approaches for enhancing the commitment and capacity for risk reduction provides an instructive contrast between traditional regulation and less coercive methods for enhancing disaster resilience, between government-led actions and government-assisted actions, and between differing mixes of public and private actions. All of this raises fundamental issues about the appropriate roles of the public and private sector in enhancing disaster resilience.

The findings about building regulation underscore what is well known among scholars who follow state and local protective regulations. There is much state-level variation in the comprehensiveness and stringency of such requirements as well as much local-level variation in the degree of effort that is put into enforcing the requirements. Of interest in this chapter is how states with larger proportions of rural residents and smaller jurisdictions fare in these regards. The short answer is not well. The data show that states with largely rural populations tend to be less proactive in mandating or encouraging local governmental building regulation than more urban-dominated states. Smaller jurisdictions tend to have weaker enforcement efforts and capacity, though the differences are not that great on average. More notable is the much weaker demands by organized interests in smaller communities for strong code enforcement in comparison to those for larger jurisdictions. All of this adds up to a picture of a highly disjunctive state and local governmental protective regulatory system in the United States for addressing risks posed by natural hazards and extreme events (May & Deyle, 1998). Rural areas seem to be at the lower end of the regulatory scale. Though the data from which these observations are drawn are dated, more recent scholarly commentary about this topic suggests little has changed (Burby, 2001; Burby et al., 2000; Zahran et al., 2008).

The findings concerning planning partnerships and the nation's Critical Infrastructure Protection Program show at best nascent communities of interest as they relate to the sectors of most relevance to rural-related interests. The Agriculture and Food Sector stands out as having trade and professional associations that are more engaged than the other sectors considered in this chapter, a wider set of rural-related associations, and relatively strong engagement among them. While this is encouraging in terms of fostering communities of interest around enhancing the resilience of agriculture and food production, the fact remains that this is an enormous challenge. The diversity of entities that make up the "farm to food" system makes it hard to develop a coherent community of interest. The findings for the Transportation, Energy, and Dams sectors show less overall engagement of relevant associations, and particularly of rural-related associations, in helping to foster relevant communities of interest.

These findings are not surprising to observers of efforts to build disaster resilience among rural areas. The lessons are not wholly negative in that some more committed and capable jurisdictions and private sector entities have taken the lead in such efforts. The challenge that is underscored by these findings is overcoming the commitment and capacity problems that undermine the willingness and abilities to take protective actions. What can be done to change the situation?

The answer to this question is not simple. One key step is to encourage states that have weaker mandates governing building codes, land use planning, and other regulatory approaches to strengthen the mandates and to better monitor local government implementation of them. The rub is that these steps fly in the face of the autonomy of states in governing these functions and run counter to the current reluctance of the federal and many state governments to advocate for stronger regulations. As well, the privatization of traditionally governmental functions concerning land use and code enforcement by outsourcing key aspects of these tasks to private entities has undermined awareness of the vulnerabilities of local communities to natural hazards and has diminished the ownership of the problems of addressing them by local governments.

Progress with respect to stronger state regulations and their enforcement at local levels of government will only come from increased demands for them. This underscores the important role of professional associations of planners, civil engineers, and the design professions along with networks of flood management, earthquake planning, and flood planning officials. As shown with the findings reported here, the advocacy of such groups is important for bringing about stronger regulations and better enforcement of them. Stronger advocacy for steps to strengthen building codes and their enforcement and for planning for natural hazards more generally will put pressure on local officials to pay more attention to vulnerabilities from natural hazards and other public risks. This, in turn, will enhance the sense of ownership of the problem and help foster needed capacity for addressing it.

Similar issues arise when considering communities of interest and the protection of critical infrastructures. The federal government cannot mandate greater participation or engagement. But, this is exactly what is required to move from nascent to more proactive communities of interest that work toward common ends in enhancing disaster resilience (Koski, 2011; May & Koski, 2012). There need to be stronger visions for each sector. And, there needs to be greater recognition of the variation in status of each sector's community of interest. Those sectors that lag in development of communities of interest need stronger oversight and incentives. For example, relevant associations can be funded to take a stronger lead in establishing meaningful sector visions, in pulling in relevant participants, and in mobilizing stronger engagement as a means for encouraging their participation. Greater funding of such planning efforts with specific targeting of key associations as evangelists could also provide justification for stronger federal efforts to monitor, assess, and encourage stronger partnership planning for the targeted sectors. This type of accountability would also provide penalties for those associations who accept funding but do not follow through with serious efforts to engage their respective members.

This discussion underscores the fact that devising strategies for addressing public risks requires rethinking public and private sector roles. The coupling of the lack of governmental resources with a philosophical shift about regulation ironically means that increasingly the primary actors involved in addressing public risks are private entities. It is the energy and resources of these entities that need to be harnessed for enhancing disaster resilience. Private entities can be effective advocates in their communities for stronger building codes, land use provisions, and other governmental actions for enhancing disaster resilience. Private sector entities are also key players along with their industry associations in devising and implementing strategies for protecting critical infrastructures that are owned and managed by them. The challenge is to activate a shared sense of responsibility among private entities for fulfilling these roles.

ACKNOWLEDGEMENTS

Two collaborators have been instrumental for the research upon which this chapter is based. Raymond J. Burby of the University of North Carolina initiated the building regulatory study and Chris Koski of Reed College has been a co-investigator for the study of critical infrastructure partnerships. A number of research assistants helped with these studies for which particular thanks is due to Jose Cabrel, Mark Donovan, T. Jens Feeley, Dan Hansen, Ashley Jochim, Barry Pump, Emma Tessier, and Breck Wilmot. Financial support for the building code study was provided by National Science Foundation Grant No. BCS-93311857 (Burby, PI; May, Co-PI). Financial support for the critical infrastructure protection study was provided by National Science Foundation Grant Nos. CMMI-0924227 (Koski, PI) and CMMI-0925306 (May, PI). Neither the NSF nor those who have offered assistance are responsible for the content of this manuscript.

NOTES

1. States for which these functions were limited to regulation of only state-owned buildings or restricted subclasses of buildings (e.g., only schools or places of public assembly) were excluded from further analyses because these states did not undertake oversight of local building regulation.
2. The population size of the cities and counties less than 10,000 residents range from 152 to 9,985 with a mean of 6,047 (median 6,034). The jurisdictions are distributed throughout the country with states with largely rural populations comprising the largest percentage of the subsample—Arkansas with 7%, Idaho 5%, Maine 2%, Montana 7%, Wyoming 2%.
3. The eighteen sectors are: Agriculture and Food; Banking and Finance; Chemical Facilities; Commercial Facilities; Communications; Critical Manufacturing; Dams; Defense Industrial Base; Emergency Services; Energy; Government Facilities; Healthcare and Public Health; Information Technology; National Monuments and Icons; Nuclear Reactors, Materials, and Waste; Postal and Shipping; Transportations Systems; and Water Systems.
4. The type of each sector's council members was coded by two researchers with initial intercoder agreement of 96% for 432 entities and after discussion 100% agreement.
5. The website entries for each organization were coded by two coders with discussion of differences in codes and eventual resolution of the codes for all of the relevant organizations. The agreement rate for a sample of fifty of these after coder discussion was 83% with eventual resolution of all but two disagreements, resulting in an overall 96% agreement.

REFERENCES

Berke, P., & Beatley, T. (1992). *Planning for earthquakes: Risk, politics and policy.* Baltimore: Johns Hopkins University Press.

Burby, R.J. (2001). Flood insurance and floodplain management: The U.S. experience. *Journal of Environmental Hazards*, 3(3), 111–122.

Burby, R.J., Deyle, R.E., Godschalk, D.R., & Olshansky, R.B. (2000). Creating hazard resilient communities through land-use planning. *Natural Hazards Review*, 1(2), 99–106.

Burby, R.J., & May, P.J. (1998). Intergovernmental environmental planning: Addressing the commitment conundrum. *Journal of Environmental Planning and Management*, 41(1), 95–110.

Burby, R.J., May, P.J., & Patterson, R.C. (1998). Improving compliance with regulations: Choices and outcomes for local governments. *Journal of the American Planning Association*, 64(3), 324–334.

Camerer, C.F., & Kunreuther, H. (1989). Decision processes for low probability risks: Policy implications. *Journal of Policy Analysis and Management*, 8(4), 565–592.

Chisholm, D. (1989). *Coordination without hierarchy*. Berkeley: University of California Press.

DHS (Department of Homeland Security). (2006). *National infrastructure protection plan*. Washington, DC: DHS.

DHS (Department of Homeland Security). (2009). *National infrastructure protection plan*. Washington, DC: DHS.

DHS (Department of Homeland Security). (2010). *Critical infrastructure partnership advisory council* Annual. Washington, DC: DHS.

Eisinger, P. (2006). Imperfect federalism: The intergovernmental partnership for homeland security. *Public Administration Review*, 66(4), 537–545.

Gerber, B.J. (2007). Disaster management in the United States: Examining key political and policy challenges. *Policy Studies Journal*, 35(2), 227–238.

Haas, P.M. (1992). Introduction: Epistemic communities and international policy coordination. *International Organization*, 46(1), 1–36.

Huber, P.W. (1986). The Bhopalization of American tort law. In National Academy of Engineering, *Hazards: Technology and fairness* (pp. 89–110). Washington, DC: The National Academies Press.

Koski, C. (2011). Committed to protection? Partnerships in critical infrastructure protection. *Journal of Homeland Security and Emergency Management*, 8(1), 1860–1879.

Kunreuther, H. (2006). Disaster mitigation and insurance: Learning from Katrina. *Annals of the American Academy of Political and Social Science*, 604, 208–227.

Linder, S.H. (2000). Coming to terms with the public-private partnership. In P.V. Rosenau (Ed.), *Public-private policy partnerships*. Cambridge, MA: MIT Press.

May, P.J. (1991a). Addressing public risks: Federal earthquake policy design. *Journal of Policy Analysis and Management*, 10(2), 263–285.

May, P.J. (1991b). Reconsidering policy design: Policies and publics. *Journal of Public Policy*, 11(2), 187–206.

May, P.J. (1997). State regulatory roles: Choices in the regulation of building safety. *State and Local Government Review*, 29(2), 70–80.

May, P.J., & Birkland, T.A. (1994). Earthquake risk reduction: An examination of local regulatory efforts. *Environmental Management*, 18(6), 923–937.

May, P.J., & Burby, R.J. (1998). Making sense out of regulatory enforcement. *Law & Policy*, 20(2), 157–182.

May, P.J., & Deyle, R. (1998). Governing land use in hazardous areas with a patchwork system. In R.J. Burby (Ed.), *Confronting natural hazards: Land-use planning for sustainable communities* (pp. 57–82). Washington, DC: The National Academies Press.

May, P.J., & Feeley, T.J. (2000). Regulatory backwaters: Earthquake risk reduction in the western United States. *State and Local Governmental Review*, 32(1), 20–33.

May, P.J., & Koski, C. (2012). *Addressing public risks: Extreme events and critical infrastructures*. Seattle: Center for American Politics and Public Policy, University of Washington.

May, P.J., & Williams, W. (1986). *Disaster policy implementation: Management strategies under shared governance*. New York: Plenum Press.

National Atlas of the United States. (2012). *Major dams of the United States: National atlas of the United States*. Reston, VA. Retrieved from http://nationalatlas.gov/atlasftp.html?openChapters=chpwater#chpwater.

NRC (National Research Council). (2011). *Building community disaster resilience through public-private collaboration*. Committee on Private-Public Sector Collaboration to Enhance Community Disaster Resilience, Geographical Science Committee. Washington, DC: The National Academies Press.

O'Toole, L.J., Jr. (2003). Interorganizational relations in implementation. In B.G. Peters & J. Pierre (Eds.), *Handbook of public administration* (pp. 234–244). Thousand Oaks, CA: Sage.

Rhodes, R.A.W. 2008. Policy network analysis. In M. Moran, M. Rein, & R.E. Godin (Eds.), *The Oxford handbook of public policy* (pp. 425–447). Oxford: Oxford University Press.

United States Office of the President. (2003). *National strategy for the protection of infrastructures and key assets*. Washington, DC: White House.

Zahran, S., Brody, S.D., Vedlitz, A., Grover, H., & Miller, C. (2008). Explaining local commitment to climate change policy in the United States. *Environment and Planning C*, 26(3), 544–562.

8 Networks, Collaborative Planning, and Perceived Quality of Comprehensive Plans

Christopher V. Hawkins

INTRODUCTION

The concept of resilience generally refers to the ability of a system to absorb, change, and still carry on. With its focus on "systems" the concept has been applied widely in the literature on ecology and environmental management (Goldstein, 2011). It has also been presented as a guiding principle in the assessment of community vulnerability and hazards mitigation (Godschalk, 2003). Much of this literature positions resilience within the context of natural ecosystems, for example keeping wetlands intact and mitigating the impact of development on the environment (Brody, Zahran et al., 2007; Brody, Highfield et al., 2007).

This chapter extends this research by investigating the quality of county comprehensive plans in improving disaster resiliency. Plan development and implementation is essential for building community resilience because planning policy can shape land development patterns and reduce the risk to populations within hazardous areas and the vulnerability to disasters. The actions of county government can be critical in improving community resilience because they are the governing units responsible for developing comprehensive plans and implementing emergency management and disaster response activities.

The development and implementation of comprehensive plans includes a wide range of local government and non-government organizations. This chapter focuses on an important but overlooked network of organizations representing Emergency Support Functions (ESFs) in Florida. The empirical analysis explains the perception these organizations have of the county comprehensive plan in improving local disaster resiliency. Having buy-in and support of organizations can improve the prospects for successful plan and project implementation.

As policy implementation studies have shown, the receptivity of target groups and the composition of actors shape the extent to which plans and programs are actually implemented (Wildavsky, 1973; Mazmanian & Sabatier, 1983). Thus knowing how the types of organizations studied in this chapter perceive plans creates the basis for the use of planning mechanisms that can build greater support for planning goals and policies. By explaining what factors increase the perceptions that comprehensive plans improve resiliency, this research will identify some initial conditions for greater success in policy implementation.

Specifically I ask, if an organization is active in sharing information with members of its planning network, will it lead to improved perceptions of plan quality? Theoretically, networks can enhance collective planning efforts by supporting information diffusion and facilitating learning among network members. I also seek to determine whether the planning process employed by the local government influences perceptions. The involvement of stakeholders commonly associated with disaster planning networks in the development of county comprehensive plans can generate policies that mitigate the effects of natural disasters on people and property and support the ability of a community to retain and improve essential functions. When local governments collaboratively plan with the types of organizations sampled in this study, it can minimize conflicting interests and generate greater consensus on plan goals and strategic public investments and actions.

I start by describing the literature on plan quality studies and discussing the concept of resiliency in relation to planning. I next discuss the literature on networks and collaborative planning, which represent the key theoretical arguments that are tested. In the analysis section I discuss the area of study, sample, survey, variable measurement, and method of analysis. I then present the results of the empirical analysis and in the conclusion discuss the implications for practice and avenues for future research.

LITERATURE REVIEW: PLAN QUALITY STUDIES

Longstaff and colleagues (2010) present a conceptual framework for assessing community resilience. Within their framework are five key community subsystems: ecological, economic, physical infrastructure, civil society, and governance. These subsystems mirror planning elements contained within comprehensive plans, such as land uses, conservation and recreation, economic development, housing, and public facilities. In their model resources are defined as objects, conditions, characteristics, and energies that people value. The comprehensive plan represents one type of local resource that is critical to a community's sustained functioning and provision of public services over the long term. A comprehensive plan is an action-oriented guide for making decisions in multiple policy areas. It expresses value judgments of the community (the manner in which a community will grow and develop among other public investments) that are identified through collaborative and open planning processes (Kent, 1964). Because they are intended to inform public resource investment and shape the character of growth and development of a community, they represent a key resource for improving community resiliency.

In the conceptual model of resilience presented by Norris et al. (2008), resilience occurs when resources are sufficiently robust to buffer or counteract a stressor in the environment. The robustness of a community resource can be evaluated based on its performance, or in other words, how well a resource accomplishes a particular function. Previous research on the quality of comprehensive plans has focused on evaluating plans in the context of urban sprawl (Brody, Carrasco, & Highfield, 2006), sustainability (Conroy & Berke, 2004), coastal management (Deyle, Chapin, & Baker, 2008), smart growth and environmental protection (Edward & Haines, 2007; Tang, 2009), as well as hazard mitigation (Berke et al., 1996; Burby & May, 1998). Much of the work on plan quality evaluation has coded goals and policy statements and evaluated fact basis through content analysis (Berke & Godshalk, 2009).

In the context of natural hazards, planning incorporates identifying and evaluating the pre-conditions of social vulnerability and the nexus with the physical nature of development. Research suggests that the risk of, and damage from, hazards is significantly reduced when appropriate land use planning and development oversight is performed, including monitoring efforts to ensure land use regulations are properly implemented (Burby, 2005, 2006; Burby, French, & Nelson, 1998). Reviews of comprehensive plans that incorporate disaster preparation and response elements indicate that they have a positive effect in fostering the development of more rigorous hazard programs (Burby et al., 1999). Moreover, communities that invest the time to specify hazard mitigation goals, objectives, and strategies are more likely to reduce loss through appropriate development and adoption of effective strategies (Burby et al., 1998; Nelson & French, 2002; Schwab et al., 1998). Evidence suggests that local plan quality is favorable in limiting development in hazardous areas (Nelson & French, 2002).

As Berke et al. (1996) note, natural hazards are connected to environmental and land use planning because natural hazards are ubiquitous in that almost all communities must contend with at least one hazard. Thus it is argued that disaster planning should be considered comprehensively by integrating mitigation policies into land use, transportation, infrastructure, environment, and other comprehensive plan components (Burby, 2003; Burby et al., 1999; Godschalk, 2003). However, it has been the traditional practice in emergency management to develop specialized stand-alone plans. Nevertheless, a number of state requirements for local comprehensive plans mandate a hazard mitigation component (Burby, 2003). With an increase in attention given to resiliency, it is likely local governments are also voluntarily integrating the concept of resiliency into multiple comprehensive plan components. This study is intended to provide one perspective to assess these efforts. In the next section I discuss the linkages between disaster resiliency and planning.

RESILIENCY AND PLANNING

Even though planners employ methods of analysis to plan for or accommodate future changes, the capacity or ability to prepare for all future events and contingencies is unrealistic. Because communities are not able to control the occurrence of natural events, there is a need to increase local adaptive capacity and plan for the generally uncertain timing and impact of disasters and hazard events. The concept of resiliency incorporates these dynamics by shifting attention from controlling change in a presumably stable environment to managing the capacity of communities to adapt to and shape change in environments that can be far from certain and predictable. From this perspective, resilience focuses on the survivability of a community's essential social, ecological, and economic functions in the long term (Cutter, Burton, & Emrich, 2010; Longstaff et al., 2010). This includes reducing vulnerabilities to hazards by developing resources that support a community's ability to bounce back and "return to equilibrium after a displacement" (Norris et al., 2008: 127).

Community resiliency is strongly linked with the policies and actions pursued at the local level and embodied in comprehensive plans that shape the character of development and guide investments in mitigation techniques to deal with unexpected events (Godschalk, 2003). Through a variety of planning processes local governments develop standards that shape the patterns of growth and development to minimize disaster impacts and institute policies that improve the ability of a community to withstand disasters and bounce back after a stressor event.

For example, development in hazard-prone areas such as coastal zones may undermine environmental resources. When a community fails to protect critical environmental areas through land use planning, resources can become excessively stressed and ultimately destroyed. This increases the vulnerability of individuals and communities to flooding, which can disrupt transportation networks, increase costs to municipal government, and impact economic conditions. Some populations may be more at risk than others. The location and intensity of development are also important factors that contribute to the loss from disasters, such as flooding (Birkland et al., 2003; Brody, Zahran et al., 2007, Brody, Highfield et al., 2007). For instance low-income neighborhoods in inferior housing and in vulnerable locations can be made worse by alterations to floodplains. To reduce vulnerability and improve the capacity to withstand a disaster, communities can develop and implement structural and non-structural mitigation techniques to protect important natural resources (Norris et al., 2008).

A challenge in developing resilient communities is not only recognizing and anticipating the scope of disasters, but integrating multiple stakeholder groups into planning efforts (Comfort, 2006; Ronan & Johnson, 2005; Pelling, 2003). The primary focus of this study is on (1) the role of planning networks that involve a wide range of individuals, organizations, and institutions (Longstaff et al., 2010; Norris et al., 2008; Bruneau et al., 2003), and (2) planning processes that focus on stakeholder involvement and collective decision making. These are discussed in the next section.

STAKEHOLDER PLANNING NETWORKS AND COLLABORATIVE PLANNING PROCESSES

Planning Networks

Stakeholder organizations that actively share information with other organizations in a disaster planning network are one key component in building resilient communities. Networks can be characterized by high levels of interdependence that involve multiple organizations (Schneider et al., 2003). The NRC (2010) argues that working with "local, state and federal agencies; small and large businesses; nonprofit and faith-based organizations; academicians, researchers, and educational institutions; the mass media; civic and neighborhood organizations; technical experts; volunteers; and other diverse community stakeholders" (39) represents the core of a collaborative network and which ought to be developed in a way that is representative of all components of a community. As Koppenjan and Klijn (2004) write, "Problem solving takes place in an arena in which mutually dependent actors mold and shape problem definitions and solutions" (45).

Planning networks are important because they provide a means by which problems that cross boundaries of agencies that are involved in a broad policy arena, such as disaster and emergency management, can be addressed (Comfort, 2006; Kapucu, 2006). Relations among these types of organizations that are identified by the NRC (2010) can facilitate access to information and resources and support opportunities for building organizational capacity to undertake policy change (Hawkins & Wang, 2012). Greater strategy coordination through task forces and coalitions is likely to emerge when organizations are embedded in information networks. In this regard, networks facilitate learning based on exchange relations. This is especially germane to disaster planning at the local level where multiple organizations interact and formulate policy that can at times be relatively complex. Networks also contribute to "institutional thickness," an interlocking web of organizations and relationships that can increase trust and stimulate the development of new activities (Amin & Thrift, 1994). When stakeholders exchange information it enriches the understanding of the issues in a particular policy domain and provides a basis for stakeholders to purposively coordinate their activities and act collectively.

Hawkins and Wang (2012) discuss this as a local support network which can improve the capacity of a community to undertake new initiatives. In their model, the local capacity to develop, implement, and manage planning initiatives is a support network of organizations that have knowledge and expertise in substantive planning foci, such as emergency management, environmental, housing, and economic issues. Information exchange among organizations within a support network improves knowledge of key policy issues, reduces conflict and facilities decision making, and thus improves the prospects for policy implementation. Moreover, a support network comprised of specialized organizations or technical experts can provide facts and knowledge for planning decision making. Such networks frame issues in a way that can not only make policy changes relevant to local conditions but can also improve the capacity to manage policy changes (Bulkeley & Betsill, 2003; Lindseth, 2004; Sharp et al., 2010).

Planning Process

The collaborative planning process spearheaded by local government is also important in developing resilient communities. Collaboration is characterized by diverse stakeholders working together to resolve a conflict or develop and advance a shared vision (Gray, 1989). Through local government comprehensive planning processes, stakeholders together develop a vision for a community and move towards specific strategic actions to meet objectives. Extant research on collaborative land use planning has emphasized stakeholder's motivation for participating, characteristics of the planning process, and the contents of resulting plans (Conroy & Berke, 2004; Koontz, 2005; Leach et al., 2002), as well as lessons for improving plan implementation (Koontz, 2005).

There are many benefits to integrating a network of organizations into collaborative comprehensive planning efforts. When key stakeholder groups are involved in the planning process it can generate a stronger collective capacity in decision making, enhance individual components of a plan, help build appropriate mitigation programs, result in stronger interorganizational coordination, and generate goals and objectives that are more inclusive and that are aligned with community needs (Berke & Campanella, 2006; Burby & May, 1998; Burby, 2003; Burby et al., 2000). If network participants are not integrated into planning processes local governments may not receive the most accurate and up-to-date information that can inform decisions on disaster mitigation. Collaboration and participation in local planning processes by stakeholders has been shown to be critical for successful plan development and project implementation (Brody, 2003; Innes & Booher, 2003). Early and ongoing involvement of stakeholders throughout plan making and implementation are also shown to be important factors in influencing better outcomes (Burby, 2003). The next section provides a design to examine the perceived quality of county comprehensive plans in improving disaster resiliency.

METHODS

The central questions of this study are (1) whether the activity of organizations in their disaster planning network and (2) the collaborative planning processes employed by local government explain the perceptions of the quality of county comprehensive plans. Perceptions are important because they are closely linked with implementation. Implementation is a particular challenge in planning, which Calkins (1979) referred to as "new plan syndrome" or why plans are continually rewritten or updated without regard to the implementation status of the originally prepared plan. As Talen (1996) observes, the analysis of planning documents (detailed assessments of what are deemed to be "model" plans) and the factors that contribute to higher standards of plan quality represent just one aspect of plan evaluation but of which a number of studies have contributed.

This study takes a different direction. Just as important are the perceptions held by organizations that can contribute to plan development and more importantly, implementation. In this next section I describe the analysis based on survey data collected from organizations representing ESFs in five Central Florida counties. I next discuss the study area, sample, variable measurement, and method of analysis.

The sample for this study is comprised of five Central Florida counties: Citrus, Orange, Osceola, Seminole, and Sumter. Florida is ranked as the state with the highest risk for flooding (FEMA, 1997), and the Central Florida region in particular is known for its vulnerability to tornados, wildfires, and hurricanes (Oxfam, 2009). These five counties were chosen because they are relatively diverse in both population and economic activities. For example, whereas Orange County has a population in excess of one million, Citrus County is relatively rural with a population of approximately 141,000 (Table 8.1). The study area was also selected because there is extensive variation in the partner organizations across the five counties. With a larger population, Orange County has more nonprofit organizations that have a responsibility for emergency management in addition to more government agencies actively involved in planning activities, such as transportation, housing, and public facilities.

Table 8.1 Study Area Characteristics

	Citrus	*Orange*	*Osceola*	*Seminole*	*Sumter*
Population	141,236	1,145,956	268,685	422,718	93,420
Percent over age 65	31.9	9.7	11.0	12.0	43.4
Percent white	93.0	63.6	71.0	78.2	86.6
Percent with bachelor degree	16.5	30.3	18.3	33.6	19.0
Percent homeowner	84.6	59.9	66.5	71.3	89.3
Median household income	$37,933	$50,138	$45,328	$58,971	43,079
Population density	242	1,268	202	1,367	170

Note: All figures are from the 2010 U.S. Census.

Because of their legal authority over development, county governments are heavily involved in decision-making processes, including setting planning goals and developing alternative strategies for land development. Florida counties represent a key local governing unit for land use decision making. Under state statute, they have the responsibility for the administration and management of planning policy in unincorporated areas and work directly with cities within their jurisdictional borders in the approval process for land development projects.

The primary source of data is a web-based survey of organizations listed in each county's Comprehensive Emergency Management Plan (CEMP). The CEMP is an operations-oriented document that Florida counties are required to adopt pursuant to state statute. The CEMP establishes the framework for an effective planning and response system to ensure that counties will be adequately prepared to deal with the occurrence of emergencies and disasters. The plan outlines the roles and responsibilities of local, state, and federal agencies and private and nonprofit organizations within each county and the region.

The CEMP unites the efforts of stakeholder organizations under ESF. Each ESF is headed by a lead or primary agency, which has been selected based on its expertise, authority, resources, and capabilities in the functional area. Within each ESF there are numerous support agencies and organizations that work cooperatively with the primary agency to provide a comprehensive approach to the mitigation, planning, response, and recovery from hazards and emergencies.

For this study representative agencies are organized into three categories: (1) *operations*, which include organizations and county departments responsible for performing emergency response, law enforcement, and health/medical operations during an emergency and/or disaster, (2) *logistics*, which include parties responsible for the logistical resources needed to support the management of an emergency and/or disaster, and (3) *planning and information* organizations that are responsible for gathering, processing, and publishing critical information for the mitigation, planning, response, and recovery from emergencies.

The survey was distributed to 318 organizations in Citrus, Orange, Osceola, Seminole, and Sumter counties. Contact information for each organization was provided by the county emergency management director in each county. Respondents were asked to provide information on their agency's involvement in the comprehensive planning process and their perceptions of the quality of the comprehensive plan in improving disaster resiliency. Fifty usable surveys were returned, representing a 15% response rate.

The quality of the county comprehensive plan is based on survey respondents' perception of how well the plan accomplishes three particular functions identified by Kaiser, Godschalk, and Chapin (1995): (1) fact basis defines local needs, (2) goals are clear and comprehensive in demonstrating commitment to address needs, and (3) policies are specific and

Table 8.2 Variable Measures and Description

Variable	*Mean*	*Std. Dev.*	*Min.*	*Max.*	*Description*
Planning process	3.82	.83	1	5	Process was participatory and incorporated innovative ways to manage risk, 1 (very poor) to 5 (excellent)
Planning network	2.46	.53	1	3	Organization exchanges information with other organizations in their disaster planning network (1 = agree, 2 = neutral, 3 = disagree)
Housing	3.28	1.31	1	5	Organization is involved in actions oriented towards housing, 1 (almost never) to 5 (almost always)
Economic development	3.07	1.26	1	5	Organization is involved in actions oriented towards economic development, 1 (almost never) to 5 (almost always)
Planning and information ESFO	.48	.50	0	1	ESFO is characterized as a planning and information organization, 1 = yes, 0 = otherwise
Orange County	.24	.43	0	1	ESFO serves primarily Orange County, 1 = yes, 0 = otherwise
Non-public organization	.29	.46	0	1	ESFO is a nonprofit organization, 1 = yes, 0 = otherwise

action-oriented in achieving plan goals. Respondents were asked to indicate on a scale of 1 (very poor) to 5 (excellent) whether the comprehensive plan for the county the organization primarily serves improves disaster resiliency by (1) presenting facts about hazards and increasing public awareness of them, (2) representing a clear vision of hazard resilience, and (3) presenting specific policy goals and objectives for achieving resiliency. These items were added to form an index that ranges from 3 to 15 of perceived plan quality with regards to disaster resiliency.

I expect that a collaborative and inclusive comprehensive planning process will improve the perception that the comprehensive plan improves resiliency. To measure this variable respondents indicated from 1 (very poor) to 5 (excellent) whether the comprehensive planning process was participatory and incorporated innovative ways to manage risk. Perceptions of the comprehensive plan in addressing resiliency are also likely to be influenced by the extent to which representing organizations are actively engaged in a disaster planning information network. I expect that when organizations share information on disaster planning activities among other organizations within their planning network they are more likely to perceive the comprehensive plan as adequately addressing disaster resiliency. This is measured using a three-point survey response to the question of the extent to which their organization exchanges information with other organizations in their disaster planning network (1=agree, 2=neutral, 3=disagree).

The study measures the involvement of representative organizations in two substantive planning areas that have implications for disaster resiliency: housing and economic development. Both planning areas can affect land development patterns and vulnerability of people and property to disasters. Respondents were asked if their organization is involved in actions oriented towards economic development and actions oriented towards housing. Both are measured on a scale from 1 (almost never) to 5 (almost always). Lastly I controlled for whether the respondent is categorized as an ESF planning and information organization, whether the organization services primarily Orange County, and is a nonprofit organization. The model is estimated using OLS regression techniques. Table 8.2 provides the descriptive statistics and a description of the independent variables.

RESULTS

The parameter estimates are reported in Table 8.2. The results provide strong support for a link between planning processes and the perception of plan quality related to resiliency. ESF organizations are more likely to perceive county comprehensive plans as improving local disaster resiliency when the comprehensive planning process is viewed as more participatory and collaborative. The analysis also focuses on the extent to which organizations share information with members of their disaster planning network and if this has a positive effect on the perception that the comprehensive plan improves disaster resiliency. The results provide support for this proposition. Whether the responding organization is in a planning and information ESF had a positive effect, but the coefficient fell short of statistical significance.

When the organizations representing ESFs are more involved in planning actions focused on housing the organizations are more likely to perceive the comprehensive plan as having higher quality with regards to disaster resiliency. However, perceptions of plan quality tend to be negative when the planning activities of the ESF organizations are oriented more towards economic development. One explanation for this result is that housing represents the greatest potential for personal injury and long-term displacement after an emergency or disaster. Planning efforts may be oriented heavily towards mitigating damage to the local housing stock rather than commercial or industrial development. Possible inequities in the focus of disaster mitigation may affect an organization's perception that protecting, maintaining, and rebuilding non-residential development is less of a priority for the community.

Table 8.3 OLS Estimates for Perceived Plan Quality

Variable	*Coefficient*	*Standard Error*
Planning process	1.77***	.261
Planning network	1.64***	.411
Housing	.46**	.221
Economic development	-.40*	.232
Planning and information ESF	.32	.428
Orange County	-.37	.494
Non-public organization	-.25	.426
Constant	.73	1.323
Number of observations	50	
Prob. > F	.000	
Adjusted R-squared	.655	

Note: *p<.10, **p<.05, ***p<.01

CONCLUSION

This study investigates how stakeholders engaged in disaster planning perceive the quality of the comprehensive plans as improving local disaster resiliency. The analysis suggests that when these organizations consider planning as more collaborative in nature they are more likely to have a positive perception of the comprehensive plan. In addition, when members of this sample of organizations are more engaged in exchanging information among organizations within their planning network they are likely to have a more favorable perception of the comprehensive plan in improving disaster resiliency. These findings are only one additional step in linking planning processes and the role of information exchange among stakeholders to the quality of a local comprehensive plan.

From a practical perspective the linkages between planning processes and plan implementation or outcomes are essential. When the perception of key stakeholders is that a plan or program is of high quality, the chances of implementation are likely to improve. Although we do not test this directly, theoretically by-in and support for local policies and programs are important for at least improving the prospects for successful plan implementation.

One caveat of this study is that the types of organizations surveyed do not represent all potential stakeholders. Rather, they represent a small and relatively specialized segment of organizations that are active in disaster planning. Nonetheless these organizations have expertise in emergency management and are important actors in planning for disasters and the implementation of disaster mitigation policy. This study also assumes that county comprehensive plans represent a key component in minimizing vulnerability and mitigating the impacts from natural disasters. Based on this premise, these stakeholders provide an additional viewpoint from which to assess and improve this local resource. Moreover, the types of organizations in this study bring to the planning process valuable knowledge of disaster planning techniques that may be overlooked by planners that are dealing with primarily land use issues.

Future studies should examine the research questions posed in this chapter under different contexts. Florida has a relatively long tradition of state-mandated land use plan development and a well-defined emergency management system that identifies key stakeholders and support systems. The extent to which these planning efforts are effectively linked remains a question in Florida as well as in other contexts.

REFERENCES

Amin, A. & Thrift, N. (eds.). (1994). *Globalisation, institutions and regional development in Europe*. Oxford University Press, Oxford.

Berke, P.R., & Campanella, T.J. (2006). Planning for postdisaster resiliency. *Annals of the American Academy of Political and Social Science*, 604, 192–207.

Berke, P., & Godschalk, D. (2009). Searching for the good plan: A meta-analysis of plan quality studies. *Journal of Planning Literature*, 23(3), 227–240.

Berke, P., Roenigk, D., Kaiser, E., & Burby, R. (1996). Enhancing plan quality: Evaluating the role of state planning mandates for natural hazard mitigation. *Journal of Environmental Planning and Management*, 39, 79–96.

Birkland, T.A., Burby, R.J., Conrad, D., Cortner, H., & Michener, W.K. (2003). River ecology and flood hazard mitigation. *Natural Hazards Review*, 4(1), 46–54.

Brody, S.D. (2003). Measuring the effects of stakeholder participation on the quality of local plans based on the principles of collaborative ecosystem management. *Journal of Planning Education and Research*, 22, 407–419.

Brody, S.D., Carrasco, V., & Highfield, W.E. (2006). Measuring the adoption of local sprawl reduction planning policies in Florida. *Journal of Planning Education and Research*, 25(3), 294–310.

Brody, S. D., Highfield, W.E., Ryu, H., & Spanel-Weber, L. (2007). Examining the relationship between wetland alteration and watershed flooding in Texas and Florida. *Natural Hazards*, 40(2), 413–428.

Brody, D.S., Zahran, J.M., Praveen, M., Grover, H., & Highfield, W.E. (2007). The rising costs of floods: Examining the impact of planning and development decisions on property damage in Florida. *Journal of the American Planning Association*, 73(3), 331–345.

Bruneau, M., Chang, S., Eguchi, R., Lee, G., O'Rourke, T., & Reinhorn, A. (2003). A framework to quantitatively assess and enhance the seismic resilience of communities. *Earthquake Spectra*, 19(4), 733–752.

Bulkeley, H., & Betsill, M.M. (2003). *Cities and climate change: Urban sustainability and global environmental governance*. New York: Routledge.

Burby, R.J. (2003). Making plans that matter: Citizen involvement and government action. *Journal of the American Planning Association*, 69(1), 33–49.

Burby, R.J. (2005). Have state comprehensive planning mandates reduced insured losses from natural disasters? *Natural Hazards Review*, 6(2), 67–81.

Burby, R.J. (2006). Hurricane Katrina and the paradoxes of government disaster policy: Bringing about wise governmental decisions for hazardous areas. *Annals of the American Academy of Political and Social Science*, 604(1), 171–191.

Burby, R. J., French, S. P., & Nelson, A.C. (1998). Plans, code enforcement, and damage reduction. *Earthquake Spectra*, 14, 59–74.

Burby, R.J., Berke, P.R., Doyle, R.E., French, S.F., Godschalk, D.R., Kaiser, E.J., Kartez, J.D., . . . & Platt, R.H. (1999). Unleashing the power of planning to create disaster-resistant communities. *Journal of the American Planning Association*, 65(3), 247–258.

Burby, R.J., Deyle, R.E., Godschalk, D.R., & Olshansky, R.B. (2000). Creating hazard resilient communities through land-use planning. *Natural Hazards Review*, 1(2), 99–106.

Burby, R.J., & May, P.J. (1998). Intergovernmental environmental planning: Addressing the commitment conundrum. *Journal of Environmental Planning and Management*, 41(1), 95–110.

Calkins, H. W. (1979). The planning monitor: An accountability theory of plan evaluation. *Environment and Planning A*. 11(7), 745–758.

Comfort, L.K. (2006). Cities at risk: Hurricane Katrina and the drowning of New Orleans. *Urban Affairs Review*, 41(4), 501–516.

Conroy, M.M., & Berke, P. 2004. What makes a good sustainable development plan? An analysis of factors that influence principles of sustainable development. *Environment and Planning*, 36, 1381–1396.

Cutter, S.L., Burton, C.G., & Emrich, C.T. (2010). Disaster resilience indicators for benchmarking baseline conditions. *Journal of Homeland Security and Emergency Management*, 7(1), 1–22.

Deyle, R.E., Chapin, T.S., & Baker, E.J. (2008). The proof of the planning is in the platting: An evaluation of Florida's hurricane exposure mitigation planning mandate. *Journal of the American Planning Association*, 74(3), 349–370.

Edwards, M.M., & Haines, A. (2007). Evaluating smart growth implications for small communities. *Journal of Planning Education & Research*, 27, 49–64.

Federal Emergency Management Agency (FEMA). (1997). *MultiHazard identification and risk assessment*. Washington. DC
: U. S. Government Printing Office.

Godschalk, D.R. (2003). Urban hazard mitigation: Creating resilient cities. *Natural Hazards Review*, 4(3), 136–143.

Goldstein, B.E. (2011). *Collaborative resilience: Moving through crisis to opportunity*. Cambridge, MA: MIT Press.

Gray, B. (1989). *Collaborating: Finding common ground for multiparty problems*. San Francisco: Jossey Bass.

Hawkins, C., & Wang, X. (2012). Sustainable development governance: Citizen participation and support networks in local sustainability initiatives. *Public Works Management & Policy*, 17(1), 7–29.

Innes, J., & Booher, D.E. (2003). *Collaborative policymaking: Governance through dialogue in deliberative policy analyses*. Cambridge: Cambridge University Press.

Kaiser, E.J., Godschalk, D.R., & Chapin, F.S. (1995). *Urban land use planning*. 4th edition. Chicago, IL: University of Illinois Press.

Kapucu, N. (2006). Interagency communication networks during emergencies: Boundary spanners in multi-agency coordination. *American Review of Public Administration*, 36(2), 207–225.

Kent, T. J. (1964). *The urban general plan*. San Francisco: Chandler Publishing Co.

Koontz, T. M. (2005). We finished the plan, so now what? Impacts of collaborative stakeholder participation on land use policy. *Policy Studies Journal*, 33(3), 459–481.

Koppenjan, J., & Klijn, E.J. (2004). *Managing uncertainties in networks.* London, UK: Routledge.

Leach, William D., Pelkey, N. W. & Sabatier, P. A. (2002). Stakeholder partnerships as collaborative policymaking: Evaluation criteria applied to watershed management in California and Washington. *Journal of Policy Analysis and Management,* 21(4), 645–70.

Lindseth, G. (2004). The cities for climate protection campaign (CCPC) and the framing of local climate policy. *Local Environment*, 9, 325–336.

Longstaff, P.H., Armstrong, N.J., Perrin, K., Parker, W.M., & Hidek, M.A. (2010). Building resilient communities: A preliminary framework for assessment. *Homeland Security Affairs*, 6(3), 1–23.

Mazmanian, D. & Sabatier, P. (1983). *Implementation and public policy.* Glenview, IL: Scott, Foresman.

Nelson, A., & French, S. (2002). Plan quality and mitigating damage from natural disasters: A case study of the Northridge earthquake with planning policy considerations. *Journal of the American Planning Association*, 68(2), 194–207.

Norris, F.H., Stevens, S.P., Pfefferbaum, B., Wyche, K.F., & Pfefferbaum, R.L. (2008). Community resilience as a metaphor, theory, set of capacities and strategy for disaster readiness. *American Journal of Community Psychology*, 41(1–2), 127–150.

NRC (National Research Council). (2010). *Private-public sector collaboration to enhance community disaster resilience: A workshop report.* Washington, DC: The National Academies Press.

Oxfam. (2009). *Exposed: Social vulnerability and climate change in the US Southeast.* Washington, DC: Oxfam.

Pelling, M. (2003). *The vulnerability of cities: Natural disasters and social resilience.* London: Earthscan.

Ronan, R.K., & Johnson, D.M. (2005). *Promoting community resiliency in disasters: The role for schools, youth, and families.* New York: Springer.

Schneider, M., Scholz, J.T., Lubell, M., Mindruta, D., & Edwardsen, M. (2003). Building consensual institutions: Networks and the national estuary program. *American Journal of Political Science*, 47(1), 143–158.

Schwab, J., Topping, K.C., Eadie, C.C., Deyle, R.E., & Smith, R.A. (1998). *Planning for post disaster recovery and reconstruction.* Planning Advisory Service report no. 483/484. Chicago: American Planning Association Publication.

Sharp, E.B., Daley, D.M., & Lynch, M.S. (2010). Understanding local adoption and implementation of climate change mitigation policy. *Urban Affairs Review*, 47(3), 433–457.

Talen, E. (1996). Do plans get implemented? A review of evaluation in planning. *Journal of Planning Literature.* 10(3), 248–259.

Tang, Z. (2009). How are California local jurisdictions incorporating a strategic environmental assessment in local comprehensive land use plans? *Local Environment*, 14(4), 313–328.

Wildavsky, A. (1973). If planning is everything, maybe it's nothing. *Policy Sciences*, 4(2), 127–153.

9 Examining Environmental Factors Contributing to Community Resilience along the Gulf of Mexico Coast

Samuel D. Brody and Joshua R. Gunn

INTRODUCTION

Scholars and practitioners are placing increased emphasis on the concept of resilient coastal communities (Beatley, 2009). Rapid population growth and development in low-lying areas combined with persistent storm events have placed millions of residents in the U.S. in harm's way. Nowhere is this problem more apparent than along the Gulf of Mexico coast (Brody et al., 2008). From 2000 to 2005, Gulf coast county/parish floods caused over $52 billion in overall flood damage with $19 billion reported through the National Flood Insurance Program (NFIP). With mounting losses accruing from both acute and chronic flood events, attention is being placed on how to develop more resilient coastal communities over the long term. However, little empirical research has been conducted on the causes and consequences of environmental factors contributing to resiliency that may provide guidance to policy makers on the Gulf coast. Even less work has been done to assess the differences between rural and urban coastal communities.

We address these issues by examining environmental indicators of resiliency across 144 counties bordering the Gulf of Mexico. First, we measure and describe the statistical pattern of four selected indicators across the study area (floodplain area, soil porosity, naturally occurring wetlands, and pervious surface coverage), with special attention given to the differences in rural versus urban communities. Second, we use explanatory statistical models to isolate the impact of these indicators on observed flood damages from 2001 to 2005 along an urban/rural spectrum. Results provide guidance to local decision makers on the role of community resiliency in mitigating the adverse impacts of floods and the degree to which rural localities may be more or less at risk from future flooding events.

In the following sections, we outline the core concepts of community resiliency and then provide a justification for our indicator selections. Next, we describe the research methods, including study area selection, concept measurement, and analytical techniques. Results are then reported in two phases. First, we present maps and descriptive statistics for each indicator. Second, we report the results of a multiple regression analysis predicting actual flood losses over a five-year period. Finally, we interpret the results from a policy/planning perspective and offer guidance for future research.

COMMUNITY RESILIENCY

Resiliency has been defined in a variety of ways depending on the discipline in which it is being discussed. The modern concept of resiliency is rooted in the literature on ecological theory. Holling (1973) is often cited as the first to introduce the concept over thirty-five years ago. In his seminal piece, Holling defines ecological resilience as a measure of the ability of a system to absorb change in the face of extreme perturbation and yet continue to persist. Since the publication of this work, resilience has become a much written about topic in several fields, including ecology, disaster studies, organizational design, and planning (Walker & Salt, 2006).

Most definitions of resiliency involve a system's ability to: a) resist or absorb impacts, and b) maintain or return, more or less, to the same form, function, structure, or qualitative state. A more resilient system (ecological or human) is one that can absorb larger shocks and rapidly bounce back to some desired state (Adger, 2000). The most recent work on resilience considers interactions within coupled socio-ecological systems (Berkes, Colding, & Folke, 2003). That is, resiliency is governed by the interplay and dependence of ecological and socio-political systems (Folke et al., 2005). Human systems have the ability to learn from new experiences, adapt to changing ecological conditions, and modify management systems in response to negative impacts (Lebel et al., 2006). In this sense, communities can modify themselves in response to impacts, becoming more resistant to future adverse events. The goal is not necessarily to bounce back to the pre-disturbance state, but rather reach a more viable or sustainable equilibrium.

The concept of resiliency as a socio-ecological system is also taken up by scholars writing in the field of natural hazard mitigation and disaster risk reduction (Paton & Johnston, 2006). Earlier work in this field recognizes community vulnerability as a place-based concept (Longhurst, 1995). In this sense, susceptibility to the adverse effects of natural hazards is based on the interaction of biophysical risk and socio-economic conditions within a specific geographic domain (Cutter, 1996). When conditions associated with ecological risk coincide with human settlements, social and economic functions are disrupted (Smith & Petley, 2001). Recognizing that the social component critically interacts with the ecological component in socio-ecological systems provides the basis for building more hazard-resilient communities (Berkes et al., 2003; Folke et al., 2005).

The concept of an integrated socio-ecological system provides insights into how to facilitate community development while reducing risk (Lebel et al., 2006; Paton & Johnston, 2006). A central theme in this area is the importance of minimizing pre-existing vulnerabilities by reducing exposure to hazards, such as flooding and storm surge (Mileti, 1999; Godschalk, 2003). Ecological components essentially moderate the impacts of natural hazards on human populations and the built environment (Beatley, 2009). Maintaining ecological functions and processes thus becomes a critical aspect of community resiliency. Also, precisely where human communities persist within ecological and geophysical landscapes is a central aspect in determining the degree of risk and vulnerability. Jurisdictions that locate or build in risk-prone areas will be more likely to incur adverse impacts associated with natural hazards unless they can adopt appropriate coping strategies or possess strong socio-economic resiliency.

Community resiliency thus implies robustness, rapidity, and enhancement in response to disasters. A system is robust with respect to its ability to absorb, deflect, and resist the impacts of a hazard agent. After experiencing a disaster event, a system is able to rapidly bounce back or reach desired levels of restoration. Finally, as part of the recovery process, a resilient system enhances its capacities by improving its mitigation status, reducing pre-existing vulnerabilities, and improving its overall level of sustainability (Peacock et al., 2010). It is important to note that simply returning or bouncing back to the pre-disaster state is not necessarily resilient or adaptive, but could instead provide the seeds for future disaster preparations.

While measuring and assessing disaster resiliency is a growing endeavor, little empirical work has been done on the differences among urban and rural communities. This distinction may be important given the continued shift of population from rural to urban areas, the widespread impacts of coastal hazards across all communities, the concentration of economic resources in major cities, and sprawling residential development into rural settings. The ecological-social systems may behave differently in more rural than urban settings given the variations in contextual characteristics.

For example, Cutter, Burton, and Emrich (2010) evaluated resiliency from a purely sociological perspective across 736 counties within eight states in the southeastern U.S. The authors found a statistical and spatial "divide," where rural counties across the study area displayed a markedly lower level of resiliency compared to urban counties, but intentionally did not include natural environmental indicators in their analysis. The conclusions may have been different if ecological components of resiliency were considered as buffers of hazard impacts. Rural communities may have less economic resources, social cohesion, and infrastructure with which to cope with hazards than their urban counterparts (Rolfe, 2006; Cuotter, 2004). But, the environmental systems in these areas may be less compromised by human development and better able to absorb or minimize the degree of impact caused by a severe storm. For example, the loss of natural capital resulting from the intrusion of impervious surfaces and other aspects of the built environment may be less of an issue in a rural setting, which can provide an advantage in the context of community resiliency. Research identifying specific characteristics of the natural environment that may buffer the impacts of coastal hazards, and then assessing the performance of these landscape features in the face of specific events, is needed to better understand community resiliency from a more holistic perspective.

IDENTIFICATION OF SPECIFIC ENVIRONMENTAL INDICATORS

While scholars recognize the importance of location-based factors when seeking to develop communities in vulnerable areas, they often stop short when it comes to identifying and measuring specific features of the natural environment that provide a foundation for minimizing flood hazard impacts. In response to this lack of study, we propose as a starting point the following four environmental/ecological indicators that serve as essential moderators between flood events and their degree of economic impact on local communities. These indicators receive strong support in the academic and practitioner-oriented literature for their ability to reduce the adverse impacts of flood hazards, and thus serve as likely empirical test beds for this study.

Floodplain Area

The 100-year floodplain (where there is a 1% chance of flooding every year) is a longstanding marker for determining the possibility of an area being inundated by a rainfall event (Freitag et al., 2009). Communities developing within designated floodplain boundaries are at a greater risk of flood impacts unless mitigation measures are in place. Filling in areas of the floodplain or raising sites above flood elevation may be a protective measure for a specific development, but flooding risks are often increased downstream. The cumulative impacts of seemingly minor alterations can compromise the ability of hydrologic systems to store runoff and literally alter the boundary of a floodplain. For example, Brody, Zahran et al. (2007) found that, on average, increasing areas of floodplain in Florida were correlated with larger amounts of property damage from floods. We expect that jurisdictions along the Gulf of Mexico coast with larger percentages of floodplain area will experience significantly greater amounts of flood losses. Conversely, localities with larger areas outside of the floodplain will experience lower amounts of losses from flooding events.

Soil Porosity

Soil porosity is another important indicator of resiliency to flooding because it helps determine the rate of surface water infiltration (Tollan, 2002; Chang & Franczyk, 2008). The amount of water that any given soil will infiltrate and retain depends primarily upon its texture and current moisture condition (Saxton & Shiau, 1990). Porous soils, such as those with high sand content drain much more quickly than low-porosity soils, making them a potentially more suitable substrate for development. The probability of higher peak and mean annual flows from basins with low soil permeability is greater than that for basins with higher permeability soils, as higher permeability allows greater infiltration, more storage, and less runoff (Rasmussen & Perry, 2000). For these reasons, we expect that counties/parishes containing soils with higher levels of porosity will incur significantly lower amounts of property damage from floods.

Naturally Occurring Wetlands

Naturally occurring wetlands are also key ecological indicators when considering the risk to flood damage because of their ability to attenuate floods caused by precipitation (Bullock & Acreman, 2003). Both anecdotal and empirical research suggest that wetlands may reduce or slow precipitation-based flooding (Lewis, 2001; Mitch & Gosselink, 2000). Some research also suggests wetlands reduce the impact of storm surge caused by hurricanes (Costanza et al., 2008).

For example, empirical research in Texas and Florida indicates the value of naturally occurring wetlands in reducing the adverse impacts of floods. Brody, Highfield et al. (2007) found that the development of wetlands significantly increased the number of exceedances in stream flow across eighty-five watersheds in Texas and Florida. When controlling for multiple socio-economic and geophysical contextual characteristics, Brody et al. (2008) found that the loss of wetlands across thirty-seven coastal counties in Texas from 1997 to 2001 significantly increased the observed amount of property damage from floods. On average, wetland alteration added over $38,000 in property damage per flood (Brody et al., 2008). A companion analysis for every county in Florida showed similar results (Brody, Zahran et al., 2007). In this case, the alteration of wetlands increased the average property damage per flood by over $400,000. Given the evidence, we expect that wetland loss will exacerbate property damage from floods even over larger study areas, such as the Gulf of Mexico coast.

Pervious Surfaces

Another key environmental indicator moderating flood impacts is the amount of area within a jurisdiction not covered by impervious surfaces, such as roads, rooftops, and parking lots (Arnold & Gibbons, 1996). Pervious surfaces (e.g., green space, protected areas, and coastal prairies) serve important hydrological functions because they absorb, store, and slowly release water (Tourbier & Westmacott, 1981). Green infrastructure, such as swales, open space, retention ponds, etc. can reduce stormwater runoff, particularly when embedded into an urban context. Conversely, large areas of impervious surface coverage correspond to a decrease in rainfall infiltration and an increase in surface runoff (Paul & Meyer, 2001). Impervious surfaces are especially implicated in increased peak discharge (Brezonik & Stadelmann, 2002). Under compromised hydrological conditions, the lag time between the center of precipitation volume and runoff volume is compressed so that floods peak more rapidly (Hirsch et al., 1990). This reduced lag time occurs because runoff reaches water bodies more quickly when rainfall is unable to infiltrate into the soil (Hey, 2002; Hsu, Chen, & Chang, 2000).

Overall, there is a substantial body of evidence to support the notion that development-based impervious surface increases runoff volume, peak discharges, and associated flood magnitudes. These conditions set the stage for more regular flood events and increasing human losses. For example, Brody, Highfield et al. (2007) found that an increase in impervious surfaces (as measured through remote sensing imagery) correlated with a significant increase in stream flow exceedences over a twelve-year period across eighty-five coastal watersheds in Texas and Florida. A follow-up study in coastal Texas examined the influence of the built environment from 1997 to 2001. The authors observed that over thirty-seven counties, every square meter of additional impervious surface translated into approximately $3,602 of added property damage caused by floods per year (Brody et al., 2008). Undeveloped or pervious areas indicate a certain level of resiliency because they act to moderate flooding events and the degree of impact in terms of property damage and human casualties. Based on this assumption, we expect that increasing areas of pervious surface will result in decreasing amounts of flood loss.

After selecting the four environmental indicators of resiliency based on supporting evidence from the literature, we tested their effectiveness in reducing damage caused by floods along the Gulf of Mexico coast. The following section outlines the research methods, including study area selection, concept measurement, and analytical procedures used to generate the results.

METHOD

We selected as our study area 144 coastal counties and parishes along the Gulf of Mexico as identified by the National Oceanic and Atmospheric Administration (NOAA) (Crossett et al., 2004). This region extends from the Florida Keys westward to the southern tip of Texas, and includes counties from the following six states: Florida, Georgia, Alabama, Mississippi, Louisiana, and Texas (Figure 9.1). The NOAA special projects office defines a county (or parish if in LA) as coastal if one of the following two criteria is met: (1) at a minimum, 15% of the county's total land area is located within a coastal watershed, or (2) a portion of or an entire county accounts for at least 15% of a coastal cataloging unit.

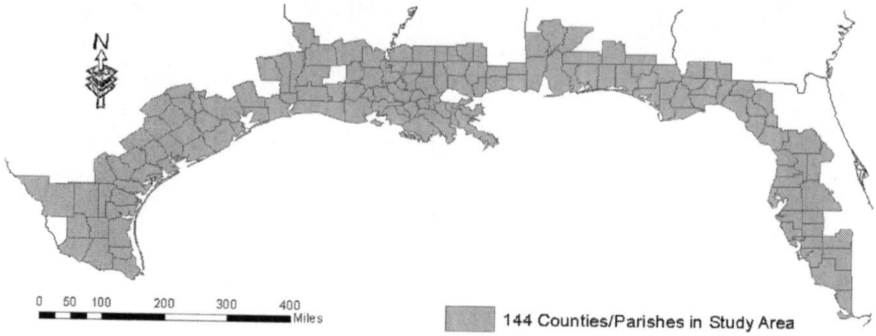

Figure 9.1 Gulf of Mexico coastal study area.

Coastal counties associated with the Gulf of Mexico provide an ideal area in which to study resiliency and flooding for several reasons. First, this low-lying coastal margin is extremely vulnerable to the adverse effects of rainfall and surge-based flooding events, resulting in a long history of losses to examine. For example, data on insured losses from 1996 to 2007 show that jurisdictions bordering the Gulf coast experienced the largest amount of property damage in the U.S. (Brody, Highfield, & Kang, 2011). Louisiana, which suffered extensively from Hurricane Katrina in 2005, reported the highest amount of property damage nationwide with over $13.8 billion over a twelve-year period, followed by Mississippi (over $2 billion), Florida (over $2 billion), Texas (over $1.8 billion), and Alabama (over $793 million).

Second, this study area contains a wide range of environmental conditions that may act as moderators between storm events and impacts to human communities. For example, some of the largest areas of naturally occurring wetlands and coastal floodplain are located within the study area, providing an ideal test bed for examining their influence on flood losses. There is also a wide variation in soil types, ranging from poorly drained clays to sand.

Third, these ecological features have been heavily impacted by human settlement. Localities fringing the Gulf of Mexico have a legacy of rapid population growth and a diversity of development patterns in flood-prone areas. For example, a NOAA survey of land use change along U.S. coasts from 1996 to 2001 revealed that 53% of new development occurred in the southeast between Texas and North Carolina. The interaction of human communities and ecological systems enables us to quantitatively examine the issue of flood resiliency.

Fourth, the study area contains a mix of urban and rural settings within which to examine the differences in the performance of environmental indicators with respect to flood losses. Jurisdictions range from agriculturally based to large metropolitan areas, such as New Orleans, Tampa, and

Houston. Finally, the study area and unit of analysis is a principle target for NOAA policy and planning for coastal resiliency. The results of our study can therefore be directly used by decision makers as they work to reduce the adverse impacts of floods and other natural hazards in the Gulf coast.

Flood loss was measured using National Flood Insurance Program (NFIP) property damage claims from 2001 to 2005. The NFIP was established in 1968 and is administered through FEMA to provide flood insurance to residents and businesses. In participating communities, residential structures within the 100-year floodplain with a mortgage are required to purchase flood insurance. While insured loss does not cover all damage incurred from a flood event, it is a reliable source of data that best represents flood impacts to residential areas. Also, because the market penetration of NFIP policies is so high within the study area, these losses more accurately reflect overall flood damage than in regions where the NFIP is less prevalent. The dollar value of individual insured claims for both contents and structures was aggregated to the county or parish level and log-transformed to better approximate a normal distribution (Table 9.1).

Urban versus rural jurisdictions were measured based on the majority of households in each category based on the U.S. Census classification. The Census Bureau denotes an urban area as housing units located within an urbanized area (UA) or urban cluster (UC). These densely settled areas consist of census blocks that have a population density of at least 1,000 people per square mile, and surrounding census blocks that have a density of at least 500 people per square mile. Territory outside of UAs or UCs is classified as rural. We measured this variable dichotomously, so that if there were more rural than urban households, the variable was scored a 0; jurisdictions with more urban than rural households were scored a 1. Using this method, 70 of the 144 jurisdictions in the sample were considered urban.

Four environmental indicators of resiliency were measured using a variety of data sources. Non-floodplain area was derived by first calculating the percentage of a county/parish occupied by FEMA-defined 100-year floodplains. We then subtracted this percentage from 1 to calculate the percent area for a locality outside of the floodplain. Information was acquired from Oxfam America and the Hazards Vulnerability Research Institute at University of South Carolina, as well as FEMA Q3 datasets. On average, 35% of the study area is within the 100-year floodplain. The indicator of soil porosity and potential surface runoff was measured using the State Soil Geographic Database (STATSGO). A measure of average permeability, based on inches per hour of water infiltration, was generated for each jurisdiction. Higher averages of porosity indicate a greater degree of soil infiltration for rainwater.

The loss of naturally occurring wetlands across the study area was measured as a percent change score over a five-year time period. We chose this approach because it is easier to detect a statistical signal representing the effect of wetlands on flood damages as these areas change over time (loss of ecological services) rather than examining only one time period. Wetland loss was measured by calculating the percent change in wetland landcover classified by NOAA C-CAP using Landsat Thematic Mapper data at thirty-meter resolution from 2001 to 2005. The pervious surface indicator was measured by subtracting the percent of impervious surface within a jurisdiction from 1. Impervious surface was calculated using NOAA C-CAP landcover classifications derived from Landsat Thematic Mapper data at thirty-meter resolution. The area of all impervious surface types (low, medium, and high intensity) were aggregated and divided by the area of each jurisdiction to derive a percentage.

Several contextual control variables were also measured and included in the explanatory model component of the data analysis. Precipitation is an essential control variable as it is usually considered the most important factor contributing to local flooding and associated property damage. Generally, more rainfall increases the likelihood streams and rivers will overflow due to excessive runoff. We calculated precipitation based on annual rainfall amounts from the PRISM dataset. Annual precipitation data for the five-year period was mapped at a scale of thirty arc-second normals and then averaged annually in millimeters for each jurisdiction. To capture precipitation events most likely to result in flooding, we totaled the number of times per year in a jurisdiction rainfall amounts exceeded the seventy-fifth percentile for the study period. This variable was thus measured as a count variable, where higher values represent a greater number of extreme rainfall events.

Two socio-economic controls were also included in the statistical model. First, we measured the number of housing units in each county/parish based on 2000 U.S. Census data. Because our dependent variable is structural flood damage, the number of housing units across the study area should be a strong determinant of the amount of losses observed. We expect that jurisdictions with more housing units are more likely to report higher amounts of flood-related property loss. Second, we controlled for relative wealth of the population using per capita income levels for each county or parish in the sample based on 2000 U.S. Census data. This variable is another important control because wealthy residents are more likely to have flood insurance and higher priced homes. We expect that higher incomes are associated with the presence of more expensive structures and resulting flood damage.

Finally, we measured the total number of points each jurisdiction accrued under the FEMA Community Rating System (CRS) for the year 2000 as a proxy for flood mitigation and planning (Brody et al., 2009). Local flood mitigation activities should be accounted for in the statistical model because they usually have a significant effect on reducing flood losses when properly implemented. It should be noted that due to the small sample size and limited statistical power, we were constrained in the number of variables we could include in the statistical models. This is a recognized limitation of our study that should be addressed in future research.

Table 9.1 Concept Measurement

Variable	*Measurement*	*Source*	*Range*	*Mean*	*Standard Deviation*
Flood Loss	Logged NFIP paid claims for flood losses from 2001–2005.	FEMA	0–22.6	13.66	4.47
Urban/Rural Jurisdiction	Dummy-coded variable where jurisdictions with more rural than urban homes were assigned a 0 and jurisdictions with more urban than rural homes were assigned a 1.	U.S. Census	0–1	.486	.502
Non-floodplain Area	Proportion area outside of the FEMA-defined 100-year floodplain.	Hazards Vulnerability Research Institute at University of South Carolina	0.03–1	.658	.245
Soil Permeability	Average soil permeability in inches per hour.	State Soil Geographic Database	0.7–13	5.35	4.33
Wetland Loss	Losses from the wetland alteration variable were made positive and wetland gains were removed.	NOAA, Coastal Change & Analysis Program	0–.038	.005	.005
Pervious Surface	Proportion area not covered by impervious surfaces based on summing 30 sq. meter pixels from Landsat Thematic Mapper remote sensing imagery.	NOAA, Coastal Change & Analysis Program	0.40–.99	.943	.082
Precipitation	Number of times precipitation exceeded the seventy-fifth percentile during the study period.	PRISM Climate Group	0–5	1.65	1.44
Number of Housing Units	U.S. Census 2000 estimates for the number of housing units in each jurisdiction.	U.S. Census	281–1,298,130	54,396	126,431
Per Capita Household Income	U.S. Census 2000 estimates for the median household income for each jurisdiction.	U.S. Census	7,069–31,195	16,194	3,593
CRS Activity	Number of credit points received through the FEMA CRS.	FEMA	0–1825	262.52	455.965
CRS Used	Dummy-coded variable where counties that don't use a CRS were assigned a 0 and counties that do use a CRS were assigned a 1.	FEMA	0–1	.285	.453

Data on environmental indicators of resiliency were analyzed in two stages. First, we examined the differences in the environmental resiliency indicators between urban and rural settings using two-sample *t*-tests. Second, we used robust regression analysis to explain the performance of these indicators in terms of insured flood losses for both urban and rural jurisdictions. Diagnostics for model misspecification, variance inflation, and spatial-autocorrelation did not yield any major violations. We did, however, find heteroskedacticity in the data, leading us to estimate the model with robust standard errors. No other violations of regression assumptions were detected.

RESULTS

The spatial distribution of urban and rural jurisdictions is mapped in Figure 9.2. As expected, jurisdictions with a majority of urban households concentrate in and around major metropolitan areas adjacent to the coastline, such as Houston, New Orleans, Tampa, etc. By contrast, rural counties/parishes persist in inland portions of the study area. Rural gaps are also evident along the Florida Panhandle, eastern Louisiana, and the southern Texas coast.

As shown in Table 9.2, the intensity of several resiliency variables is significantly different in rural versus urban jurisdictions. For example, soil porosity is on average significantly higher (p=.04) in predominantly urban residential settings than in rural. As expected, the differences are more pronounced when it comes to impervious surfaces. Rural jurisdictions have significantly more pervious surface area (p<.001) which can be explained by the likelihood that these jurisdictions contain less high-intensity or concentrated development. Rural jurisdictions also perform better when it comes to the floodplain area resiliency indicator. On average, rural

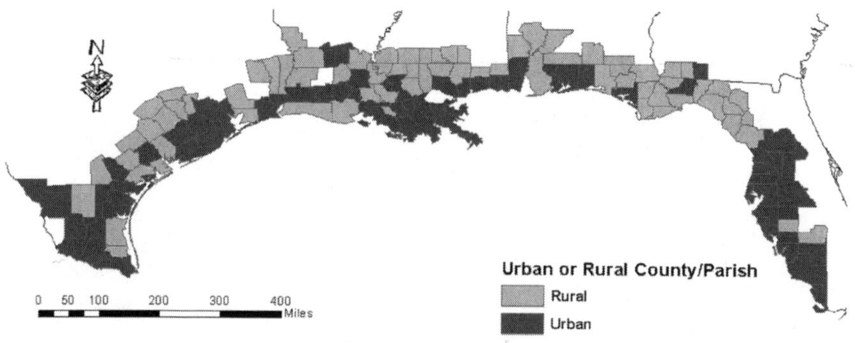

Figure 9.2 Gulf of Mexico coastal study area.

Table 9.2 Differences between Rural and Urban Counties on Resiliency Variables

	Rural (N = 74)		*Urban (N = 70)*		
Variable	*Mean*	*SD*	*Mean*	*SD*	*t*
Soil Permeability †	4.73	3.804	6.009	4.733	-1.888**
Pervious Surface †	0.98	0.011	0.904	0.103	6.132***
Wetland Loss †	0.005	0.007	0.004	0.003	1.943**
Non-floodplain Area	0.69	0.234	0.606	0.251	2.079**
CRS Activity †	100.405	246.4965	433.9	555.5247	-4.6116***
CRS Used ‡	0.1486	n/a	0.42857	n/a	-3.7203***

† Unequal variances. ‡ Test of proportions.
* p<.1 ** p<.05 ***p<.01

counties/parishes contain significantly larger areas outside of the 100-year floodplain (p<.05), despite the fact that more people and structures occupy urban areas. Interestingly, our data show that jurisdictions with mostly rural housing have a significantly greater percent loss of naturally occurring wetlands (p=.05) from 2001 to 2005 compared to urban settings. This result is most likely caused by new, sprawling development outside of city areas that intrudes upon wetlands previously left undisturbed. Lastly, in terms of flood mitigation activities (both structural and non-structural), it is evident that urban jurisdictions have implemented a far greater number of policies (p<.000) than their rural counterparts. Assuming local flood mitigation strategies reduce the degree of disturbance from flood events, urban counties/parishes along the gulf coast should be, from a policy perspective, more resilient in the face of repetitive storm events.

Regression models isolate the impact of selected environmental resiliency indicators on observed flood losses in both rural and urban residential areas. As illustrated in Table 9.3, several indicators play an important role in modifying the amount of insured property damage caused by flood at the county/parish level for rural jurisdictions. For example, increasing amounts of land outside of the 100-year floodplain significantly reduces the economic impact of floods (p<.000). Percent wetland loss from 2001 to 2005 in rural jurisdictions also mitigates the amount of flood damage (p<.05). In contrast, while soil permeability and pervious surfaces are negatively associated with flood loss, these indicators are not statistically significant predictors in a rural setting. Control variables in the model generate mixed results. As expected, the number of housing units significantly increases property damage caused by floods. But, severe rainfall, usually a strong predictor of flood loss, is not statistically significant. Also, the amount of effort in the FEMA CRS program has no discernible effect on minimizing the amount of flood damage in rural areas.

Table 9.3 Resiliency and Flooding in Rural Jurisdictions

Variable	*Coefficient*	*Standard Error*	*t*
Non-floodplain Area	1.069409	-4.196613	-3.92***
Pervious Surface	25.90644	-10.6351	-0.41
Soil Permeability	0.0726252	-0.0284128	-0.39
Wetland Loss	38.59709	-99.4276	-2.58**
Per Capita Household Income	0.0001229	0.0001067	0.87
Number of Household Units	0.0000321	0.0000848	2.64**
CRS Activity	0.0011769	0.0000692	0.06
Precipitation	0.624371	0.9524996	1.53
Constant	25.06808	23.92113	0.95
N		74	
F(8, 65)		6.57	
Significance		0.000	

*p<.1 ** p<.05 ***p<.01

Table 9.4 Resiliency and Flooding in Urban Jurisdictions

Variable	*Coefficient*	*Standard Error*	*t*
Non-floodplain Area	1.052693	-0.3758551	-0.36
Pervious Surface	4.360135	3.280465	0.75
Soil Permeability	0.0619574	-0.0545438	-0.88
Wetland Loss	79.98825	297.1578	3.72***
Per Capita Household Income	0.0000632	0.0002859	4.52***
Number of Household Units	4.74E-06	6.83E-06	1.44
CRS Activity	0.0005579	-0.0011416	-2.05**
Precipitation	0.5635158	2.673874	4.74***
Constant	4.29738	6.615991	1.54
N		69	
F(8, 60)		13.12	
Significance		0.000	

*p<.1 ** p<.05 ***p<.01

Environmental indicators of flood resiliency behave differently in predominantly urban-oriented jurisdictions (Table 9.4). In this setting, increasing areas outside of the floodplain do not, on average, significantly reduce observed residential flood loss. The outcome is the same for both pervious surfaces and soil porosity. Percent wetland loss, however, is statistically significant ($p<.000$), where increasing alteration leads to greater amounts of property damage from floods. Contextual control variables also behave differently in urban settings. For example, per capita income emerges as a significant positive predictor ($p<.000$). Large amounts of precipitation also drive a marked increase in property damage ($p<.000$). Finally, flood mitigation makes a significant difference, where higher overall CRS scores per jurisdiction reduce the amount of subsequent losses ($p<.05$).

DISCUSSIONS

Our statistical results indicate that the presence and performance of environmental indicators associated with resiliency differ across rural and urban jurisdictions. Furthermore, it is impossible to conclude that one of these settings is markedly more resilient than the other from an ecological perspective. For example, two-sample *t*-tests show that urban jurisdictions contain better drained soils, which is an important aspect of reducing surface water runoff. If urban communities are building on more porous soils, they may be less apt to experience flooding. On the other hand, these same urban counties/parishes have significantly greater floodplain area, making it more difficult to place structures outside of the most vulnerable locations. Also, while urban jurisdictions may contain better drained soils that can reduce runoff, they seem to be negating the positive resiliency effects by covering the surface with pavement, buildings, and other impervious surfaces. In contrast, rural areas have significantly more pervious areas with which to absorb heavy rainfall events.

Despite having less total areas of pavement, rural jurisdictions in our sample experienced significantly higher rates of wetland loss. At first, this result may seem counterintuitive. But, considering our wetland change measure focused on the period 2001–2005, it is likely that we detected more recent sprawling development outside of urban centers that intrude upon existing wetland systems. The majority of wetland conversion in urban-oriented counties/parishes probably occurred prior to our study period, although both settings experienced major declines. Finally, from a preparedness and flood mitigation standpoint, our research suggests that urban jurisdictions are far better prepared than their rural counterparts. This result is expected considering the majority of housing units and NFIP policies are located in cities and metro areas where flood reduction strategies may have the largest effect.

From a policy/planning perspective, it may be more important to understand the degree to which these indicators of resiliency help reduce the adverse impacts of a natural disturbance (i.e., a flood). The statistical models in our chapter isolate the effects of each indicator across the rural/urban spectrum and provide guidance to local decision makers on how to enhance their community's level of resilience to future flood hazards. First, having more area outside of the 100-year floodplain appears to pay off for rural communities by significantly reducing property damage. We speculate that in urban jurisdictions, structures are more apt to be located in the floodplain, making this variable an insignificant factor in contributing to flood resiliency. Also, the intrusion of the built environment in urban areas often alters the boundaries of the floodplain to the extent that they become irrelevant predictors of flooding and associated damage. We recommend that localities make a concerted effort to avoid the floodplain as well as understand the implications of surrounding development on changing these boundaries.

Second, percent wetland loss in rural and urban settings is of critical importance from a flood resiliency standpoint. In rural jurisdictions, wetland alteration significantly reduced flood losses even though the overall rate of loss was higher than in urban areas. We believe this result is due to the fact that: a) there is still more total wetland area remaining in rural jurisdictions to absorb heavy precipitation; b) the low-density pattern of development typical of rural areas has less of an obliterating effect on wetland systems so they are able to better retain their core function; and c) wetlands are being altered in both settings, but in rural jurisdictions, there is less "crowding" of these areas by new homes that can exacerbate flood impacts. The opposite may be true in urban settings. In these locales, wetlands are more likely to be more completely removed, voiding their function as naturally occurring flood mitigation devices. Also, high-density residential development often takes the place of a wetland, setting the stage for significant damage during a flood event, even though structural drainage measures may have been installed. Based on these results, we recommend that even remnant wetlands occurring in urban areas are vital forms of resiliency with respect to floods. Protecting these areas should be a priority given the degree to which their alteration greatly increases property damage from flooding events.

Interestingly, in our study, neither soil porosity nor pervious surfaces have a significant impact on flood losses. Much of the study area contains soils so poorly drained that many consider them as ineffective as pavement in absorbing runoff. However, we note that a heavy rainfall event has a much greater impact on losses in urban versus rural jurisdictions. This outcome is most likely the result of increased runoff from paved surfaces, inadequate drainage infrastructure, and lack of recognition from planners of downstream impacts. We urge local development officials to be more cognizant of the water catchment/hydrological system (as opposed to just the development site) and the effects of development on downstream communities.

Lastly, our results show that mostly non-structural mitigation techniques under the FEMA CRS program significantly reduce flood losses in urban settings. When mitigation measures are put in place, they appear to be working, even to the extent to which they are overcoming the weaknesses revealed in environmental indicators of resiliency. We recommend expansion of the CRS program or implementation of similar techniques into rural jurisdictions even though these areas are not major population centers. Mitigation strategies, such as protected areas, relocation of vulnerable structures, higher regulatory standards, etc. appear to effectively reduce observed flood damages and should be implemented beyond major metropolitan areas.

CONCLUSION

This chapter suggests that specific environmental components of the landscape play an important role in enhancing community resiliency to flood hazards. Furthermore, it may not be simply the presence of a landscape component, but the degree to which it is functional and intact that may make the difference in protecting communities from future flood loss. While the chapter makes some valuable contributions to understanding the impact of the natural landscape on community resilience, particularly with regard to rural and urban settings, it should only be considered an initial step in examining the overall topic.

First, we consider resiliency as the ability to withstand a flood disturbance without catastrophic impacts to humans during one time frame. Future research should take a more longitudinal approach and assess the degree to which communities respond to hazard events over multiple time periods. Second, many of our indicators lack precision in terms of their interaction with rural or urban households. For example, while we can measure the area outside of the 100-year floodplain for each jurisdiction, we do not know how many rural or urban households are located within their actual boundaries. Such a measure would be computationally intensive, but should be addressed in the future if data is available. Third, due to the relatively small sample size, we were limited in the number and scope of control variables we could include in the regression models.

Future research should assess larger samples and include more control variables, including surge risk, repetitive loss estimates, the number of NFIP policies, and specific mitigation activities. Fourth, our study was conducted at a large spatial scale and makes "high altitude" statistical conclusions. Additional qualitative work should be done focusing on individual localities to provide a more detailed context for understanding the influence of environmental characteristics on flood-resilient communities. Finally, we only assess the performance of four indicators. Future projects should expand the number and scope of resiliency measures to form a more complete overall picture of the way environmental components interact with the well-being of coastal communities.

REFERENCES

Adger, W.N. (2000). Social and ecological resilience—are they related? *Progress in Human Geography*, 24(3), 347–364.

Arnold, C.L., & Gibbons, J.C. (1996). Impervious surface coverage: The emergence of a key environmental indicator. *Journal of the American Planning Association*, 62(2), 243–258.

Beatley, T. (2009). *Planning for coastal resilience: Best practices for calamitous times*. Washington, DC: Island Press.

Berkes, F., Colding, J., & Folke, C. (2003). *Navigating social ecological systems: Building resilience for complexity and change*. Cambridge: Cambridge University Press.

Brezonik, P.L., & Stadelmann, T.H. (2002). Analysis and predictive models of stormwater runoff volumes, loads and pollutant concentrations from watersheds in the twin cities metropolitan area, Minnesota, USA. *Water Resources*, 36, 1743–1757.

Brody, S.D., Highfield, W.E., & Kang, J.E. (2011). *Rising waters: Causes and consequences of flooding in the United States*. Cambridge: Cambridge University Press.

Brody, S.D., Highfield, W., Ryu, H.C., & Spanel-Weber, L. (2007). Examining the relationship between wetland alteration and watershed flooding in Texas and Florida. *Natural Hazards*, 40(2), 413–428.

Brody, S.D., Zahran, S., Highfield, W., Bernhardt, S., & Vedlitz, A. (2009). Policy learning for flood mitigation: A longitudinal assessment of CRS activities in Florida. *Risk Analysis*, 29(6), 912–929.

Brody, S.D., Zahran, S., Highfield, Grover, H., & Vedlitz, A. (2008). Identifying the impact of the built environment on flood damage in Texas. *Disasters*, 32(1), 1–18.

Brody, S.D., Zahran, S., Maghelal, P., Grover, H., & Highfield, W. (2007). The rising costs of floods: Examining the impact of planning and development decisions on property damage in Florida. *Journal of the American Planning Association*, 73(3), 331–345.

Bullock, A., & Acreman, M. (2003). The role of wetlands in the hydrological cycle. *Hydrology and Earth System Sciences*, 7(3), 358–389.

Chang, H., & Franczyk, J. (2008). Climate change, land-use change, and floods: Toward an integrated assessment. *Geography Compass*, 2(5), 1549–1579.

Costanza, R., Pe´rez-Maqueo, O., Luisa, M., Sutton, P., Anderson, S.J., & Mulder, K. (2008). The value of coastal wetlands for hurricane protection. *Ambio*, 37(4), 241–247.

Crossett, K.M., Culliton, T.J., Wiley, P.C., & Goodspeed, T.R. (2004). Population trend along the coastal United States: 1980–2008 [electronic version]. Retrieved October 20, 2006 from www.oceanservice.noaa.gov.

Cuotter, D.A. (2004). Communities of work: Rural restructuring in local and global contexts. *Rural Sociology*, 69, 581–584.

Cutter, S.L. (1996). Vulnerability to environmental hazards. *Progress in Human Geography*, 20(4), 529–539.

Cutter, S.L., Burton, C., & Emrich, C. (2010). Disaster resilience indicators for benchmarking baseline conditions. *Journal of Homeland Security and Emergency Management*, 7(1), article 51, 1–22.

Folke, C., Hahn, T., Olsson, P., & Norberg, J. (2005). Adaptive governance of social-ecological systems. *Annual Review of Environmental Resources*, 30, 441–473.

Freitag, B., Bolton, S., Westerlund, F., & Clark, J. (2009). *Floodplain management: A new approach for a new era.* Washington, DC: Island Press.

Godschalk, D. (2003). Urban hazard mitigation: Creating resilient cities. *Natural Hazards Review*, 4(3), 136–142.

Hey, D.L. (2002). Modern drainage design: The pros, the cons, and the future. *Hydrological Science and Technology*, 18(14), 89–99.

Hirsch, R.M., Walker, J.F., Day, J.C., & Kallio, R. (1990). The influence of main on hydrological systems. In M.G. Wolman & H.C. Riggs (Eds.), *Surface water hydrology* (pp. 329–359). Boulder, CO: Geological Society of America.

Holling, C.S. (1973). Resilience and stability of ecological systems. *Annual Review of Ecology and Systematics*, 4, 2–23.

Hsu, M.H., Chen, S.H., & Chang, T.J. (2000). Inundation simulation for urban drainage basin with storm sewer system. *Journal of Hydrology*, 234, 21–37.

Lebel, L., Anderies, J.M., Campbell, B., Folke, C., Hatfield-Dodds, S., Hughes, T.P., & Wilson, J. (2006). Governance and the capacity to manage resilience in regional social-ecological systems. *Ecology and Society*, 11(1), 19, 1–15.

Lewis, W.M. (2001). *Wetlands explained: Wetland science, policy, and politics in America.* New York: Oxford University Press.

Longhurst, R. (1995). The assessment of community vulnerability in hazard prone areas. *Disasters*, 19, 269–270.

Mileti, D.S. (1999). *Disaster by design: A reassessment of natural hazards in the United States.* Washington, DC: Joseph Henry Press.

Mitch, W.J., & Gosselink, J.G. (2000). *Wetlands.* 3rd edition. New York: John Wiley and Sons.

Paton, D., & Johnston, D. (Eds.) (2006). *Disaster resilience: An integrated approach.* Springfield, IL: Charles Thomas.

Paul, M.J., & Meyer, J.L. (2001). Streams in the urban landscape. *Annual Review of Ecological Systems*, 32, 333–365.

Peacock, W.G., Brody, S.D., Seitz, W.A., Merrell, W.J., Vedlitz, A., Zahran, S., Harriss, R.C., & Stickney, R. (2010). Advancing the resilience of coastal localities: Developing, implementing and sustaining the use of coastal resilience indicators. Final report for NOAA CSC grant no. NA07NOS4730147.

Rasmussen, P.P., & Perry, C.A. (2000). Estimation of peak streamflows for unregulated rural streams in Kansas. U.S. Geological Survey, Water-Resources Investigations, report no. 00–4079.

Rolfe, R.E. (2006). *Social cohesion and community resilience: A multi-disciplinary review of literature for rural health research.* Halifax, NS: Rural Center.

Saxton, K.E., & Shiau, S.Y. (1990). Surface waters of North America: Influence of land and vegetation on streamflow. In M.G. Wolman & H.C. Riggs (Eds.), *The geology of North America*, vol. 0–1, *Surface water hydrology* (pp. 55–80). Washington, DC: The Geologic Society of America.

Smith, K., & Petley, D.N. (2001). *Environmental hazards: Assessing risk and reducing disaster.* New York: Routledge.

Tollan, A. (2002). Land-use change and floods: What do we need most, research or management? *Water Science and Technology*, 45(8), 183–190.

Tourbier, J.T., & Westmacott, R. (1981). *Water resources protection technology: A handbook of measures to protect water resources in land development.* Washington, DC: The Urban Land Institute.

Walker, B., & Salt, D. (2006). *Resilience thinking: Sustaining ecosystems and people in a changing world.* Washington, DC: Island Press.

10 Resilience Planning in the Face of Uncertainty

Adapting to Climate Change Effects on Coastal Hazards

Robert E. Deyle and William H. Butler

INTRODUCTION

The resilience of coastal communities has long been precarious as the coastal environment is a highly dynamic system, prone to shocks from the wind, waves, and surges of coastal storms, as well as daily and seasonal cycles of sediment erosion and accretion. Many of the strategies used to mitigate these impacts on human development along the coast, and the conventional methods for evaluating and choosing among alternatives, are grounded on assumptions of stationarity and predictability of coastal system conditions and trends and the external forces that affect them. The unfolding narrative of climate change poses challenges to using conventional methods of assessing alternatives for maintaining and enhancing community resilience to such hazards. Assumptions of stationarity are eroding, and predictability is elusive. The rate of global sea level rise has increased in recent decades and is very likely accelerating. Scientists also have determined that the intensity of tropical cyclones has increased in several of the world's oceans and theorize that this might be a result of global warming. Still, very large uncertainties remain about how these changes will translate into altered local exposure and associated vulnerability. Scientists are unable to estimate objective probabilities for how high sea level will get, where, and when, or how more frequently higher intensity hurricanes will occur.

The capacity to adapt in the face of changing system dynamics is essential for community resilience to coastal hazards (Adger et al., 2005; Beatley, 2009; U.S. Indian Ocean Tsunami Warning System Program, 2007; Walker et al., 2004). Conventional methods inadequately address uncertainty in both the ecological and social domains and are data-intensive and complex which makes them inaccessible to low-resource coastal communities. Such coastal communities need a different approach for analyzing alternatives and choosing responses where uncertainty is indeterminate and where they wish to apply broader goals to assessing alternatives in a transparent and participatory manner.

In this chapter, we explore how small and low-resource coastal communities can effectively address the challenges that confront them as they attempt to build the resilience with limited economic and human capital. We argue that the practices of adaptive management and scenario analysis can provide the means to identify, select, and adopt robust and flexible mitigation measures that promote resilience while coping with uncertainty. We introduce a multi-criteria decision tool based on Hill's (1968) goals achievement matrix (GAM) and demonstrate how it can be used to support complex alternatives analysis under indeterminate uncertainty in low-resource communities. We also show how communities can build their adaptive capacity by using such a tool in an adaptive governance framework that facilitates social learning and collaborative management.

We begin this chapter by contrasting conventional views of resilience with the concept of adaptive resilience that recognizes that most socio-ecological systems do not exist in a state of dynamic equilibrium. Then, we explore how the context of coastal management is changing in the face of accelerating sea level rise and the possibility of increasing intensities of tropical cyclones. Next, we elaborate on the shortcomings of conventional approaches to alternatives analysis, and make our case for adaptive management and scenario analysis as more effective for dealing with complexity and uncertainty. We then introduce Hill's GAM and show how it can be applied to the particulars of adapting to the effects of climate change on coastal hazards. We conclude by outlining the elements of an adaptive governance process through which collaborative, social learning can be incorporated into iterative and ongoing use of a GAM as new knowledge and changing understandings, values, and perceptions are incorporated into climate change adaptation responses.

RESILIENCE AND COASTAL HAZARDS MANAGEMENT

The conventional approach to hazards management conceives of resilience as the ability of a physical or social system to recover from the major disturbance of a natural disaster. This so-called "engineering resilience" presumes stability of system structure and function centered on some optimal equilibrium. It is measured by the time necessary to repair or rebuild the impacted elements of the system and to restore its normal structure and functions (Folke, 2006; Zellner, Hock, & Welch, 2011), hence the conventional goal of post-disaster recovery to "return to normal as quickly as possible" (Schwab et al., 1998).

Scholars have begun to understand that stability is an elusive state for complex socio-ecological systems that are subject to periodic acute perturbations, such as natural hazards and economic shocks, and slow-moving chronic changes, such as shifting demographic patterns and climate change which repeatedly and continuously alter system conditions and dynamics (Walker & Salt, 2006). In this view, the system's resilience is not its ability to bounce back to a previous equilibrium point, rather it is the ability of system relationships and functions to persist, perhaps in an altered state, in a new context (Boyd & Folke, 2012; Holling, 1973). This mode of resilience, which has been dubbed by some as "adaptive resilience" (Ali, Angelov, & Hutchison, 2011; Robinson, 2010), depends upon a system's "adaptive capacity," that is, its ability to "modify or change its characteristics or behaviour so as to cope better with existing or anticipated external stresses" (Adger et al., 2004: 34).

Adger et al. (2005: 1036), citing Carpenter et al. (2001) and Folke (2002), argue that such resilience is in part a function of the ability of the system to "build capacity for learning." Such learning is especially critical where planners and managers face uncertainty about how socio-ecological systems work, how they are affected by external perturbations and management interventions, and how external drivers are changing.

Scholars who have examined the capacity of communities to adapt to climate change have determined that this capacity is a function of human and social capital, as well as economic capital, and is uneven across communities, even within developed countries (Adger et al., 2007). Important factors include education, income, and health, local economic development and access to technology, as well as government institutions, community organizations, and social networks (Adger & Vincent, 2005; Adger et al., 2004; Yohe & Tol, 2002). Adaptive capacity in small and low-resource coastal communities is constrained by limited resources on which to draw to undertake planning and implement management actions. Thus, smaller coastal cities and low-resource communities are likely to have less capacity to adapt to the added complexities and uncertainties that climate change brings to managing coastal hazards resiliently. It is to these uncertainties and complexities in coastal climate change adaptation that we now turn our attention.

UNCERTAIN COMPLEX HAZARDS

The equilibrium resilience assumptions of conventional coastal hazard mitigation along sedimentary coasts reflect in part implicit or explicit assumptions that the external drivers of chronic erosion and storm forces are stationary phenomena: chronic erosion rates are driven by a constant rate of sea level rise and coastal storms have return frequencies that can be ascertained by analyzing historic data. Thus coastal communities have sought to reduce their vulnerability to these hazards through such strategies as constructing soft (beach nourishment) and hard (bulkheads, revetments, seawalls) engineered structures designed to withstand storms with specified return frequencies. They similarly adopt building elevation standards and policies for restricting development of the most hazardous areas based on maps of wind zones, flood hazard areas, and storm surge flood zones with specified return frequencies. And some communities define fixed-distance development setbacks based on a reference feature such as the mean high tide line or some multiple of the average annual erosion rate.

Climate change, however, confronts coastal planners with two dimensions of uncertainty as they attempt to maintain or enhance the resilience of their communities over planning horizons of multiple decades:

1. knowledge of the future trends of the external drivers of physical and ecological coastal systems and how those systems will adjust autonomously, and
2. knowledge of the current and future goals and priorities of the stakeholders who comprise the social systems for which they plan.

The uncertainties about future trends of the external drivers of sea level rise rates and tropical cyclone intensity are currently indeterminate, a function of all three dimensions of what the U.S. Army Corps of Engineers defines as "knowledge uncertainty": "Incomplete understanding of . . . [the] system, modeling limitations, and . . . limited data" (USACE, 2011: 10). Until recently, the rate of global eustatic sea level rise has been quite modest,[1] approximately 1.8 mm/year (Meehl et al., 2007). As sea level rises, sedimentary shorelines recede from the combined effects of erosion and inundation. Receding shorelines and higher sea levels also shift coastal flood zones further inland thereby increasing the return frequencies of floods of a given magnitude at any given location (Ning et al., 2012). Coastal planners have accommodated rising sea level either by ignoring it because the rate has been so slow or by accounting for average erosion rates in the design of beach renourishment projects, shore protection structures, and erosion-based development setbacks.

More recent observations indicate, however, that the rate of global eustatic sea level rise has been higher since at least the early 1990s, averaging 3.3 mm/yr between 1993 and 2003 (Meehl et al., 2007). Both theory and evidence suggest that the rate is accelerating due to an array of positive global warming and sea level rise feedbacks, the most prominent of which is accelerated melting of the Greenland and West Antarctic ice sheets (Grinsted, Moore, & Jevrejeva, 2010; Hansen & Sato, 2011; Pfeffer, Harper, & O'Neel, 2008; Steffen et al., 2008; Vermeer & Rahmstorf, 2009). At present, however, scientists are unable to fully explain how and why the ice sheets are melting more rapidly based on models derived from geologic time scales of glacial changes (Steffen et al., 2008). Adding to these uncertainties are the complexities of predicting rates of sea level rise in particular locations. This local "relative sea level" rise is a function of differential thermal expansion of sea water, salinity, winds and currents, proximity to melting ice, and the direction and rate of land surface movement (Meehl et al., 2007).

Climate change also is likely to affect tropical cyclone activity, adding a second dimension of uncertainty to managing the resilience of coastal communities. There is some evidence that increases in tropical cyclone frequency may have accompanied rising sea surface temperatures in the northern Atlantic since the 1970s (Webster et al., 2005), but not elsewhere. The percentage of higher intensity tropical cyclones, i.e., Category 4 and 5 hurricanes, has apparently increased in the northern Atlantic and in other oceans (Elsner, Kossin, & Jagger, 2008; Emanuel, 2005, 2007; USEPA, 2010). Uncertainty remains, however, about the extent to which recent trends are a direct reflection of climate change as opposed to other temporal patterns in factors that influence tropical cyclone intensity (Karl et al., 2008; USEPA, 2010).

Coastal planners and managers also may face knowledge uncertainty about social goals and priorities because of the diversity of perceptions and value-based perspectives held by stakeholders about the causes of observed changes in coastal hazard phenomena and the appropriate management responses (Pahl-Wostl et al., 2007). Uncertainty in the social system may arise not only from a lack of data or information to understand these values under current conditions, but also from changing stakeholder values and perceptions over time about the nature and importance of the threats posed by climate change and coastal hazards (NRC, 2004).

CONVENTIONAL DECISION MAKING IN UNCERTAIN ENVIRONMENTS

Environmental planners and managers have developed a variety of multi-criteria decision analysis tools to assess alternative strategies (Kiker et al., 2005). Cost-benefit analysis (CBA) is the bedrock of public sector alternatives evaluation in the U.S. (USACE, 1983, 2011, 2012) and is widely embraced for assessing climate change adaptation options (see for example Callaway, 2004; Fankhauser et al., 1999; Smith, 1997; Toman, 2006). The methods for doing so are complex and data intensive, however, and likely well beyond the means of small local governments with limited staff and fiscal resources. CBA also presents a number of limitations to fully assessing the effects that alternative strategies will have on the resilience of coastal communities to the effects of climate change on natural hazards.

Where the costs and benefits of alternative adaptive strategies can be fully specified with certainty, planners maximize efficiency by choosing the alternative with the highest net present value. However, scholars and practitioners have long recognized the challenges of monetizing non-market costs and benefits and the limitations of standard approaches for contending with this problem (Bendick et al., 2010; Hill, 1968; Kneese, 1984; Rothman, 2000). Analysts estimate shadow prices for some costs or benefits that are not traded in private markets, but other "intangibles" may not even be susceptible to quantification and often are given no more than "lip service" (Hill, 1968: 20).

Where the timing and/or magnitude of costs and benefits are uncertain but can be assigned probabilities, planners employ so-called risk-benefit analysis (RBA) in which probability values for individual costs and benefits are incorporated in net present value calculations. For coastal hazards adaptation projects, they estimate probabilities from historic frequency data across the array of possible storm scenarios or by employing Monte Carlo simulation (USACE, 2011). RBA adds additional layers of complexity that require further technical sophistication. But more important is that this approach cannot be applied where uncertainties are indeterminate (Fussel, 2007).

In such situations, planners often apply a heuristic approach that reflects their level of risk averseness. Some commentators, concerned about the opportunity costs of unnecessary overprotection, have advocated a "wait and see" approach, until uncertainties are resolved. Others suggest initiating adaptation strategies with near-term benefits that are more certain, or following the so-called "low regrets" or "no regrets" approach of taking initiatives that have net benefits even in the absence of climate change (Fankhauser, 2010; Stern, 2006; Titus & Neumann, 2009). Relying on such heuristics runs the risk of taking insufficient adaptation measures to forestall significant long-term climate change impact costs and irreversible damages (Adger et al., 2007; Fussel, 2007). The consequences of waiting for greater certainty may be substantial because of the long-term commitments embodied in land use development and infrastructure investment (Deyle, Bailey, & Matheny, 2007; Parry et al., 2007; Smith, 1997) and the fact that the burden of anthropogenic greenhouse gases already in the atmosphere makes accelerated sea level rise a certainty for centuries (Nicholls et al., 2007; Wigley, 2005). Where concerned with underprotection planners may incorporate a "safety factor" in engineering designs (USACE, 2011) or invoke the so-called "precautionary principle," under which lack of full scientific certainty is not used as a reason for postponing actions to address "threats of serious or irreversible damage" (UN, 1992).

Two additional concerns with the conventional approaches of CBA and RBA are the frequent lack of transparency of value-based and empirical assumptions, and the limited involvement of stakeholders in defining evaluation criteria and methods (Toman, 2006). While lack of transparency is not inherent to the methods themselves, because of the technical nature of these analyses, many explicit and implicit assumptions are not always made known to, much less validated with, parties who have a stake in the outcomes.

ADAPTIVE MANAGEMENT AND SCENARIO ANALYSIS

Adaptive management and scenario analysis offer alternatives to conventional approaches for assessing strategies for mitigating coastal hazards that can contend with the indeterminate uncertainties of climate change while avoiding the oversimplification of resorting to heuristics. They also can facilitate transparency and stakeholder involvement in the selection and implementation of actions. Used in concert, scenario analysis and adaptive management can maximize the resilience of the chosen adaptive strategies by ensuring that flexibility and robustness criteria are incorporated into the evaluation of alternatives.

Adaptive Management

When uncertainty is high and the future is unpredictable, a system's adaptability becomes paramount to its resilience. Adaptive management offers an approach to both adjust to and resolve uncertainty. The classic approach, first described by Holling (1978), offers the means to inform iterative management decisions using an experimental, hypothesis-testing approach in which management actions are taken on the landscape, results are monitored and analyzed, information is fed back into the planning process, and management strategies are altered to account for a revised understanding of ecological dynamics (Anderson et al., 2003; Holling, 1978; Lee, 1993). While the practice of monitoring project implementation and making adjustments, so-called "passive" adaptive management (Anderson et al., 2003; NRC, 2004), is not new (Hill, 1968), this explicitly experimental, "active" form of adaptive management (Anderson et al., 2003; NRC, 2004; Pahl-Wostl et al., 2007) was a departure from an emphasis on identifying the best alternative and attempting to implement it as successfully as possible.

Adaptive management assumes that the principal uncertainties to be resolved concern how the system will react to management interventions. Climate change, however, presents coastal planners and managers with uncertainties about the future trends of external drivers that affect the systems with which they are concerned as well as the very social systems for which they are trying to enhance resilience. Changes in the rate of sea level rise or the intensity of tropical cyclones may influence how effective management interventions are over time and even render some infeasible or untenable. But these uncertainties cannot be resolved through experimentation. Coastal planners and managers must, therefore, rely on passive adaptive management while they gain better understanding of how external drivers are changing and how the vulnerabilities of the socio-ecological systems with which they are concerned are being affected (Adger et al., 2005; Folke, 2006; Walker et al., 2004).

Two important attributes of the adaptation strategies chosen by communities will have a significant effect on the community's adaptive capacity under these circumstances: flexibility and robustness (Adger & Vincent, 2005; Fankhauser et al., 1999; NRC, 2004; Pahl-Wostl et al., 2007; Quay, 2010). Fankhauser et al. (1999) define flexibility in terms of "accelerated capital turnover," i.e., capital projects with relatively rapid depreciation and short design lives. Viewed more broadly, flexible adaptive responses are susceptible to adjustment or substitution over time as planners and managers gain knowledge that helps to resolve uncertainty about systems and their drivers and as they contend with changing perceptions and priorities of stakeholders. Quay (2010: 499), citing Hallegatte (2009) and Easterling et al. (2004), suggests "creating flexible actions that can be broken into modules and implemented as needed as the future unfolds." Costs can then be distributed across time and losses minimized in the event that certain investments are no longer needed. Selection of flexible measures can also help to minimize "decision-making gridlock by making it clear that decisions are provisional, that there is . . . no "right" or "wrong" decision, and that modifications are [possible and] expected" (NRC, 2004: 20).

Robust actions are those that are capable of remaining effective over a range of possible system changes or multiple futures (Fankhauser et al., 2009; Pahl-Wostl et al., 2007; Quay, 2010). Such actions have the potential to achieve desired outcomes under various scenarios thereby reducing the need to shift strategies as future conditions unfold. Strategies that are both robust and flexible will optimize a community's adaptive capacity.

Scenario Analysis

Scenario analysis provides a way to assess the robustness of adaptive response alternatives as well as a means for engaging stakeholders in the process of identifying and analyzing those alternatives.[2] It involves defining an array of possible future states of the world that, in the absence of interventions, represent "alternative plausible combinations of the values of the key uncertainties" (USACE, 2011: 88). This process seeks to characterize multiple plausible futures that might unfold in response to foreseeable trends and/or disturbances over time and assessing each alternative against each scenario.

Planners can use the scenario development process as an opportunity to engage stakeholders in collaboratively learning about the uncertainties the community faces and the vulnerabilities for which they want to plan (Marcy et al., 2012; Yoe, 2004). How far into the future should we look when assessing land use and infrastructure investment policies that will commit us to fifty to one hundred years or more of urban development in areas that may become progressively more exposed to coastal hazards? What sea level rise scenarios should we consider at what points in time across that planning horizon? For what tropical storm intensities should we assess future vulnerabilities and the effects of alternative adaptive strategies on mitigating those vulnerabilities?

Once planners and stakeholders have specified the scenarios against which they wish to test their adaptive response alternatives, robustness can be defined as the extent to which an alternative performs satisfactorily across the array of plausible future conditions (Yoe, 2004). Planners and stakeholders can apply this criterion, along with adaptive flexibility, to identify those coastal hazard mitigation alternatives that are likely to maximize the community's capacity to adapt as future climate conditions and their effects are manifest.

Structuring Uncertainty with a Goals Achievement Matrix

Hill (1968) presents the goals achievement matrix (GAM) as a tool for structuring complex decisions where the decision maker wishes to apply multiple, non-commensurable evaluation criteria. He advocates using the GAM as a means to contend with the exclusion of non-market commodities in conventional cost-benefit analysis and with the practice of relying predominantly or exclusively on economic efficiency as the basis for evaluating public sector policy and program alternatives. The GAM also provides transparency about underlying goals and their importance and can accommodate various levels of analytic sophistication.

We illustrate here the use of a simplified GAM design suggested by Hill that is well suited to the needs and capabilities of low-resource communities. This approach minimizes analytic complexity and facilitates comparison of alternatives across all goals in the aggregate while retaining transparency about the goals used and their weights. Rather than estimating attainment of each goal using an individualized scale and measurement unit, the planner converts the scales for all goals to a common, unitless, ordinal scale, e.g., high, medium, and low (3, 2, 1). Doing so allows for less precise estimation of goal achievement levels and permits calculation of a weighted sums score for each alternative, thus simplifying the comparison of alternatives. This approach also permits straightforward sensitivity analyses in which planners and decision makers can assess the effects of alternative weights for individual goals and/or stakeholders.

Coastal planners can use the GAM in several ways to assess the adaptive potential of alternatives, i.e., how effectively alternatives will contribute to a community's adaptive capacity for managing the uncertain effects of climate change on coastal hazards. They can include flexibility of the strategy itself, and the resulting coastal development that it supports, as a goal in the matrix. They can operationalize flexibility of the strategy in terms of short economic and/or design life (Fankhauser et al., 1999) or more broadly as the ability to adjust or replace the intervention to accommodate changing conditions. Planners also can include a goal to minimize interference with autonomous adaptation by coastal ecosystems. They can evaluate the robustness of alternative strategies by constructing GAMs for two or more climate change scenarios.

ASSESSING ALTERNATIVE RESPONSES TO CLIMATE CHANGE IMPACTS ON COASTAL HAZARDS

In this section, we demonstrate how the GAM could be used for assessing a selection of alternative strategies for mitigating acute and chronic erosion, storm surge flooding, and wave damage from coastal storms while accounting for the uncertainties of climate change impacts on

Table 10.1 Sample Adaptive Response Options for Coastal Hazards with Accelerated Sea Level Rise and Increased Tropical Cyclone Intensity

Adaptive Response Strategy	*Response Option*	*Description*
Protect	Shore armoring	Seawalls, bulkheads, and revetments constructed parallel to the shore to protect upland embankments or structures from flooding, waves, and erosion.
	Beach nourishment	Artificial replenishment of beach (and dune) sediments to counteract the effects of chronic and/or acute erosion.
Accommodate	Elevate structures	Require elevating of habitable structures above a specified flood level using fill, or pilings (FEMA, 2009).
	Erosion-based setback	Require setback of habitable structures based on some multiple of the average annual erosion rate, e.g., thirty years
Avoid/ Retreat	Prohibit development plus transfer of development rights (TDR)	Zone vacant land vulnerable to coastal erosion and flooding so as to prohibit development and facilitate transfer of development rights from those properties to others determined to be more suitable for development.
	Prohibit development plus acquisition	Zone of vacant land vulnerable as above, and acquire those properties in fee-simple.
	Post-disaster down-zoning plus TDR	Down-zone built-out land so as to prohibit redevelopment of properties damaged by storm-induced erosion and/or flooding and facilitate transfer of development rights from those properties to other less vulnerable properties.
	Post-disaster down-zoning plus acquisition	Down-zone built-out land as above, and acquire those properties in fee-simple.
	Rolling easement	Prohibit shoreline armoring and require that structures be moved landward or removed altogether when the mean high water line reaches the seaward toe of the structure or some similar threshold condition.

rates of sea level rise and intensity of tropical cyclones. We define a set of goals that could be used to assess those alternatives, and we describe a coastal community vulnerability and assets typology that can be used to set weights for the goals that reflect local conditions. We simplify the method for this demonstration by not analyzing costs and benefits for different types of stakeholders.

Table 10.1 briefly describes a limited set of common strategies for mitigating damages from coastal erosion, storm surge flooding, and waves, plus a strategy specifically designed to accommodate accelerating sea level rise, rolling easements. We have classified these strategies into three categories of adaptive response to sea level rise generally following Dronkers et al. (1990): protect, accommodate, and avoid/retreat. While a number of other strategies can be considered (Grannis 2011; Titus & Craghan, 2009; van Raalten et al., 2009), this abbreviated set suffices for our demonstration.

Drawing from the criteria that an array of scholars and practitioners have used to assess climate change adaptation and hazard mitigation alternatives (Bendick et al., 2010; Fankhauser et al., 1999; Fussel, 2007; Godschalk et al., 1998; Hill, 1968; Smith, 1997; Titus & Neumann, 2009; USACE, 2011), we apply eight goals for evaluating these strategies: minimize risk to human development; minimize public sector capital and operating costs; minimize opportunity costs of regulation; minimize interference with ecological adaptation; maximize flexibility to adapt as conditions change and new knowledge is gained; minimize legal challenges, i.e., account for federal and state constitutional and state statutory constraints; maximize political feasibility vis-à-vis costs, property rights, etc.; and mitigate environmental stresses including climate change.

Several authors suggest relatively simple typologies for assessing the suitability of different adaptive response strategies for sea level rise. Titus et al. (2009) apply a heuristic typology based on existing and planned development intensity to prioritize the alternative responses of protection or retreat. They posit that communities are more likely to opt for protection against sea level rise as the mix of existing and planned land use shifts from areas dedicated to conservation and "vacant" land (e.g., agriculture or forestry) to higher levels of development intensity. Van Raalten et al. (2009) follow a similar approach in which they reduce the adaptive response decision framework to a four-cell "Strategy Development Method" matrix based on two primary values ascribed to a specific geographic area: (1) economic importance and intensity of existing development (high/low) and (2) natural hydrological and ecosystem dynamics (high-natural/low-altered).

We use a three-dimensional coastal community vulnerability and assets typology informed by these two approaches to illustrate how goal weights can be defined to reflect local conditions (see Table 10.2) based on coastal shoreline vulnerability (CSV), urban system value (USV), and natural system value (NSV). Table 10.3 shows how weights on a 100-point scale might be assigned to reflect those conditions and the priorities of stakeholders involved in formulating the GAM.

Table 10.4 illustrates how the weighted GAM might be applied to a rural coastal community on beach-dune coastal shoreline. We assign the ordinal scores for each adaptive response strategy based on our knowledge of these systems and pertinent literature (Titus & Craghan, 2009). These scores can be adjusted as new knowledge becomes available, but they should be

Table 10.2 Coastal Community Vulnerability and Assets Typology

	Scale		
Dimension	*High*	*Moderate*	*Low*
Coastal Shoreline Vulnerability (CSV)	High wave energy + high erosion vulnerability	High wave energy + low erosion vulnerability OR low wave energy + high to moderate erosion vulnerability	Low wave energy + low or no erosion vulnerability
AND/OR high inundation/flood exposure	AND/OR moderate inundation/flood exposure	AND/OR low inundation/flood exposure	
Urban System Value (USV)	Significant private and/or public sunk assets and important to current local and regional economies and culture	Substantial private and/or public sunk assets, important to local economy and culture and/or with substantial potential to become important to local and regional economies and culture	Few or no private and/or public sunk assets, lower economic potential, and low cultural value
Natural System Value (NSV)	Threatened or endangered species habitat and/or high recreational or economic value ecosystems, e.g., beaches, coastal wetlands that support fisheries	Other valuable coastal natural systems	Altered coastal ecosystems that do not give substantial support to recreational or commercial interests

High wave energy = National Flood Insurance Program (NFIP) V-zone
Low wave energy = National Flood Insurance Program A-zone
High erosion vulnerability = Sedimentary substrate (sand beach and dune system)
Moderate erosion vulnerability = Wetland (salt marsh or mangrove) substrate
Low erosion vulnerability = Consolidated substrate, e.g., Florida Keys
No erosion vulnerability = Armored shoreline (seawalls, bulkheads, revetments, etc.)
High inundation/flood exposure = Areas <3 ft above mean high water line (MHWL), or other factors*
Moderate inundation/flood exposure = Areas >3 ft and <5 ft above MHWL, or other factors*
Low inundation/flood exposure = Areas >5 ft above MHWL*

*Based on updated NFIP maps, storm surge projections, and local sea level rise inundation maps and time-frame projections. These may vary based on site-specific characteristics. For example, lands particularly subject to storm surge may be "high inundation" even though they are more than 3 ft above MHWL. Similarly, areas not at great risk of storm surge may be "low inundation" at elevations lower than 5 ft above MHWL.

Table 10.3 Example Assignment of Weights to Sea Level Rise Adaptive Response Goals

	Goal Weights								
Community Vulnerability and Assets Scenario	*Minimize Risk to Human Development*	*Minimize Capital and Operating Costs*	*Minimize Opportunity Costs of Regulation*	*Minimize Environmental Stresses*	*Minimize Interference with Ecological Adaptation*	*Maximize Adaptive Flexibility*	*Maximize Political Feasibility*	*Minimize Legal Challenges*	*Totals*
High-high-low									
CSV = high USV = high NSV = low	20.0	12.5	20.0	5.0	5.0	12.5	12.5	12.5	100.0
High-low-high									
CSV = high USV = low NSV = high	5.0	12.5	5.0	20.0	20.0	12.5	12.5	12.5	100.0
Low-high-low									
CSV = low USV = high NSV = low	20.0	20.0	20.0	5.0	5.0	5.0	12.5	12.5	100.0
Low-low-high									
CSV = low USV = low NSV = high	5.0	20.0	5.0	20.0	20.0	5.0	12.5	12.5	100.0

CSV = Coastal Shoreline Vulnerability
USV = Urban System Value
NSV = Natural System Value

Table 10.4 Adaptive Response Goals Achievement Matrix for a Rural Coastal Community on Beach-Dune Coastal Shoreline

Scenario CSV = high USV = low NSV = high	*Goals*								
	Minimize Risk to Human Development	*Minimize Capital and Operating Costs*	*Minimize Opportunity Costs of Regulation*	*Minimize Environmental Stresses*	*Minimize Interference with Ecological Adaptation*	*Maximize Adaptive Flexibility*	*Maximize Political Feasibility*	*Minimize Legal Challenges*	*Total Weighted Score*
Goal weights	*5.0*	*12.5*	*5.0*	*20.0*	*20.0*	*12.5*	*12.5*	*12.5*	
Adaptive Response Strategies									
Protect									
Shore armoring	3	-3	3	-2	-3	-2	0	2	-107.5
Beach nourishment	2	-2	1	1	-1	2	0	1	27.5
Accommodate									
Elevate structures	3	-2	2	1	-1	1	0	2	37.5
Erosion-based setback	1	2	2	2	1	1	0	2	137.5
Avoid/Retreat									
Prohibit development plus TDR	3	2	-2	3	3	2	0	2	200.0
Prohibit development plus acquisition	3	-2	-3	3	3	2	0	0	120.0
Post-disaster down-zoning plus TDR	3	2	-2	3	3	2	-3	2	162.5
Post-disaster down-zoning plus acquisition	3	-2	-3	3	3	2	-3	0	82.5
Rolling easement	2	1	1	2	3	2	0	-1	152.5

Rating scale:
+ + + highly positive impact on goal = 3
+ + moderately positive impact = 2
+ somewhat positive impact = 1
+/–neutral or uncertain impact on goal = 0
–somewhat negative impact = -1
—moderately negative impact = -2
——highly negative impact = -3

held constant across community vulnerability and asset conditions. In the following paragraphs we illustrate the manner in which stakeholders might score two protection alternatives: shore armoring and beach nourishment.

We rate hard-engineered structures, such as shore armoring (Nordstrom, 2000) highly for minimizing risk to human development because they can be designed to withstand high waves and storm surges. They are, however, very expensive to construct (Beever et al., 2009) and will eventually incur costs for repair and/or replacement due to advancing shore erosion and reduced effectiveness as sea level rises. Shore armoring minimizes opportunity costs of regulation by allowing development of vulnerable lands for some periods of time. However, shore armoring can prevent landward migration of beach and dune systems and/or coastal wetlands as sea level rises and can exacerbate erosion both in front of and adjacent to the structure. These structures therefore increase environmental stress and present impediments to autonomous landward migration of beach-dune ecosystems. The sunk capital costs in hard-engineered structures, coupled with design lives of fifty years, acording to American Society of Civil Engineers (ASCE), limit flexibility to adapt as better knowledge is gained about future trends in sea level rise and tropical cyclone intensity. Political opposition and legal challenges are likely minimal except where environmental advocates oppose shore armoring or where the structures will adversely affect neighboring properties.

"Soft" engineered approaches such as beach nourishment often provide lower design levels of protection, may have lower capital costs, with fewer direct adverse impacts on natural systems and greater long-term flexibility for both natural and human systems to adapt to changing tropical cyclone intensity as well as sea level rise and changing shorelines (NRC, 1995). Beach nourishment benefits can be short-lived if a project area is hit by a severe storm, potentially requiring expensive subsequent renourishment. Nonetheless, nourishment is often sought because it helps preserve the recreation, tourism, and ecosystem values of the beach. Beach nourishment has historically been viewed favorably by property owners and the public and thus has experienced little political opposition and few legal challenges. While property owners recently challenged the constitutionality of publicly funded beach nourishment in Florida (*Stop the Beach Renourishment, Inc. v. Florida Department of Environmental Protection*, 130 S. Ct. 2592 [2010]), the state prevailed.

The total weighted scores for the alternative strategies shown in Table 10.4 illustrate application of the weights for the high-low-high community vulnerability and assets scenario from Table 10.3: the ordinal score in each cell is multiplied by the goal weight and the products are summed across the row for each strategy. Table 10.5 shows how the total weighted scores of the alternative strategies would change for the low-high-low scenario shown in Table 10.3, e.g., a small coastal city facing the bay with high urban density and limited remaining natural system assets. Comparison of

Table 10.5 Adaptive Response Goals Achievement Matrix for a Small Coastal Bay-Front City

Scenario CSV = high USV = low NSV = high	*Goals*								
	Minimize Risk to Human Development	*Minimize Capital and Operating Costs*	*Minimize Opportunity Costs of Regulation*	*Minimize Environmental Stresses*	*Minimize Interference with Ecological Adaptation*	*Maximize Adaptive Flexibility*	*Maximize Political Feasibility*	*Minimize Legal Challenges*	*Total Weighted Score*
Goal weights	*20.0*	*20.0*	*20.0*	*5.0*	*5.0*	*5.0*	*12.5*	*12.5*	
Adaptive Response Strategies									
Protect									
Shore armoring	3	-3	3	-2	-3	-2	0	2	65.0
Beach nourishment	2	-2	1	1	-1	2	0	1	42.5
Accommodate									
Elevate structures	3	-2	2	1	-1	1	0	2	90.0
Erosion-based setback	1	2	2	2	1	1	0	2	145
Avoid/Retreat									
Prohibit development plus TDR	3	2	-2	3	3	2	0	2	125
Prohibit development plus acquisition	3	-2	-3	3	3	2	0	0	10
Post-disaster down-zoning plus TDR	3	2	-2	3	3	2	-3	2	87.5
Post-disaster down-zoning plus acquisition	3	-2	-3	3	3	2	-3	0	-37.5
Rolling easement	2	1	1	2	3	2	0	-1	102.5

Rating scale:
+ + + highly positive impact on goal = 3
+ + moderately positive impact = 2
+ somewhat positive impact = 1
+/–neutral or uncertain impact on goal = 0
–somewhat negative impact = -1
——moderately negative impact = -2
———highly negative impact = -3

Tables 10.4 and 10.5 reveals the sensitivity of the strategy of total weighted scores to the goal weights and the importance of devising decision frameworks that are sensitive to variations in shoreline vulnerability, urban system value, and natural system value.

To assess the robustness of different adaptive response strategies, one could evaluate their performance for some combination of possible sea level rise scenarios coupled with storm surge associated with tropical cyclones of different intensities. States and local governments are using the projected amount of sea level rise by 2100 relative to 1990 as a common basis for defining future sea level rise scenarios. A number of scholars and practitioners have defined 1.0 meter of sea level rise relative to 1990 by 2100 as a plausible adaptive response scenario (Miami-Dade County Climate Change Task Force, 2008; Mitchum, 2011; Pfeffer et al., 2008; Vermeer & Rahmstorf, 2009). Several scientists have defined an upper bound by 2100 on the order of 2.0 meters (Pfeffer et al., 2008; Vermeer & Rahmstorf, 2009), while Hansen (2007) indicates that with a ten-year doubling acceleration rate, sea level could reach 5.0 meters by 2100. Marcy et al. (2012) report that the average 2100 sea level rise scenario ranges for ten state and local governments range from 0.6 to 1.1 meters. Intermediate scenarios might be set at half the 2100 value for 2050. To incorporate the effects of tropical cyclone intensity, one could construct scenarios using a Category 2 or Category 4 hurricane with the different sea level rise scenarios.

Marcy et al. (2012) lay out a process for developing local sea level rise scenarios, but the data gathering and processing involved require significant time and technical expertise. Formal vulnerability assessments will require even more advanced technology and associated skills such as processing LiDAR data to construct digital elevation maps and conducting sophisticated damage estimates using such software as HAZUS-MH. The National Oceanic and Atmospheric Administration's Coastal Services Center provides a variety of analytic tools that smaller communities can use to produce less sophisticated simulations of sea level rise and its impacts that can be used for preliminary awareness building and alternatives identification (NOAA CSC, 2012). However, most small communities will need to rely on technical assistance from states to produce the detailed information and analyses they need or they will have to pay consultants to prepare sea level rise scenarios and vulnerability assessments for them. Fortunately, this is a role already played by many states in the context of coastal hazards. The state of Florida, for example, financed LiDAR mapping of its entire coast and preparation of SLOSH storm surge maps and regional evacuation plans through its regional planning councils (ARPC, 2010). A similar effort could produce sea level rise inundation and storm-surge-plus-sea-level-rise scenario mapping for use by local governments.

Adaptive Use of Goals Achievement Matrices

The GAM as we have described it provides a decision support tool that can be employed by a resource-constrained community in order to take action in the face of uncertainty. In this section, we argue that the GAM must not only incorporate measures of the adaptive potential of alternative strategies, but also must be used in a way that fosters greater adaptive capacity for the community through ongoing learning and adaptation of the tool itself.

Whatever decision-making process is used in resource-constrained coastal communities, planning for complex dynamic systems when faced with uncertainty requires learning and feedback (Diduck, 2010; Pahl-Wostl et al., 2007). As internal or external system conditions change, planners and managers need to have the ability to assess changes, take into account new conditions and new information, and alter management strategies to reflect those changes. There are several sources of new information that should be reflected in adaptive use of the GAM. First, new information can be obtained from experience by monitoring the effectiveness of the implemented strategy. Second, information can be shared with other communities that are attempting to adapt to the same environmental change phenomena and that have similar coastal hazard vulnerabilities. Third, ongoing scientific research may reveal new knowledge about the underlying environmental change phenomena which may reduce (or increase) uncertainty about future change, associated impacts, and the likely effectiveness of strategies being implemented. In addition, stakeholder perceptions may shift in response to new understandings of environmental change phenomena, the onset of disruptions to the community's socio-ecological system (e.g., from a natural disaster or a socio-economic crisis such as a national or global recession), or changes in values that accompany demographic shifts in the community. These changes in perceptions or values may alter how new information is interpreted and applied.

Coastal planners and stakeholders may choose to reassess and revise elements of the GAM to reflect what they have learned and to incorporate new perspectives about the challenges to community resiliency and the suitability of different management alternatives. For example, the community learns that enhanced models of ice sheet melting now indicate that sea level rise is more likely to reach 1 meter above 1990 levels by 2050 rather than 0.5 meters and that the rate is doubling every ten years. This means that erosion-based setbacks, seawalls, beach renourishment projects, and elevation standards will provide considerably less protection than anticipated over the next forty years. It also means that such measures will need to incorporate greater margins of safety or risk becoming obsolete before they are fully depreciated. Coastal planners and stakeholders may want to reassess their risk averseness, i.e., how willing they are to risk under- or overadaptation, and in that light, to reconsider their options for mitigating coastal hazards as climate change continues to alter the environment for which they are planning.

Adaptive Governance for Resilience Planning

Tools for decision making about mitigating vulnerability and enhancing resilience in the face of uncertainty are most effective when deployed through a collaborative process that engenders social learning and fosters social capital through networks to build adaptive capacity in the system (Adger & Vincent, 2005; Boyd & Folke, 2011; Cutter et al., 2012). Thus, coastal communities will need to develop an adaptive governance approach for planning in the face of uncertainty, to enable learning and to build a greater capacity to act on what has been learned in their management efforts.

Knowledge and Social Learning

Sorting through relevant knowledge and information to assess alternative adaptive strategies is essential to successfully adapting management strategies over time as conditions, social values, and perspectives change. Knowledge and information will arise from diverse sources, be incomplete, and will be a source of uncertainty unto itself. Furthermore, coastal communities will be constantly challenged to stay abreast of the rapid advances being made in climate change science.

Joint fact finding can be an effective way for communities to engage in adaptive learning in the face of uncertainty. Participants, functioning as a "community of inquiry," seek new information, sort through and test it, and educate themselves to build a common new understanding about the planning challenges they face (Innes & Booher, 2010; Karl, Susskind, & Wallace, 2007). Scientists and other experts can play an important role by informing community understanding based on research and explicit knowledge. Planners can help to translate complex scientific information into lay terms and work with stakeholders to construct common understanding by applying new knowledge and information to local contexts and their experiences *in situ*. Joint fact finding can help communities uncover "truth" as "deeply embedded in context, a time and a place" not simply by revealing facts, but by exploring "relationships, causes, and predictions" (Innes & Booher, 2010: 160).

Participation of Diverse Stakeholders

No individual or organization will be able to gather, comprehend, and apply all of the relevant information about the complex layers of a coastal socio-ecological system in a context of high levels of uncertainty about external climate drivers and their effects (Innes, 1996; Innes & Booher, 2010). Joint fact finding will most effectively foster social learning for iterative climate change adaptation when coastal planners engage experts from multiple disciplines and organizations, as well as lay and political participants, in open communication to generate knowledge about facts, values, problems, and opportunities; areas of agreement and disagreement; alternative actions; and possibilities for working together (Diduck 2010; Pahl-Wostl et al., 2007; Schusler, Decker, & Pfeffer, 2003; Toman, 2006). Communities can maximize creativity and innovation if they engage experts and stakeholders through an approach in which "community members are empowered to take control of the processes involved" (Cutter et al., 2012: 300).

Climate change and its associated impacts on coastal hazards will affect a wide array of individuals and organizations including landowners, government agents, business owners, tourists, seasonal visitors, etc. Participation of these diverse parties in deciding how to adapt, with their multiple interests, perspectives, values, and ways of understanding the world, will foster greater opportunities for learning and creativity (Innes & Booher 2010; Lowe, Foster, & Winkelman, 2009; Schusler et al., 2003; Walker & Salt, 2006). Coastal planners should engage an equally wide array of empirical observers, including knowledgeable locals as well as experts, environmentalists, advocates of social equity, land use professionals, etc., to inform efforts to adapt to these changes (Cutter et al., 2012).

But how do planners motivate stakeholders to participate? Natural hazards issues have notoriously low salience among local officials and their constituents (see for example Burby et al., 1999; May & Williams, 1986; Rossi, Wright, & Weber-Burdin, 1982) and sea level rise is more so because of its gradual onset. A common recommendation is to increase awareness and understanding of the threats posed and the benefits to be gained through mitigation (Burby & May, 1998; Fussel, 2007; Lowe et al., 2009).

In its Urban Leaders Adaptation Initiative, the Center for Clean Air Policy (CCAP) recommends documenting the economic benefits of adaptive action versus inaction while stressing that adaptation is a necessity because of climate change that already has occurred (Foster, Winkelman, & Lowe, 2011). CCAP emphasizes the importance of engaging policy elites such as "the professional class of developers, planners, financiers, insurers, and lawyers who work with elected leaders and the city or county managers" to serve as "prime motivators" for adaptation (Foster et al., 2011: 11). CCAP also recommends packaging climate change impact information to show its relevance to local concerns.

Public concern with sea level rise can best be engaged by emphasizing its effects on tangible problems with which communities are already concerned such as coastal flooding (Foster et al., 2011). While the effects of periodic large coastal storms will be aggravated by sea level rise, a more immediate and regularly occurring impact is the increased height of monthly spring high tides and associated local flooding. Several authorities have thus recommended that planners map sea level rise vulnerability based on spring high tide elevations rather than mean high tide. Gregg et al. (2011: 57–58) describe an array of techniques that communities have successfully used to engage stakeholders in dialogue about climate change adaptation including "meetings and workshops, one-on-one interactions, . . . websites, . . . interactive, participatory discussions, problem solving sessions, and role-playing exercises." Although ways to increase the participation of stakeholders will vary from community to community, planners and managers will benefit from raising the salience of climate change risk and adaptation in their communities.

Monitoring and Continuous Learning

Ongoing monitoring of changes in the external environment, and of the responses of a community's socio-ecological system to those changes and to management initiatives intended to enhance resilience, can be a valuable source of input to the community's continuous process of adaptive learning. Such monitoring can help redress uncertainties or highlight the effectiveness (or lack thereof) of management alternatives. Engaging the multiple parties of the adaptive management "community of inquiry" in the monitoring effort can facilitate collective learning and trust building (Moote, 2011).

Coastal communities can gather a variety of data to keep track of changing conditions as well as evaluate the effectiveness of interventions. For example, they can deduce local relative sea level rise based on information collected from the nearest tide gauges and scientific studies about global or regional eustatic sea level rise rates as well as draw on updated sea level rise estimates developed by the Intergovernmental Panel on Climate Change (IPCC) and others and the mapping and analytic tools available from the NOAA Coastal Services Center (NOAA CSC, 2012). Some states (e.g., Florida) monitor erosion rates and the U.S. Geological Survey publishes subsidence and rebound rates (Marcy et al., 2012), information that can be combined with local observations of changes in coastal shoreline systems. To monitor the effectiveness of interventions, stakeholders will need to observe how well strategies perform in achieving multiple goals in their GAM, particularly following coastal storms and in relation to chronic erosion and inundation associated with rising sea levels. As participants monitor these changing conditions and the performance of management strategies, they will need to revisit and reassess the GAM to incorporate new information and knowledge. Stakeholders may choose to revise goals, weights, strategies, or scoring based on the ongoing collection of data and information over time.

CONCLUSIONS

Innes and Booher (2010: 206) suggest that "resilience thinking shifts the purpose of governance from the questions 'where do we want to be and how do we get there' to 'how do we move in a desirable direction in the face of uncertainty?'" We have sought to offer guidance to planners and managers working with low-resource and small coastal communities as they seek to maintain and enhance their resilience in the face of indeterminate uncertainty about climate change impacts on coastal hazards. Such communities face significant challenges as they attempt to adapt to accelerating sea level rise and the possibility of increasing intensity of tropical cyclones. Knowledge and information requirements are high, uncertainty and complexity are largely irreducible, and social perceptions, priorities, and values are likely to change over the decades that adaptive strategies must function. Resilience of these communities will depend upon increasing adaptive capacity by implementing robust and flexible strategies while learning and acting on new understandings about the changing internal and external drivers of change.

In response to these challenges, planners and public managers have drawn on several decision-making tools, techniques, and approaches as they seek to manage "within their means" while building new capacities for adaptation. We argue that conventional, resource-intensive tools such as cost-benefit analysis and risk-benefit analysis are poorly suited to account for the diverse social and ecological goals of resilience planning and the complexities and indeterminate uncertainties of climate change effects on coastal hazards. We suggest that passive adaptive management coupled with scenario analysis offers an alternative approach that may more effectively address uncertainty and complexity while providing ample opportunity to identify robust and flexible strategies going forward. As the latest IPCC report on risk management clarifies, "Management strategies need to consider adaptation as a process rather than measures and actions for a particular event or time period" (Cutter et al., 2012: 323).

To aid in this endeavor, we draw inspiration from Hill's (1968) GAM to develop an approach that planners and stakeholders can employ to identify adaptive strategies that are scientifically defensible, socially acceptable, politically viable, and economically feasible. We argue that this tool is most aptly used within an adaptive governance framework that fosters collaborative fact finding and learning. We provide guidance for how coastal communities can design this governance approach, suggesting that it involve multiple stakeholders and experts in ongoing inquiry as they engage new knowledge from scientific research, experiments, and monitoring. This approach to using a GAM can provide the flexibility to assess and modify management strategies in response to changes in internal and external conditions and to changes in stakeholder values, perceptions, and goals and to build community adaptive capacity through social learning.

Undertaking this work will require technical assistance from states to enable low-resource coastal communities to develop future sea level rise scenarios against which they can assess alternative adaptation strategies. Building community resilience also will require a level of humility on the part of planners and managers, a clear recognition of uncertainty and complexity and the challenges that they bring. It also will require a level of openness and flexibility in decision making and relinquishing control to a broader base of stakeholders to enable greater creativity and innovation in the face of unpredictable and, in many ways, unthinkable changes ahead.

ACKNOWLEDGEMENTS

The authors wish to acknowledge the contributions of Thomas Ruppert and Michael Volk to the conceptualization of the shoreline typology and adaptive response goals achievement matrix during an earlier collaboration with Robert Deyle.

NOTES

1. Eustatic sea level rise is the increase in the volume of the oceans that results primarily from the thermal expansion of sea water as heat is transferred from the atmosphere and from the melting of glaciers, ice caps, and the Greenland and West Antarctic ice sheets. Sea level rise experienced by an observer on land is referred to as "relative" sea level rise. This is a function of changes in eustatic sea level as well as shifts in the elevation of the land.
2. A number of scholars also describe and advocate what they refer to as "scenario planning" (see for example Peterson, Cumming, & Carpenter, 2003). While closely related methodologically, scenario planning typically refers to the construction of complex, multi-dimensional alternative futures that represent whole systems and the effects of changes in multiple variables and drivers. We focus here on the narrower concept of scenario analysis as it can be applied to assessing the effectiveness of a set of more discrete alternative actions.

REFERENCES

Adger, W.N., Agrawala, S., Mirza, M.M.Q., Conde, C., O'Brien, K., Pulhin, J., Pulwarty, R., Smit, B., & Takahashi, K. (2007). Assessment of adaptation practices, options, constraints and capacity. In M.L. Parry, O.F. Canziani, J.P. Palutikof, P.J. van der Linden, & C.E. Hanson (Eds.), *Climate change 2007: Impacts, adaptation and vulnerability; Contribution of Working Group II to the fourth assessment report of the Intergovernmental Panel on Climate Change*. New York: Cambridge University Press.

Adger, W.N., Brooks, N., Bentham, G., Agnew, M., & Eriksen, S. (2004). *New indicators of vulnerability and adaptive capacity*. Technical report no. 7. Tyndall Centre for Climate Change Research. Norwich, UK: East Anglia University.

Adger, W.N., Hughes, T.P., Folke, C., Carpenter, S.R., & Rockstrom, J. (2005) Social-ecological resilience to coastal disasters. *Science*, 309, 1036–1039.

Adger, W.N., & Vincent, K. (2005). Uncertainty in adaptive capacity. *Comptes Rendues Geoscience*, 337, 399–410.

Ali, A., Angelov, P., & Hutchison, D. (2011). Towards an adaptive resilience strategy for future computer networks. In *Proceedings of the 11th UK workshop on computational intelligence* (pp. 201–206). Manchester, UK: University of Manchester School of Computer Science. Retrieved February 10, 2012 from http://ukci.cs.manchester.ac.uk/files/Proceedings.pdf.

Anderson, J.L., Hilborn, R.W., Lackey, R.T., & Ludwig, D. (2003). Watershed restoration: Adaptive decision making in the face of uncertainty. In R.C. Wissmar & P.A. Bisson (Eds.), *Strategies for restoring river ecosystems: Sources of variability and uncertainty in natural and managed systems* (pp. 203–232). Bethesda, MD: American Fisheries Society.

ARPC (Apalachee Regional Planning Council). (2010). *Statewide regional evacuation study program*, vol. 1–2, *Apalachee region technical data report*. Blountstown, FL: ARPC.

Beatley, T. (2009). *Planning for coastal resilience: Best practices for calamitous times*. Washington, DC: Island Press.

Beever, J.W., Gray, W., Trescott, D., Cobb, D., Utley, J., Hutchinson, D., Gibbons, J., . . . & Ott, J. (2009). *City of Punta Gorda adaptation plan*. Technical report 09–4. Ft. Myers, FL: Southwest Florida Regional Planning Council and Charlotte Harbor National Estuary Program.

Bendick, R., Dahlin, K.M., Smoliak, B.V., Kumler, L., Jones, S.J., Aktipis, A., . . . & Scott, D. (2010). Choosing carbon mitigation strategies using ethical deliberation. *Weather, Climate and Society*, 2, 140–147.

Boyd, E., & Folke, C. (2011). Conclusions: Adapting Institutions and Resilience. In: Boyd, E. and C. Folke (eds.). *Adapting Institutions: Governance, Complexity and Social-Ecological Resilience* pp 264–280. Cambridge University Press, Cambridge, UK.

Boyd, E., & Folke, C. (2012). *Adapting institutions: Governance, complexity and social-ecological resilience*. Cambridge: Cambridge University Press.

Burby, R.J., Beatley, T., Berke, P.R., Deyle, R.E., French, S.P., Godschalk, D.R., Kaiser, E.J., . . . & Platt, R.H. (1999). Unleashing the power of planning to create disaster resistant communities. *Journal of the American Planning Association*, 65(3), 247–258.

Burby, R.J., & May, P.J. (1998). Intergovernmental environmental planning: Addressing the commitment conundrum. *Journal of Environmental Planning and Management*, 41(1), 95–110.

Callaway, J.M. (2004). Adaptation benefits and costs: Are they important in the global policy picture and how can we estimate them? *Global Environmental Change*, 14, 273–282.

Carpenter, S., Walker. B., Anderies, M., & Abel, N. (2001). From Methapor to Measurement: Resilience of What to What? *Ecosystems*, 4, 765–781.

Cutter, S., Osman-Elasha, B., Campbell, J., Cheong, S.-M., McCormick, S., Pulwarty, R., Supratid, S., & Ziervogel, G. (2012). Managing the risks from climate extremes at the local level. In C.B. Field, V. Barros, T.F. Stocker, D. Qin, D.J. Dokken, K.L. Ebi, M.D. Mastrandrea, . . . & P.M. Midgley (Eds.), *Managing the risks of extreme events and disasters to advance climate change adaptation: A special report of Working Groups I and II of the Intergovernmental Panel on Climate Change (IPCC)* (pp. 291–338). Cambridge: Cambridge University Press.

Deyle, R.E., Bailey, K.C., & Matheny, A. (2007). *Adaptive response planning to sea level rise in Florida and implications for comprehensive and public-facilities planning*. Tallahassee: Florida Planning and Development Lab, Florida State University.

Diduck, A. (2010). The learning dimension of adaptive capacity: Untangling the multi-level connections. In D. Armitage & R. Plummer (Eds.), *Adaptive capacity and environmental governance*. New York: Springer-Verlag.

Dronkers, J., Misdorp, R., Schröder, P.C., Carey, J.J., Spradley, J.R., Vallianos, L., . . . & James, E. (1990). *Strategies for adaptation to sea level rise: Report of the Coastal Zone Management Subgroup to the Intergovernmental Panel on Climate Change response strategies working group*. The Netherlands: Ministry of Transport and Public Works. Retrieved from http://papers.risingsea.net/federal_reports/IPCC-1990-adaption-to-sea-level-rise.pdf

Elsner, J.B., Kossin, J.P., & Jagger, T.H. (2008). The increasing intensity of the strongest tropical cyclones. *Nature*, 455, 92–95.

Emanuel, K.A. (2005). Increasing destructiveness of tropical cyclones over the past 30 years. *Nature*, 436, 686–688.

Emanuel, K.A. (2007). Environmental factors affecting tropical cyclone power dissipation. *Journal of Climate*, 20(22), 5497–5509.

Easterling, W. E., Hurd, B. H., & Smith, J. B. (2004). *Coping with global climate change: The role of adaptation in the United States*. Arlington, VA: The Pew Center on Global Climate Change.

Fankhauser, S. (2010). The costs of adaptation. *WIRES Climate Change*, 1(1), 23–30.

Fankhauser, S., Smith, J.B., & Tol, R.S.J. (1999). Weathering climate change: Some simple rules to guide adaptation decisions. *Ecological Economics*, 30, 67–78.

Folke C. (2001) Building social-ecological resilience for sustainability. in Svedin U., (ed), *Human Dimensions of Global Environmental Change: Swedish Perspectives 2001*. Swedish Council for Planning and Coordination of Research. Stockholm.

Folke, C. (2006). Resilience: The emergence of a perspective for social–ecological systems analyses. *Global Environmental Change*, 16, 253–267.

Foster, J., Winkelman, S., & Lowe, A. (2011). *Lessons learned on local climate adaptation from the Urban Leaders Adaptation Initiative*. Washington, DC: Center for Clean Air Policy.

Fussel, H.M. (2007). Adaptation planning for climate change: Concepts, assessment approaches, and key lessons. *Sustainability Science*, 2, 265–275.

Godschalk, D.R., Kaiser, E.J., & Berke, P.R. (1998). Integrating hazard mitigation and local land use planning. In R.J. Burby (Ed.), *Cooperating with nature: Confronting natural hazards with land-use planning for sustainable communities* (pp. 85–118). Washington, DC: Joseph Henry Press.

Grannis, J. (2011). *Adaptation tool kit: Sea-level rise and coastal land use*. Washington, DC: Georgetown Climate Center. Retrieved from http://www.georgetownclimate.org/sites/default/files/Adaptation_Tool_Kit_SLR.pdf.

Gregg, R.M., Hansen, L.J., Feifel, K.M., Hitt, J.L., Kershner, J.M., Score, A., & Hoffman, J.R. (2011). *The state of marine and coastal adaptation in North America: A synthesis of emerging ideas*. Bainbridge Island, WA: EcoAdapt.

Grinsted, A., Moore, J.C., & Jevrejeva, S. (2010). Reconstructing sea level from paleo and projected temperatures 200 to 2100 AD. *Climate Dynamics*, 34, 461–472.

Hallegatte, S. (2009). Strategies to adapt to an uncertain climate change. *Global Environmental Change*, 19(2), 240–247.

Hansen, J.E. (2007). Scientific reticence and sea level rise. *Environmental Research Letters*, 2(2), 024002 (6 pp.).

Hansen, J.E., & Sato, M. (2011). Paleoclimate implications for human-made climate change. Retrieved December 29, 2011 from http://arxiv.org/ftp/arxiv/papers/1105/1105.0968.pdf.

Hill, M. (1968). A goals-achievement matrix for evaluating alternative plans. *Journal of the American Planning Association*, 34(1), 19–29.

Holling, C.S. (1973). Resilience and Stability of Ecological Systems. *Annual Review of Ecology and Systematics*, 4, 1–23.

Holling, C.S. (1978). *Adaptive environmental assessment and management*. New York: John Wiley and Sons.

Innes, J.E. (1996). Planning through consensus building: A new view of the comprehensive planning ideal. *Journal of the American Planning Association*, 62(4), 460–472.

Innes, J.E., & Booher, D.E. (2010). *Planning with complexity: An introduction to collaborative rationality for public policy*. New York: Routledge Press.

Karl, T.R., Meehl, G.A., Miller, C.D., Hassol, S.J., Waple, A.M., & Murray, W.L. (2008). *Weather and climate extremes in a changing climate; Regions of focus: North America, Hawaii, Caribbean, and U.S. Pacific Islands*. Synthesis and Assessment Product 3.3. Washington, DC: U.S. Climate Change Science Program and the Subcommittee on Global Change Research.

Karl, H., Susskind, L., & Wallace, K. (2007). A dialogue, not a diatribe: Effective integration of science and policy through joint fact finding. *Environment*, 49(1), 20–34.

Kiker, G.A., Bridges, T., Varghese, A., Seager, T., & Linkov, I. (2005). Application of multicriteria decision analysis in environmental decision making. *Integrated Environmental Assessment and Management*, 1(2), 95–108.

Kneese, A.V. (1984). *Measuring the benefits of clean air and water*. Washington, DC: Resources for the Future.

Lee, K.N. (1993). *Compass and gyroscope: Integrating politics and science for the environment*. Washington, DC: Island Press.

Lowe, A., Foster, J., & Winkelman, S. (2009). *Ask the climate question: Adapting to climate change impacts in urban regions*. Washington, DC: Center for Clean Air Policy.

Marcy, D., Allen, A., Sweet, W., Gill, S., Luscher-Aissaoui, A., Myers, E., & Zervas, C. (2012). Incorporating sea level change scenarios at the local level. National Oceanic and Atmospheric Administration. Retrieved July 25, 2012 http://csc.noaa.gov/digitalcoast/_/pdf/slcscenarios.pdf

May, P.J., & Williams, W. (1986). *Disaster policy implementation: Managing programs under shared governance*. New York: Plenum Press.

Meehl, G.A., Stocker, T.F., Collins, W.D., Friedlingstein, P., Gaye, A.T., Gregory, J.M., . . . & Zhao, Z.-C. (2007). Global climate projections. In S. Solomon, D. Qin, M. Manning, Z. Chen, M. Marquis, K.B. Averyt, M. Tignor, & H.L. Miller (Eds.), *Climate change 2007: The physical science basis; Contribution of Working Group I to the fourth assessment report of the Intergovernmental Panel on Climate Change* (pp. 747–845). New York: Cambridge University Press.

Miami-Dade County Climate Change Advisory Task Force. (2008). *Second report and initial recommendations*. Miami: Miami-Dade County Climate Change Advisory Task Force.

Mitchum, G.T. (2011). Sea level changes in the southeastern United States: Past, present, and future. Southeast Climate Consortium and the Florida Climate Institute. Retrieved from http://www.seclimate.org/pdfpubs/201108mitchum_sealevel.pdf.

Moote, A. (2011). *Multi-party monitoring and stewardship contracting: A tool for adaptive management*. Portland, OR: Sustainable Northwest. Retrieved February 9, 2012 from http://www.sustainablenorthwest.org/resources/publications/Multiparty%20Monitoring%20Guidebook%202012_finalV3_links.pdf.

Nicholls, R.J., Wong, P.P., Burkett, V.R., Codignotto, J.O., Hay, J.E., McLean, R.F., Ragoonaden, S., & Woodroffe, C.D. (2007). Coastal systems and low-lying areas. In M. Parry, O.F. Canziani, J.P. Palutikof, P. van der Linden, & C.E. Hanson (Eds.), *Climate change 2007: Impacts, adaptation and vulnerability*. Cambridge: Cambridge University Press.

Ning L., Emanuel, K., Oppenheimer, M., & Vanmarcke, E. (2012). Physically based assessment of hurricane surge threat under climate change. *Nature Climate Change*, 2, 426–467.

NOAA CSC (National Oceanic and Atmospheric Administration Coastal Services Center). (2012). Sea level rise and coastal flooding impacts viewer. Retrieved from http://www.csc.noaa.gov/tools/.

Nordstrom, K.F. (2000). *Beaches and dunes of developed coasts*. Cambridge: Cambridge University Press.

NRC (National Research Council). (1995). *Beach nourishment and protection*. Washington, DC: The National Academies Press.

NRC (National Research Council). (2004). *Adaptive management for water resources project planning*. Washington, DC: The National Academies Press. Retrieved January 23, 2012 from http://www.nap.edu/catalog/10972.html.

Pahl-Wostl, C., Sendzimir, J., Jeffrey, P., Aerts, J., Berkamp, G., & Cross, K. (2007). Managing change toward adaptive water management through social learning. *Ecology and Society*, 12(2), 30. Retrieved January 26, 2012 from http://www.ecologyandsociety.org/vol12/iss2/art30/.

Parry, M., Canziani, O., Palutikof, J., van der Linden, P., & Hanson, C. (2007). *Climate change 2007: Impacts, adaptation, and vulnerability. Contribution of Working Group II to the fourth assessment report of the Intergovernmental Panel on Climate Change* [electronic version]. Retrieved May 23, 2007 from http://www.ipcc.ch.

Peterson, G., Cumming, G., & Carpenter, S. (2003). Scenario planning: A tool for conservation in an uncertain world. *Conservation Biology*, 17(2), 358–366.

Pfeffer, W.T., Harper, J.T., & O'Neel, S. (2008). Kinematic constraints on glacier contributions to 21st-century sea-level rise. *Science*, 321, 1340–1343.

Quay, R. (2010). Anticipatory governance: A tool for climate change adaptation. *Journal of the American Planning Association*, 76(4), 496–511.

Robinson, M. (2010). *Making adaptive resilience real*. London: Arts Council of England. Retrieved February 10, 2012 from http://www.britishcouncil.org/making_adaptive_resilience_real-2.pdf.

Rothman, D.S. (2000). Measuring environmental values and environmental impacts: Going from the local to the global. *Climatic Change*, 44, 351–376.

Rossi, P., Wright, J.D., & Weber-Burdin, E. (1982). *Natural hazards and public choice: The state and local politics of hazard mitigation*. New York: Academic Press.

Schusler, T.M., Decker, D., & Pfeffer, M. (2003). Social learning for collaborative natural resource management. *Society and Natural Resources*, 16(4), 309–326.

Schwab, J., Topping, K.C., Eadie, C.C., Deyle, R.E., & Smith, R.A. (1998). *Planning for post-disaster recovery and reconstruction*. Planning Advisory Service report no. 483/484. Chicago: American Planning Association.

Smith, J.B. (1997). Setting priorities for adapting to climate change. *Global Environmental Change*, 7, 251–264.

Steffen, K., Clark, P.U., Cogley, J.G., Holland, D., Marshall, S., Rignot, E., & Thomas, R. (2008). Rapid changes in glaciers and ice sheets and their impacts on sea level. In J.P. McGeehin (Ed.), *Abrupt climate change*, Synthesis and Assessment Product 3.4 (pp. 29–66). Washington, DC: U.S. Climate Change Science Program and the Subcommittee on Global Change Research.

Stern, N. (2006). *Stern review on the economics of climate change: Report to HM Treasury*. London: HM Treasury.

Titus, J.G., & Craghan, M. (2009). Shore protection and retreat. In J.G. Titus (Ed.), *Coastal sensitivity to sea-level rise: A focus on the mid-Atlantic region*, Synthesis and Assessment Product 4.1 (pp. 87–103). Washington, DC: U.S. Climate Change Science Program and the Subcommittee on Global Change Research.

Titus, J.G., Hudgens, D.E., Trescott, D.L., Craghan, M . . . Tanski, J., & Wang, J. (2009). State and local governments plan for development of most land vulnerable to rising sea level along the US Atlantic coast. *Environmental Research Letters*, 4, 1–7.Titus, J.G., & Neumann, J.E. (2009). Implications for decisions. In J.G. Titus (Ed.), *Coastal sensitivity to sea-level rise: A focus on the mid-Atlantic region*, Synthesis and Assessment Product 4.1 (pp. 141–156). Washington, DC: U.S. Climate Change Science Program and the Subcommittee on Global Change Research.

Toman, M. (2006). Values in the economics of climate change. *Environmental Values*, 15, 365–379.

UN (United Nations). (1992). Proceedings of the United Nations framework convention on climate change. Retrieved from http://unfccc.int/resource/docs/convkp/conveng.pdf.

United States Indian Ocean Tsunami Warning System Program. (2007). *How resilient is your coastal community? A guide for evaluating coastal community resilience to tsunamis and other coastal hazards*. Bangkok, Thailand: U.S. Indian Ocean Tsunami Warning System Program.

USACE (United States Army Corps of Engineers). (1983). Economic and environmental principles for water and related land resources implementation studies. Retrieved February 8, 2012 from http://planning.usace.army.mil/toolbox/library%5CGuidance%5CPrinciples_Guidelines.pdf.

USACE (United States Army Corps of Engineers). (2011). Coastal storm risk management national economic development manual. Retrieved February 9, 2012 from http://www.corpsnedmanuals.us/Includes/PDFs/2011-R-09.pdf.

USACE (United States Army Corps of Engineers). (2012). National economic development (NED) procedures manual for flood risk management. Retrieved February 13, 2012 from http://www.corpsnedmanuals.us/FloodDamageReduction/FDRID001Welcome.asp.

USEPA (United States Environmental Protection Agency). (2010). Climate change indicators in the United States. EPA no. 430-R-10–007. Retrieved July 2010 from http://epa.gov/climatechange/indicators.html.

Van Raalten, D., van der Laan, T., Wijsman, P., Boeije, L., Schellekens, E., Dircke, P., . . . & Schoots, M. (2009). San Francisco Bay: Preparing for the next level. San Francisco Bay Conservation and Development Commission. Retrieved from www.bcdc.ca.gov/planning/climate_change/SFBay_preparing_%20for_the_next_Level.pdf.

Vermeer, M., & Rahmstorf, S. (2009). Global sea level linked to global temperature. *Proceedings of the National Academy of Sciences of the United States of America*, 106(51), 21527–21532. doi:10.1073/pnas.0907765106.

Walker, B., Holling, C.S., Carpenter, S.R., & Kinzig, A. (2004). Resilience, adaptability and transformability in socialecological systems. *Ecology and Society*, 9(2), 5.

Walker, B., & Salt, D. (2006). *Resilience thinking: Sustaining ecosystems and people in a changing world*. Washington, DC: Island Press.

Webster, P.J., Holland, G.J., Curry, J.A., & Chang, H.-R. (2005). Changes in tropical cyclone number, duration, and intensity in a warming environment. *Science*, 309, 1844–1846.

Wigley, T.M.L. (2005). The climate change commitment. *Science*, 307(5716), 1766–1769.

Yoe, C. (2004). Scenario planning literature review. Retrieved February 13, 2012 from http://www.corpsnedmanuals.us/FloodDamageReduction/FDRIncludes/FDRYoeLitReviewSept2204PG102504.pdf.

Yohe, G., & Tol, R. (2002). Indicators for social and economic coping capacity: Moving toward a working definition of adaptive capacity. *Global Environmental Change*, 12(1), 25–40.

Zellner, M., Hock, C.J., & Welch, E.W. (2011). Leaping forward: Building resilience by communicating vulnerability. In Bruce Goldstein (Ed.), *Collaborative resilience: Moving through crisis to opportunity*. Cambridge, MA: MIT Press.

11 Incorporating Resilience of Rural Communities for Proactive Risk Reduction in Shikoku, Japan

Rajib Shaw

INTRODUCTION

Japan is located on the eastern edge of the monsoon region of Asia and its climate varies according to the seasonal and regional conditions. Typically, heavy rains occur in various parts of the country both during the rainy season in June and July and during the typhoon season from August to October. This precipitation is predominantly in the form of locally specific temporary downpours. In winter, the northern part of the country usually receives heavy snowfall, which causes prolonged floods in spring caused by the melting of snow.

Owing to these meteorological characteristics, flood disasters caused by heavy rains occur frequently in Japan. Topographically, about 70% of the land is mountainous and covered with forests. Additionally, rivers in Japan are generally very short and steep, causing flash flooding with high concentrated peak discharges soon after an intense rainfall. The remaining 30% of the land is mostly alluvial plains where housing, farming, and industries are densely concentrated, consequently increasing the vulnerability to flood disasters.

In addition to these climate challenges, the aging society is a major issue in Japan, especially for rural communities. The population of elderly people compromises 23% of the total population in the country (Statistics Bureau, 2008). The national government expects this to rise to 30% by 2025 (Shaw & Takeuchi, 2011). Recognizing these trends the Japanese government has declared that people over sixty-five years old are easily affected by disasters. An older population has implications for disaster resiliency. For example, housing for elderly tends to be relatively large and requires large parcels of land for new construction. Hence, it is often the case that facilities for the elderly are located on hills and in the mountains where land is relatively cheaper, but which are prone to landslides.

At the same time, the elderly population in Japan has shown significant levels of resilience due to higher social capital and community linkages during times of disasters. Thus the opportunities and challenges for policy makers are to utilize the social resilience of rural communities during normal times to prepare for disaster response and recovery. This chapter underlines the community resilience in rural communities in Japan while exemplifying the disaster risks and community-based risk reduction activities in three contrasting communities on Shikoku Island.

CHARACTERIZING MOUNTAIN DISASTERS IN RURAL COMMUNITIES

Japan has suffered from frequent erosion and sediment movement such as mountain collapses, debris flows, and pyroclastic flows due to the steep mountainous terrain. Because many of the gentle slopes have been developed, Japan is considered a "sediment disaster archipelago" (Ohta, 2006). Therefore, the country has a long history to cope with a range of natural disasters.

Forest Management Practices as Root Cause

The torrential rains that damaged the city of Hiroshima in 1999 triggered the Government of Japan to establish the Act on Promotion of Sediment Disaster Countermeasures for Sediment Disaster Prone Areas. The act aims to promote non-structural measures such as developing an early evacuation system. The damaged area is a typical area with sediment-related disasters, where housing land development has been promoted in mountain areas (Watanabe & Fukuoka, 2003). Building housings in high hazard risk areas, in addition to the time gap between occurrence of hazards (that makes people feel more complacent about them) and the lack of forest management often escalate the impacts of hazards and turns them into disasters (Mimaki, Takeuchi, & Shaw, 2009).

Forest resources are protective in reducing disaster risk as they can influence climate protection and are essential for wind erosion and coastal protection. They also contribute to healthy water resources and prevent soil erosion. In Japan, developed forests have no possibility to increase the risk of soil collapse, but it is effective for preventing mountain disasters (Fujita & Shaw, 2010). Thus forest management and disaster risk reduction are closely linked. However, protecting forested land from development has not been a priority issue for Japan.

Private forest land amounts to 16% of the world forest resources. In Japan around 58% is owned by private owners, 31% by national entities, and 11% are public forests. Artificial (or planted) forest amounts to 41% of the total forest, while 53% is natural forest. This has serious consequences for forest management, as a result of population decreases, increases in elderly population, and absentee forest owners (Fujita, Takeuchi & Shaw, 2011). A review of the forest resources of Japan has identified an increased need for protective functions of forest ecosystems, instead of need for productive functions. The analysis pointed to the five elements for sustainable forest management: extent of forest resources, biological diversity, forest health and vitality, productive functions of forest resources, protective and socio-economic functions (Fujita, Takeuchi & Shaw, 2011).

Climatic Conditions as Trigger of Hazard

Climate change affects the forest environment. The average precipitation in Japan is about 1,800 mm, which is twice the world average precipitation of about 970 mm. Most of the precipitation happens from early June to mid-July (rainy season), from July to October as torrential rain (typhoon season), and during the winter (snow season). Data from 1901 to 2006 show a long-term trend of increasing days with daily precipitation of more than 100 mm (JMA, 2010). In addition, it compares annual precipitation of thirty years from 1931 to 2006. The national average annual precipitation is rather decreasing long term, but the average annual standard deviation is increasing every year. The trend of occurrence of localized torrential rain is increasing, and local areas tend to suffer enormous damage. The gap between high-precipitation and low-precipitation years tends to be wide; big-scale local floods and droughts tend to occur. Moreover, owing to global warming, the temperature of Japan has increased from the late 1980s and the highest temperature was recorded in 1990 for the first time in the last one hundred years. Meteorological Agency observed that the temperature increased 1.07°C in the last one hundred years (MOA, 2006). The damage caused by mountain disasters for five years from 2002 to 2006 effected 3,500 places and the cost of damage is about 120 billion yen on annual average. Under the circumstances, average occurrence of sediment-related disasters, which are debris flow, landslides, and collapse, is 1,000/year; the average number of deaths is 60 per year (Fujita & Shaw, 2010).

Social Conditions as Trigger of Vulnerability

Elderly population in mountain areas is an important issue for forest management. Depopulation in mountainous areas after the period of high economic growth has escalated. Although the total population increased 30% from 1955 to 2005, the necessary population in the mountain villages nominated by Sanson-Fukko Law decreased by 40%. In addition, according to a report of population trend in Japan, the national population is estimated to decline 30% between 2000 and 2030, but the population in restoration necessity mountain villages is estimated to decrease by 36% ((NIPSSR2003). Moreover, in the restoration necessity villages, the percentage of the population fifteen to sixty-four years of age is estimated to be less than 50%, but the elderly population (age sixty-five and over) is estimated to be more than 40%. As a result of this trend, it is predicted that the function of the village will decline, and absentee forest owners increase more and more. The aging population of foresters is also increasing. The number of foresters has been decreasing in the long term and the ratio of foresters who are more than sixty-five years has been increasing rapidly since 1990.

People who are vulnerable to disasters due to their age constitute 20% of the population. Furthermore, they account for about half of the victims of sediment-related disasters. The percentage of elderly people in mountainous areas is higher than the cities. Another reason for the increasing percentage of disaster-vulnerable people is that many special elderly nursing homes were built in sediment-related disaster warning areas.

ISSUES OF RURAL COASTAL AREAS IN JAPAN

The coastal areas are physically vulnerable to both geological and hydrological disasters. For example, in 1993 the Hokkaido-Nanseioki earthquake (magnitude 7.8) struck Okushiri Island, Hokkaido. Significant damages were reported, including widespread fires and extensive damages of fishery harbors by debris. In 2003, typhoon 23, which was the largest recorded event in the observational history at that time, made its way across central Japan and caused significant damages particularly along coastal areas (the number of missing and dead is reported to be 98). The 2011 Great East Japan Earthquake and Tsunami have also reminded us of the vulnerability of the coastal regions in the country (Shaw & Takeuchi, 2012).

The period of high economic growth (1960–1973) resulted in a population flow from rural areas to cities. In the Shikoku region, the ratio of population outflow was the most massive (Shinohara, 1991). Ono (2005), for this region, defines "the community [where] the population aging rate is over fifty percent" as a marginal village, "the community [where] the population over fifty-five years old occupies over fifty percent" as a quasi-marginal village, and "the community [where] the population under fifty-five years old occupies over fifty percent" as a survival village. This is also evident in the different coastal villages in Shikoku.

For Japan, being an archipelago country and the rivers being of short length, the upstream and downstream relationship is quite close. Therefore, the forest management is closely related to the fishing activities in the coastal areas. The "forest fisher relationship," as it is commonly known, is important for sustaining the ecosystem of the river basin, and reducing the impacts of disaster and securing livelihoods.

Since the administrative reform (trinity reform) the subsidy from the national government has been reduced. Moreover, business recession has been continuing in the rural areas. Therefore, the local government faced a revenue shortfall and it became very difficult to allocate a sufficient budget for necessary countermeasures regarding disaster preparedness.

COMMUNITY ORGANIZATION AS THE CORE OF COMMUNITY DISASTER RESPONSES

In order to minimize the impact of frequently occurring mountain and coastal disasters, Japan has developed several structural and non-structural mitigation mechanism preparedness strategies. The disaster management system in Japan has been developed and strengthened following the bitter experiences of large-scale natural disasters over the past twenty years (CAO, 2008). Although structural and non-structural mitigation strategies reduce the impact of hazards, they are inadequate without individual and community-level disaster awareness and preparation.

The Great Hanshin-Awaji Earthquake hit an urban city, Kobe, in 1995. The damage was unprecedented and it gave an opportunity to raise awareness of risk management, and disaster prevention plans have been revised all over Japan. Preparedness strategies such as voluntary self-help organizations need to be developed as soon as possible mainly by local residents, fire departments, and fire volunteers. Neighbors played an important role to help each other when modern heavy machineries as an emergency measure did not function well because roads were blocked and staff of disaster risk reduction also became victims (Iwami, 1997). Though as it has long been known that "mountain conservation and flood control is the work of government," people still believe that national and local government should be involved in disaster risk reduction. The fundamental system for disaster risk reduction is protecting my life by myself, protecting local safety by ourselves, and promoting local infrastructure by local government and municipality. Mimaki (2009) also pointed out that establishment of community-based organizations leads to both the promoting of self-help for possible disasters and the understanding of the community's role in disaster preparedness. Thus a community-based organization is considered essential for promoting the notion of self-help.

STUDY LOCATIONS AND THEIR CHARACTERISTICS

This chapter focuses on the three brief case studies from the rural areas of Shikoku Island, which are: Saijo in Ehime Prefecture and Reihoku and Tosashimizu in Kochi Prefecture. Figure 11.1 shows the locations of the study areas, and Table 11.1 shows the demographic characters of the study locations. The city of Saijo is located in the eastern part of Ehime Prefecture. It has an area of 509.04 sq km, with a population of 116,059 (October 2006). On November 1, 2004, Saijo City, Toyo City, the town of Tanbara, and the town of Komatsu from Shuso County merged to form the new "Saijo City."

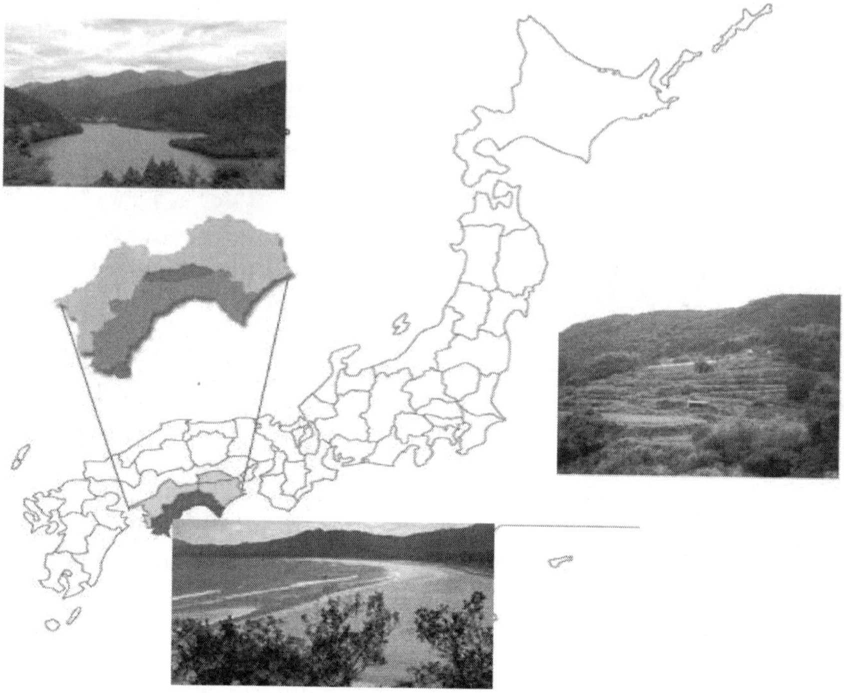

Figure 11.1 Characteristics of the case study areas.

The geography of Saijo City is classified broadly into four characterisitcs: plain area along the coast; hilly area between Saijo City and Nihama City; hilly terrain ranging in the north side of the median tectonic line along the south side of the plain; and precipitous mountains in the southern side of the median tectonic line. Mount Ishiduchi, the highest mountain in the western part of Japan, is located in the area. Two big rivers, Kamo River

Table 11.1 Characteristics of the Case Study Areas

City/Area	*Prefecture*	*Population*	*Elderly population (%)*	*Mountain cover (%)*	*Hazards*
Saijo	Ehime	116,000	30	47	Flood, typhoon
Reihoku	Kochi	15,270	45	88	Landslide, flash flood
Tosashimizu	Kochi	16,00	37	55	Typhoon, flood

and Nakayama River, flow in the center of the city. Saijo City is famous for its spring water called "Uchinuki." The river water percolates underground and in pools, then spouts above ground under pressure. Just by driving a pipe into the ground, water comes out. The amount of flowing water is around 90,000 m^3 per day. Due to the little temperature change through all seasons, it is used as drinking water, agricultural water, and industrial water. "Uchinuki" is one of the one hundred best waters in Japan. There is a traditional annual festival called "Saijo Matsuri" in October when almost all people, both young and old, get wildly excited. Each "Jichikai" (resident association) has its own "Danjiri" (portable shrine) and the relationship among people in the communities is strongly felt. A "Jichikai" is a neighborhood association which is organized in each area in the municipality at the people's own initiative. There are 540 Jichikai in Saijo City and they make twenty-eight Jichikai unions (Yoshida, Takeuchi, & Shaw, 2009).

Kochi Prefecture is located in the south of Shikoku with a high ratio of forests and ocean. North of this prefecture is mountainous and touches Ehime and Tokushima prefectures. It touches the Pacific Ocean on the south. The coastal line is long and the western part is saw-toothed coastline and the east part is upheaval coastline with flat beaches. The weather is warm and humid. The weather, the complex topography, and frequent typhoons are typical of Kochi Prefecture. Because of the weather and topography, Kochi suffers from landslide disasters frequently (Fujita & Shaw, 2010). The Reihoku area is located in the north of Kochi and in the center of Shikoku. The area is about 757 km^2 with three towns (Otoyo, Motoyama, Tosa) and one village (Okawa).The headwaters of Yoshino River are located in the Reihoku area. It is mountainous with steep mountains 200 to 1,700 meters high in the north. Forest shares 87.9% of Reihoku, and it is much higher than Japan's forest rate of 67%. Agricultural land shares 1.7% and residential area shares only 0.6%, therefore it is a typical mountain village. Landslides in Yoshino river basin are some of the largest landslides in the fracture zone. The Reihoku area is located in the upper Yoshino river basin. The population more than sixty-five years old is more than 30% of the total population in the Reihoku area. The total population has been decreasing from 34,801 in 1965 to 15,270 in 2000 and the ratio of population more than sixty-five years old has been increasing from 11.6% in 1965 to 38.7% in 2000. It is estimated that the population will decrease to 12,623 and the ratio of population more than sixty-five years old will increase to 44.7% in 2015 (Masanori et al., 2006).

In addition to the aging and decreasing population, importing cheap timber has accelerated weakening the forest industry. As a result of import liberalization, the import volume of lumber had increased sharply from about 1955 to 1975. Though the key industry in the Reihoku area is agriculture and forestry, the number of workers in the primary sector of industry decreased from 8,308 to 1,913. The decrease has been less sharp in the secondary sector, and less in the tertiary sector. Because of the declining agriculture and forest industry, people have no other choice but to find other jobs in the area or to move out of the area to find jobs. This causes the forest to be untreated and forest roads unused. Therefore the owners and foresters have more difficulties to enter untreated forest for thinning, cutting, and carrying lumber.

The third and the final case study area is also located in Kochi Prefecture, in a small coastal city in Tosashimizu. Because Tosashimizu City is on the Pacific coast, it is frequently attacked by typhoons. Additionally, the Kochi prefectural government has warned about the possibility of a Nankai earthquake within the next thirty years. In the event of an earthquake it is estimated that nearly 70% of the Tosashimizuarea will be damaged by tsunamis. The Kochi Prefecture government, the target area in this study, set the rate of participation by Voluntary Disaster Preparedness Organization (VDPO) in preparpedness activities at 100% while also providing signficant financial support for VDPOs in collaboration with local governments. The network of VDPOs is expected to play a role to educate residents about disaster preparedness, including the implementation of evacuation drills.

Recently, besides the physical condition, the risk due to social factors has increased. The local government had promoted tree-planting programs in the 1950s and 1960s. However, due to the stagnation of forest industry, the number of the forestry labors declined and their forest management became poor. The lack of forest management has caused the floods in the downstream basin area, and increased the number of abandoned farmland and the risk of the possible disasters brought by the abandoned farmland. Actually, in 2001 Tosashimizu City was heavily affected by flooding. The cost of the resulting damage was 18.8 billion yen. Lack of forest management triggered the flooding.

The population of the city is 17,000 (2005 census) and among which senior citizens (over sixty-five years old) are 35%. The national average population aging rate is 20% (CAO, 2007). Local government predicts that as of 2020 senior citizens will come close to the rate of 50% of the entire population. As for the Nakanohama community, the population is 733, the rate of senior citizens is 34%. It is coincident with the data of the city and is selected as a typical community of the city. The residents over fifty-five years old in the Nakanohama community occupy 57%. Therefore, it is a "quasi-marginal village."

RISK REDUCTION INITIATIVES

School-Based Disaster Education in Saijo City

In 2004, ten typhoons hit Japan. This frequency is approximately four times higher than the average number of two to three typhoons per year. As a result, the mountainous area of Saijo City was heavily damaged. In the mountainous area, elderly people had difficulty evacuating due to lack of appropriate assistance and functional warning system (Shaw, Takeuchi, & Badaoui, 2008; Yoshida, Takeuchi, & Shaw, 2009).

In the aftermath of the disaster, the local government, under the leadership of the mayor, undertook recovery process, including mitigation and preparedness strategies. Among these, a significant one is the town-watching and mountain-watching programs, conducted by the education department of the Saijo City government with guidance from Kyoto University. The plain area in Saijo City is rather urban and there are many young people. So, it is necessary to make a "disaster prevention network" between the plain area and the mountainous area, so as to help elderly people in the mountainous area in case of a disaster. As the driftwood stuck with bridge pier caused flooding to the plain area, disaster in the mountainous area have bearings with that in the plain area. Both residents have to know about each other's circumstances. For these reasons, mountain watching is proposed in Saijo City. Mountain watching is just like town watching and it is conducted in the mountainous area. The main target is children, and also residents in the mountain, teachers, municipal officials, and forest workers are involved. The working field is the upper area of a river along a school. Participants watch the site damaged by the typhoon in 2004 and hear the story from victims. At the same time, town watching was proposed in the plain area. The main target was students, teachers, parents, Jichikai, and municipal officers. They walk around the school zone and search for dangerous places, useful facilities in case of disasters, and favorite places, which they do not notice otherwise in daily life. This time, town watching is to be implemented in five elementary schools and mountain watching in three junior high schools as a "disaster education program," which is an activity of a twelve-year-old education project.

Town watching or mountain watching is considered to be a suitable tool to understand and resolve the community problems. It involves many stakeholders, such as pupils in elementary schools and junior high schools, teachers, parents, Jichikai, residents in mountains, forest workers, the Citizens' Safety Department, and the Education Board. So, it provides a good opportunity for them to communicate with each other. In case of a disaster, such a relationship is very important. Through town watching, participants get interested in the local area and also get knowledge about disaster prevention. But town/mountain watching should not end up only a one-time event. Through a series of continuous actions, it will become better and develop a disaster-resilient area. So, a clear implementing body and guideline is necessary for continuing town/mountain watching.

In Saijo City, a "teachers' association of disaster education" has been set up. This association consists of teachers who have incentive to promote disaster education. They study on the way of disaster education and share information. Now they have several meetings and make a guideline of town/mountain watching, making use of the past experience. In addition,

"Kids Disaster Prevention Club" is set up at the city level. It is different from the boys and girls fire prevention club. It consists of students, teachers, parents, community people, who get interested in disaster prevention through town/mountain watching. Students suggest what they want to know more about or questions arise in their minds through town/mountain watching in the club activity. Also with daily learning, they study about disaster prevention. A forum of disaster prevention such as Kids Summit is held once or twice in a year and students from each school in Saijo City make presentations about what they have learned. In this way, sustainable disaster prevention will start from school and involve the entire city (Shaw & Takeuchi, 2011).

Fire Volunteers as the Core DRR Activities in the Reihoku Area

In 2004, Kochi Prefecture had been hit by typhoons and localized torrential rains. It is the year that a record number of five typhoons hit this region. The village of Okawa and the town of Tosa was also hit by torrential rain ("Sameura rain") in August 2004. The precipitation of Sameurarain is more than 1,000 mm in three days, which is more than one-third of the annual precipitation. Because of the high precipitation, mountain collapses occurred mostly in the east part of Okawa, and the roads were cut off; as a result all of Okawa and the Minami-kawa settlement in Tosa were isolated. However, there were no casualties. Then Okawa emphasized that governments need to take the responsibility of providing and maintaining structural measures and self, mutual, and public help is needed for saving lives by themselves in addition to maintaining disaster prevention equipment. The mayor of Tosa also emphasized the necessity of self-help organization. In addition, establishing an information collection system is needed, especially in case a settlement becomes isolated (Fujita & Shaw, 2010).

A survey was implemented to Reihoku local residents with the cooperation of Reihoku NPOs (Fujita, Takeuchi, & Shaw, 2011). The purpose was to know what the people think about local organizations for disaster risk reduction, how they relate to the local bodies like voluntary self-help organizations, and how to improve their activities. To know the reason of the gap, the result was counted by districts and by age bracket. When the difference was compared by district, the rate of recognition in Okawa was rather low compared to the other three towns. Because the old-age rate in Okawa is higher than the rate in the other three towns, there is a possibility to have a relation with age. The analysis shows that the number of respondents with "yes" is higher in younger age brackets. One of the reasons of the gap of perception between young and old generations is because the voluntary help-self organization is a relatively recently spread idea and the young generation has more opportunity to know it through public information and education.

Then which organization respondents rely on was answered in the questionnaire survey. Out of the respondents, 67% answered "fire volunteer," 8% answered "voluntary self-help organization," and 14% answered "nothing special," which means they do not trust any organization. Fire volunteering is a traditional public system under the fire department and it has a close and more daily relationship with residents in addition to the activity in an emergency case of a disaster. Because fire volunteering has a long history, it is considered that many respondents answered that fire volunteering is the most reliable organization in case of disaster. To know more in detail, the result was counted by districts and by age bracket. There was no big difference by districts. Little difference was seen by age bracket as the X-axis. Though the young generation recognized the self-help organizations more than the old generation, most of the young generation still rely on fire volunteering as the key local organization for DRR.

Because it is said that fire volunteers have a close and more daily relationship with residents in addition to the activity in an emergency case of a disaster, daily communication seems a key point of reliable relationship between the residents and fire volunteers. Seventy-six percent of the respondents have community meetings and 86% of them have group work. These percentages were high, because the activities are necessary to sustain the community. Their main group activities are cutting grass, and cleaning streets and shrines. Meetings and group works are community-level activities and the results show a high rate in communication. Because the result of communicating with neighbors is individual-level activity, they

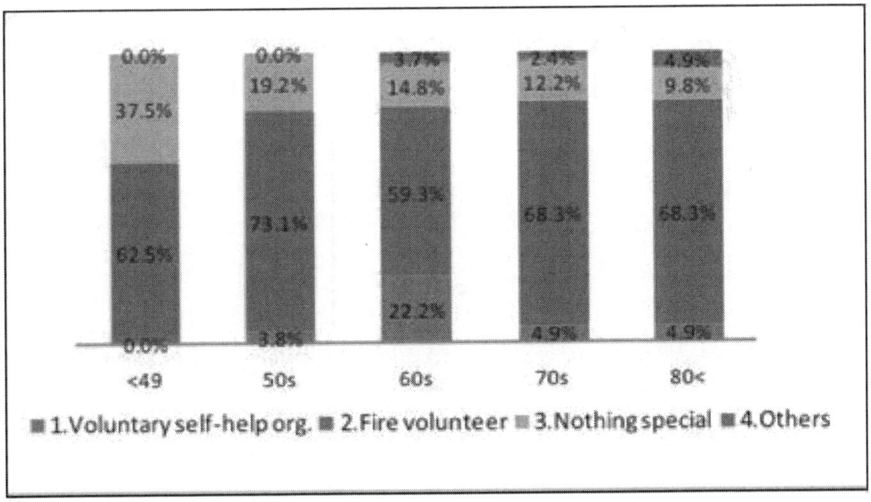

Figure 11.2 Reliable organizations in case of disasters by age bracket.

have more choice whether they communicate or not. As the results show, they have close and daily communication in individual-level activities.

These high rates in communication support why 67% of respondents rely on fire volunteers. The fire volunteers are also residents in the community, so they have been helped people in daily life also. Based on the close relationship in daily life, it is possible for fire volunteers to help residents in case of disasters and to gain the higher degree of trust.

In this context, a local nonprofit organization (NPO) called Reihoku NPO started a strong initiative of community preparedness by organizing community groups. The key idea was to ensure people's participation in the local activities, and to link the local activities to external actors. The NPO had good collaboration with the local business sector (timber-related local industry), which aimed to enhance the use of the local timber and thereby revitalize the local economy. The NPO was the key facilitator for organizing eco-tours of people outside the region. The key activities of eco-tours include: visiting local forests, enjoying seasonal flowers and local food (prepared by using local resources by the residents), visits to the local timber industry (to understand the use of forest products and timbers). The local university students also facilitated this process through participation as interns in the local NPO. This process had multiplying effects: in several cases, outside people wanted to come and live in the Reihoku area, in some cases, outside people wanted to use Reihoku timber, and in some cases, they wanted to contribute to the local economy. The key of this process is to revitalize the local economy, the timber industry, and thereby take good care of the mountain, and in turn reduce the mountain disaster losses.

Voluntary Disaster Preparedness Organization in Tosashimizu City

The Nakanohama community in Tosashimizu City is located on the coast and has a high risk of damages caused by tsunamis and storm surges. In 2004, the leader participated in the seminar about disaster preparedness at the local government's request. By attending the seminar, the leader recognized the importance of communal preparedness for the possible disasters and existence of the local government's financial support system for establishment of a VDPO. After the seminar he called the meetings to consult with residents about establishment of a VDPO and they established a VDPO in 2005 on the advice of local government. At that time, the community was facing a problem concerning community communication caused by the political fight in the 1990s. There was no community event. However, the board members began VDPO activities as communal activities (Mimaki, Takeuchi, & Shaw, 2009).

Figure 11.3 shows the process of the disaster preparedness in the Nakanohama community. At first, they began to hold periodic meetings every five districts to hear the residents' requests regarding disaster preparedness. In

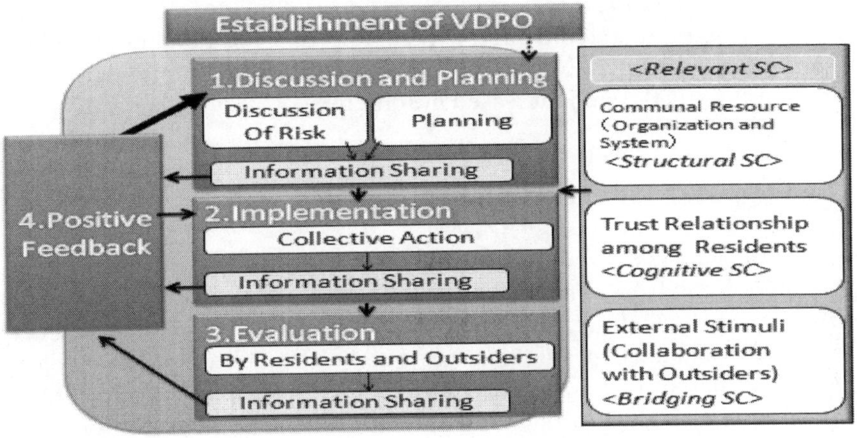

Figure 11.3 Process of promoting disaster preparedness through VDPO (SC: Social Capital).

parallel, they started periodic board meetings not only to study about disaster risk, but also to discuss residents' request regarding disaster preparedness in order to decide the next action. After those meetings they pasted up a notice in the assembly house to report to the residents about the progress and results of discussions. Moreover, with the support of the residents, events were held periodically to build evacuation routes by themselves. The community incurred the cost related to construction and reused waste materials to reduce the cost. Originally, the community had a longstanding convention of a volunteer work system called "deyaku"; each family was required to contribute to the community volunteer work. Therefore the events were organized under the system. After the construction of the road, they put seals written that, "The road was constructed by Nakanohama VDPO" on the handrails and pasted up a report and pictures in the assembly house to make residents aware of the existence of the VDPO and some residents' efforts. Then, the board members began to make efforts to involve various generations into VDPO activities. They added some easy work (taking care of flower beds, etc.) to volunteer work so that senior citizens also could participate in it. In 2006 and 2007, they had athletic meetings and evacuation drills to enlighten children as well as senior citizens on disaster preparedness. Such activities had a tremendous impact on residents' disaster preparedness. The residents came to notice the risk and the number of the residents who requested to the board members to take necessary countermeasures for "new" community risks had increased.

In the beginning, the number of the residents who participated in the volunteer work was limited (almost residents did not take the board members' plan seriously). However, the residents gradually came to notice other residents' and the board members' efforts to enhance communal disaster preparedness and thanked them for their supports. The number of the participants not only in volunteer work but also in other communal events (communal festivals, etc.) increased gradually.

TOWARDS PROACTIVE RISK REDUCTION

Each study area discussed has its characteristic features. In case of Saijo, local school children were involved along with local residents for a collective regional watching which includes local residents, local government officials, school teachers, students, and parents. In the case of Reihoku, the local NGO has started different activities to revitalize the local economy and link to the resident association, and local private companies. It is noted that fire volunteers are the most trusted community organizations. In case of Tosashimizu, the local residents have taken active involvement, and social capital analysis shows that leaders play an important role in sustaining the initiatives in the local level.

Different countries and different regions have specific perspectives of local intuition's role in disaster risk reduction (DRR). It is increasingly observed and agreed that a sustainable DRR activity is only possible when there is a strong involvement and commitment from the local institutions. The three case studies described here show different roles of local institutions. In the case of Saijo, it is the local government, which played an important role in enhancing community awareness through disaster education in schools, and sustaining the activities in the local government by making it a regular event in the schools. In the Tosashimizu experience, it is the VDPO, which played an important role in implementing and sustaining the initiatives with the local resources. The internal and external social capital enhanced the sustainability, as argued by Nakagawa and Shaw (2004). In the third example of Reihoku, it is the local NPO, which helped in revitalizing the local economy and developing the bridge between the local communities to outside people.

Shaw (2009) has concluded that the key point of the role of local actors is the partnership and collaboration. Each group has its own resources, knowledge base, and information. Sharing of the information is significantly important. It is required to have a proper information management system to utilize the right information in the right time and right place. Collaboration and partnership plays the key role for that. Based on my earlier discussion, no single organization or agency can perform all the roles in the disaster reduction system. Policy makers, local emergency managers, community organizers, trainers, and academic institutions have important roles. Local media can play a useful role in information dissemination. In many places, it is seen that community radio is an effective tool in engaging local communities in decision making. This is especially effective in the rural areas. Also, the local business community can provide useful resources through the community-based disaster risk management. For example, a new concept called "Corporate Community Interface" played an important role in engaging corporate sectors as the local actors in CBDRM (Miyaguchi & Shaw, 2007). The concept of resilience of the community is thus linked to different elements, starting from collaboration, policy, decision making, ownership, and partnership.

Christoplos, Mitchell, & Liljelund (2001) argued that disaster mitigation and preparedness can no longer be assumed to be a set piece collection of activities that can merely be "implemented." It is a challenge to see how a concern for risk fits into the values, incentives, theories, and interests which form the cognitive structures by which key actors, be they villagers, politicians, NGO volunteers, scientists, or insurance salesmen, construct their own policy narratives. Thus, the local institution's role has been re-emphasized in the general context. In a more specific example from the Philippines, Allen (2006) has described that the community-based disaster preparedness processes are associated with a policy trend that values the knowledge and capacities of local people and builds on local resources, including social capital. CBDP may be instrumental not only in formulating local coping and adaptation strategies, but also in situating them within wider development planning and debates. In theory, local people can be mobilized to resist unsustainable (vulnerability-increasing) forms of development or livelihood practices and to raise local concerns more effectively with political representatives. Agrawal (2008) has analyzed the role of local institutions for climate change adaptation, and concluded that institutions influence adaptation and climate vulnerability in three critical ways: a) they structure impacts and vulnerability, b) they mediate between individual and collective responses to climate impacts and thereby shape outcomes of adaptation, and c) they act as the means of delivery of external resources to facilitate adaptation, and thus govern access to such resources.

In a post-disaster analysis of the 1995 Kobe earthquake of Japan, Nakagawa and Shaw (2004) have shown that the community with higher social capital and with a tradition of community activities can proactively participate in the reconstruction program, and thereby can make a successful and speedy recovery. A model for bonding, bridging, and linking social capital can enhance total recovery. It was observed that the community with social capital recorded the highest satisfaction rate for the new town planning and had the speediest recovery rate. The role of community leaders has been prominent in utilizing social capital in the recovery process, and facilitating collective decision making. A similar observation was also made by Shaw and Goda (2004) while analyzing the community recovery in Kobe. Their study reveals that the civil societies in urban areas are sustainable if, first, the activities related to daily services are provided by the resident's associations, and second, these are linked to economic incentives. Leadership plays a crucial role in collective decision making. Creation of the support system is essential for long-term sustainability of civil-society activities.

Almost seventeen years have passed since the Kobe earthquake. Significant changes are observed in the field of disaster risk reduction globally and especially in Japan. On the one hand, the DRR field is well equipped with the Hyogo Framework for Action (HFA, 2005) as an action agenda to be implemented for ten years. On the other hand, increasing impacts of climatic hazards in different parts of the world have brought significant attention to the climate change adaptation (CCA) issue. DRR and CCA have natural overlaps, and this overlap becomes obvious when the actions need to be taken at the local or community levels. Shaw, Pulhin, & Pereira (2010) have argued the linkages of CCA and DRR. Citing Leary et al. (2008), they argued that there are nine adaptation lessons: employ adaptation action now, create conditions to enable adaptation, integrate adaptation with development, increase awareness and knowledge, strengthen institutions, protect natural resources, provide financial assistance, involve those at risk, and use place-specific strategies.

The same lessons are equally applicable to the disaster risk reduction field,. When it is seen though the lens of Hyogo Framework for Action (HFA, 2005) it ensures that disaster risk reduction is a national and a local priority with a strong institutional basis to implement (institutional issue), identify, assess and monitor disaster risks and enhance early warning (risk assessment), use knowledge, innovation and education to build a culture of safety and resilience at all levels (education), reduce the underlying risk factors (links to development), and strengthen disaster preparedness for effective response at all levels (emergency response). Among all these, the current paper focuses very much on local institutions.

The Great East Japan Earthquake and Tsunami of 2011 have shown that in the areas where there were prior activities of tsunami risk reduction with community participation, the evacuation was effective and there were less life losses (Shaw & Takeuchi, 2012). "Kamaishi miracle" is often given as a good example of disaster education. Even in the aged population, it was the old people who helped each other to evacuate effectively to safer areas.

CONCLUSION

In a disaster-prone and natural-resource-poor country like Japan, the key resource is human resources and its network. Japan, over the last forty to fifty years, through economic development emphasized the importance of infrastructures on DRR. However, the recent realization is that there is a limit to the role of infrastructure, and increasingly the focus has been on the roles of people and community. Before one hundred or two hundred years, it was the people and communities who helped each other during the disasters and developed different coping mechanisms and an indigenous knowledge base. Then the state and the government were formed, who took control on the disaster prevention issue for several decades. Now, we are again going back to the basics of nature-human relationship, and the community preparedness or role of people and the local institutions (formal and/or informal) is the core to that. The current efforts of local institutions in Japan are faced with different challenges, out of which the following are worth mentioning: the local institutions are found good to cope with the regular disasters, however when there is a catastrophic disaster, which is a low-frequency and high-consequence event, it is found difficult to cope.

The second is that most of the local institutions in Japan are dominated by an elderly population, and lack youth participation. Unlike other countries, Japan does not have an active youth club or association. To enhance the sustainability, there needs to be a strong emphasis on youth participation. Also, there are very few formal ways to incorporate youth voices in the DRR and also in the post-disaster recovery process. The third point, which needs improvement, is recognition by the local governments. Unlike the Saijo City's initiative, not all the activities are locally institutionalized in the governance system. Increasing focus is required to link it to the local governance system. The fourth point is the link to the external stimuli. While, it is important to focus on the local institutions, it is equally important to link the activities to the external group. A balanced mix of bonding and bridging social capital is important to stimulate some local activities. With these four improvements, Japan's examples of local institutional involvement can be leading examples of community-based disaster risk reduction activities in the local level.

In the current context of Japan, it is rather difficult to continue the disaster risk reduction as a stand-alone activity. To enhance the resilience of the communities, it needs to be merged with the daily needs, which are related to welfare, social protection, and health-related activities. The key argument is how well and effectively the disaster risk reduction can be embedded in the daily needs of the local communities. The evacuation drill is definitely one important step of preparedness, which is possibly not enough as a stand-alone activity. The disaster preparedness possibly needs to come through over years of experiences, practice, awareness, and social cohesion. With Japan's increasing aging population, vulnerability due to climatic changes, and limited resources of the local governments in the rural areas, it is necessary to develop collective networks, mainly based on the people and community's social capital, and link it to the institutional system.

ACKNOWLEDGEMENTS

This chapter is a product of several years of work with the local governments and stakeholders in the case study areas, which produced two doctorate theses and one master thesis. Cooperation of local stakeholders, hard work of the students, and research funding from different sources are highly acknowledged.

REFERENCES

Agrawal, A. (2008). The role of local institutions in adaptation to climate change. Paper prepared for the Social Dimensions of Climate Change, Social Development Department, World Bank, Washington, DC, March 5–6.

Allen, K.M. (2006). Community-based disaster preparedness and climate change adaptation: Local capacity building in the Philippines. *Disasters*, 30(19), 81–101.

CAO. (2007). White Paper on Disaster Management 2007 edition (Bousaihakusho). In Japanese. Retrieved June 26, 2008 from http://www.bousai.go.jp/hakusho/h19/BOUSAI_2007/html/index.htm.

CAO. (2008). Cabinet Office: Disaster management in Japan, Director General for Disaster Management. In Japanese.

Christoplos, I., Mitchell, J., & Liljelund A. (2001). Re-framing risk: The changing context of disaster mitigation and preparedness. *Disasters*, 25(3), 185–198.

Fujita, K., & Shaw, R. (2010). Environmental changes and Japan's forest management. *Asian Journal of Environment and Disaster Management*, 2(2), 153–178.

Fujita, K., Takeuchi, Y., & Shaw, R. (2011). Voluntary self-help organization and fire volunteer for mountain disaster risk reduction in Reihoku area, Japan. *Asian Journal of Environment and Disaster Management*, 3(3), 317–337.

HFA. (2005). *Hyogo framework for action 2005–2015: Building the resilience of nations and communities to disasters.* Geneva, Switzerland: UN ISDR.

Iwami, T. (1997). A study on community self-help organization for disaster prevention. *Journal of the Japan Association for Social and Economic Systems Studies*, 16, 43–49

JMA (Japan Meteorological Agency). (2010). Abnormal weather risk map. 3–1 day precipitation with more than 100mm. In Japanese. Retrieved October 31, 2011 from http://www.data.kishou.go.jp/climate/riskmap/index.html.

Leary, N., Adjuwon, J., Barros, V., Batima, P., Biagini, B., Burton, I., . . . & Webbe, M. (2008). A stitch in time: General lessons from specific cases. In N. Leary, J. Adejuwwon, V. Barroos, I. Burton, J. Kulkarni, & R. Lasco (Eds.), *Climate change and adaptation* (pp. 1–27). London: Earthscan.

Masanori, M., Sassa, K., Yokoyama, S., Hiura, H., & Nakaya, S. (2006). Estimation of the flood peak by using the trace of debris flow: Cases of debris flow disasters in Reihoku District, Kochi Prefecture, induced by the heavy rainfall of Typhoon Megi in 2004. *Journal of Japan Landslide Society*, 42(6), 478–492.

Mimaki, J. (2009). Role of community based organization and social capital in disaster preparedness: Perspective from rural areas in Japan. Doctoral thesis. Kyoto University.

Mimaki, J., Takeuchi, Y., & Shaw, R. (2009). The role of community-based organization in the promotion of disaster preparedness at the community level: A case study of a coastal town in the Kochi Prefecture of the Shikoku Region, Japan. *Journal of Coastal Conservation*, 13(4), 207–215.

Miyaguchi, T., & Shaw, R. (2007). Corporate community interface in disaster management: A preliminary study from Mumbai, India. *Risk Management*, 9, 209–222.

MOA. (2006). White paper on forest and forestry in 2006. In Japanese. Ministry of Agriculture, Forest and Fisheries .

Nakagawa, Y., & Shaw, R. (2004). Social capital: A missing link to disaster recovery. *International Journal of Mass Emergency and Disaster*, 22(1), 5–34.

(Ohta, T. (2006). Sediment-related disasters and new forest management. In Japanese. *Forest Science*, 4.

Ono, A. (2005). The present state of less-favored areas and the direct payment system. In S. Yoshizawa (Ed.), *Japanese less favored areas and regional revitalization*. Tokyo: Tsukuba-Shobo.

Shaw, R. (2009). Role of local actors in community based disaster risk reduction. In *Perspectives in disaster management* (pp. 123–145). METU, ISDR, WBI.

Shaw, R., & Goda, K. (2004). From disaster to sustainable community planning and development: The Kobe experiences. *Disaster*, 28(1), 16–40.

Shaw, R., Pulhin J., & Pereira, J. (2010). Climate change adaptation and disaster risk reduction: Overview of issues and challenges. In R. Shaw, J. Pulhin, & J. Pereira (Eds.), *Climate change adaptation and disaster risk reduction: Issues and challenges*. Bingley, UK: Emerald Publishers.

Shaw R., & Takeuchi, Y. (2011). Role of local institutions in climate related disasters in small and medium sized cities of Japan. *Asian Journal of Environment and Disaster Management*, 3(2), 137–150.

Shaw, R., & Takeuchi, Y. (2012). *East Japan earthquake and tsunami: Evacuation, communication, education and volunteerism*. Singapore: RPS Publishers.

Shaw, R., Takeuchi, Y., & Badaoui, R. (2008): Education, capacity building, public awareness for disaster reduction. In K. Sassa & P. Canuti (Eds.), *Landslides disaster risk reduction*. Berlin: Springer-Verlag.

Shinohara, S. (1991). Kasochiiki no Henbo to Sanson no Doukou [Changes in under populated region and movement in mountainous areas]. In Japanese. Tokyo: Taimeido.

Statistics Bureau. (2008). Chapter 2: Population. In *Statistical Handbook of Japan*. Tokyo: Statistics Bureau. Retrieved May 28, 2012 from http://www.stat.go.jp/english/data/handbook/c02cont.htm#cha2_2.

Watanabe, A., & Fukuoka, S. (2003). Characteristics of the heavy rain in Hiroshima disaster 1999 and the significance of information of radar rain gage. *Japan Society for Natural Disaster Science, Journal of Natural Disaster Science*, 22(2), 166–185.

Yoshida Y., Takeuchi Y., & Shaw, R. (2009). Town watching as the useful tool of urban risk reduction: Case of Saijo, Japan. In R. Shaw, H. Srinivas, & A. Sharma (Eds.), *Urban risk: An Asian perspective*. Bingley, UK: Emerald Publishers.

Part III

Improving Community Resilience through Networks

Maureen (Mary) Brown

How do communities improve their capacity to plan for and manage disasters? The following set of chapters provides focus on the concept of improving resilience, especially in the rural setting, through interpersonal and interorganizational networks of key stakeholders within and across communities.

Try as you might to seek simplicity, community resilience is inherently a networked phenomenon that never rests under the auspices of one jurisdiction or command structure (Kapucu, 2005). Complex structures of federal, state, local, nonprofit, and private organizations are typically required to manage disasters effectively and to build community resilience (Kapucu, 2006). The complex structure, an unavoidable necessity, is largely due to the need to tap a wider array of expertise than any one organization can provide. These complex structures, or networks of organizations, are especially pivotal in rural communities where resources are limited.

The following chapters examine the critical role that interpersonal and interorganizational networks play in promoting community resilience with special emphasis on the rural setting. The authors provide insights on how to improve the capacity for planning and managing disasters under a networked umbrella of organizations.

Drawing on their research on disaster response in the context of large-scale wildfires, Nowell and Steelman identify a series of network performance goals during the response phase of a disaster. They remind us that integrating the diverse array of responding agencies into a coordinated response can be fraught with challenge, especially under the stressful conditions of disasters. Relying on the work of Cutter et al. (2008) and Norris et al. (2008), they shed light on the notion that the ability to effectively navigate complex relationships is an essential element of community resilience. They call special attention to the critical process and outcome goals of the response stage of disaster management. Specifically, they identify key process goals as the need to manage distributed information, collaborate when advantage can be gained, and minimize destructive interference from the lack of coordination within the network of service providers. Their outcome goals focus on minimizing injury and loss of life and to advance broader community goals and values as opportunity allows. Finally, they call attention to the importance of addressing community needs in a respectful, just, timely, orderly, and efficient manner. Nowell and Steelman's research underscores the importance of interpersonal familiarity and trust, local knowledge, collaborative leadership, community focus, and a robust communication infrastructure to disaster management and community resilience.

Comfort and Okada, in their study of the Tohoku Earthquake, Tsunami, and Nuclear Reactor Breach, emphasize the complexity of interacting systems raised the degree of risk to a level that policy makers and citizens alike had not imagined. The functions of cognition, communication, coordination, and control are fundamental to reorganizing networks of action that require full collaboration across agencies, disciplines, and jurisdictions. They argue that developing the capacity to adapt to extreme events requires rapid integration of information from multiple sources simultaneously, which in turn requires building the technical and social infrastructure to enable the community to do so.

Lemyre and O'Sullivan highlight the research being conducted in Canada and the continuum of research from the psychological field to the system field to help demonstrate the complexity of resilience concepts. Most importantly, they emphasize the necessity to acknowledge and consider a multi-level, multi-tiered framework to the study of resilience. Their work highlights the inherent implications of interorganizational decision making and how this may differ from previous approaches. Their work reiterates the reality that while it may not be possible to predict the consequences, an increased understanding of the system and its elements increases support and provides insight to understand the possibilities.

Waugh's chapter begins the discussion by defining community resilience as a goal-driven activity that seeks to reduce reliance on federal resources. He claims that the field of community resilience seeks to encourage local responsibility for managing hazards. Waugh indicates that rural communities are particularly vulnerable because they lack the resources of urban areas and often have at-risk populations that require greater support. Resilience, he claims, is usually defined in terms of the capacity to recover quickly from catastrophic disasters.

Despite the reputation of self-reliance, Waugh claims that rural communities may be especially vulnerable to a lack of resilience. He argues that the ability of rural communities to respond to emergencies and provide medical care for victims is less certain. Of central concern is the notion that the expense of resilience is especially onerous due to small tax bases, low population density, and large service areas. As a result, tremendous attention has focused on the development of mechanisms that allow communities to share resources when major disasters occur. Waugh identifies that the acknowledgement that disaster management depends on shared resources has changed response operations. He begs the questions: How can communities increase their capacities to minimize losses and their capacities to recover quickly from catastrophe? What are the essential variables?

Celik and Corbacioglu's chapter shifts focus and examines the effectiveness of legislative changes that attempt to formalize a network approach to disaster management in Turkey. They provide the reader with a before and after case study of response operations to two earthquakes. Between the 1999 Duzce earthquake and the 2011 Van earthquake, new legislation was instituted to incorporate greater levels of intergovernmental coordination in the Turkish disaster system. The cross comparison of the two earthquake responses allowed the authors to isolate how the roles of the central actors differed as a result of the new legislation. In short, they sought to determine how the new legislative enactment influenced a coordinated disaster response.

Celik and Corbacioglu define disasters as "symmetry-breaking events." In other words, they change the regular flow and order of organizational and societal life. These changes may be temporary or permanent in nature. They provide a cogent argument that centralized approaches are not only ineffective to addressing the "symmetry-breaking event" but can be detrimental to the mission. Apparently, centralized hierarchical approaches perform slowly and lack the needed flexibility to overcome the damage and sudden disruptions of disasters.

Noting that firms do not act by themselves, Granovetter (1985) found that organizations are deeply embedded in "networks of external relationships" that influence the "exchange of resources and capabilities among them." Networks, when defined by interdependent relationship structures, are often touted for their ability to adapt to sudden changes in the environment. These complex systems are distinguished by a capacity for "self-organization." They are capable of rearranging and reforming configurations of operation in mutual adaptation to the changing needs and capacities of their components (Comfort, 1994). They are also capable of mutually adapting to the changing demands and opportunities imposed by the environment. The distinguishing characteristic of this process is that it occurs as a result of communication, selection, and adaptation processes within the system itself and between the evolving system and its environment (Comfort, 1994; Kauffman, 1993).

Past research suggests that the success of networks is contingent on *intentionally* adapting, coordinating, and safeguarding the exchanges that occur within the network (Jones, Hesterly, & Borgatti, 1997). But, how one successfully coordinates and safeguards the exchanges remains the subject of tremendous debate. Milward and Provan (2001) claim that managers must continuously deal with problems requiring negotiating, coordinating, monitoring, holding third parties accountable, and writing and enforcing contracts all in an interorganizational setting in which information asymmetries and moral hazards abound.

Nonetheless, Celik and Corbacioglu argue that the adaptive capabilities of networks, as opposed to a hierarchical rule-bound organization, is especially well suited for disaster management operations. Their chapter illustrates how a linear public policy approach coupled with a command-control-oriented bureaucratic model cannot stand up to the challenges of disaster management. They also call attention to the important role that local-level governments play in meeting the challenges of a disaster. They argue that local conditions influence the scale of the disaster. Moreover, local conditions determine the policy and programmatic choices available to the managers at the various stages of the disaster management process. Because local conditions prove pivotal, central and local actors must collaborate effectively.

Of utmost importance is Celik and Corbacioglu's finding that, despite the new legislative enactment, when confronted with the turmoil of a disaster, the actors abandoned the new process and reverted back to the well-entrenched centralized behaviors. Their finding calls attention to the need to incorporate some of the lessons learned from change management theorists. As organizations, especially in rural communities, strive to build resilience, they may be plagued by many of the challenges of organizational change.

Demiroz, Kapucu, and Dodson discuss the importance of interorganizational networks in disaster resiliency. As Florida is one of the most hazard-prone states, this study considers the impact of disasters on both urban (Orange County, Florida) and rural (Sumter County, Florida) counties aiming to identify the friendship, preparedness, and response networks during disaster response. Recommendations are presented in relation to strengthening local networks and those factors which inhibit sustained relationships in rural communities.

The chapters in this part of the book call attention to the important role of interpersonal and interorganizational relations in planning for and managing disasters. Rural communities may lack the resources to build resiliency, and in the face of a disaster, may have to depend heavily on their urban neighbors. Nonetheless, the pivotal finding is that regardless of the setting, resiliency is a function of building successful partnerships among a number of autonomous organizations. Clearly, resiliency, especially in the rural setting, is grounded in the success of collaborative networks.

REFERENCES

Comfort, L.K. (1994). Risk and resilience: Inter-organizational learning following the Northridge earthquake of 17 January 1994. *Journal of Contingencies and Crisis Management*, 2(3), 157–170.

Cutter, S.L., Barnes, L., Berry, M., Burton, C., Evans, E., Tate, E., & Webb, J. (2008). A place- based model for understanding community resilience to natural disasters. *Global Environmental Change-Human and Policy Dimensions*, 18(4), 598–606.
Granovetter, M. (1985). Economic action and social structure: The problem of embeddedness. *American Journal of Sociology*, 91(3), 481–510.
Jones, C., Hesterly, W.S., & Borgatti, S.P. (1997). A general theory of network governance: Exchange conditions and social mechanisms. *Academy of Management Review*, 22(4), 911–945.
Kapucu, N. (2006). Interagency communication networks during emergencies: Boundary spanners in multi-agency coordination. *The American Review of Public Administration*, 36(2), 207–225.
Kapucu, N. (2005). Interorganizational coordination in dynamic contexts: Networks in emergency management. *Connections: Journal of International Network for Social Network Analysis*, 26(2), 9–24.
Kauffman, S.A. (1993). *The origins of order: Self-organization and selection in evolution*. New York: Oxford University Press.
Milward, H.B., & Provan, K. (2001). Do networks really work? A framework for evaluating public sector organizational networks. *Public Administration Review*, 61(4), 414–423.
Norris, F., Stevens, S., Pfefferbaum, B., Wyche, K., & Pfefferbaum, R. (2008). Community resilience as a metaphor, theory, set of capacities, and strategy for disaster readiness. *American Journal of Community Psychology*, 41(1), 127–150.

12 The Role of Responder Networks in Promoting Community Resilience

Toward a Measurement Framework of Network Capacity

Branda Nowell and Toddi Steelman

INTRODUCTION

Population shifts and a changing climate are expected to contribute to greater vulnerability to disasters in rural America (Jensen, 2009). In an effort to build more resilient communities, a key focus in the study of disasters has been identifying local-level capacities that can improve the community's ability to respond to—and subsequently minimize hardship from—future disasters (Adger et al., 2004; Birkmann, 2006; Norris et al., 2008). Managing a large-scale disaster usually entails working within an intergovernmental structure with overlapping jurisdictions (Moynihan, 2009). This means local community agencies (e.g., law enforcement, Red Cross, utility companies) need to work collectively with federal disaster management teams to coordinate a response—usually in a networked structure. Integrating the diverse array of responding agencies into a coordinated response can be fraught with challenge, especially under the stressful conditions of disasters. Nonetheless, the ability to effectively navigate these complex relationships has been recognized as an essential element of resilience in disaster management (Cutter et al., 2008; Norris, et al., 2008).

These realizations have contributed to two major trends in the field of disaster response. First, there is increasing interest and emphasis in understanding how local-level capacities developed in advance of a disaster can contribute to better performance and thus greater community resilience during and following a disaster (Kapucu, Arslan, & Collins, 2010). Second, there has been increasing interest in shifting the level of analysis toward an examination of whole networks of responders and their interactions with one another (Comfort & Haase, 2006; Kapucu, 2006; Kapucu, 2005; Magsino, 2009).

Despite these trends, empirical research on the capacities of whole networks before a disaster and their impact on network performance during a disaster remains in its infancy (Magsino, 2009). Such a research agenda is hindered, in part, by a dearth of frameworks and tools for conceptualizing and assessing both local-level pre-disaster capacity and during-disaster performance in whole responder networks (Brower et al., 2009). While the literature on resilience abounds with works tackling resiliency frameworks, the dominant emphasis has been on frameworks that span temporal aspects of disaster preparedness, response, recovery, and mitigation as well as levels of analysis including individual, organizations, institutions, communities, and even whole ecologies (Birkman, 2006; Manyena, 2006; Norris et al., 2008). Such macrotheorizing is valuable for considering the dynamics of resilience that may transcend scale, context, and temporality. However, in trade-off, it necessitates a level of abstraction that creates challenges in empirical operationalization and can obscure important elements of context that can advance our theories of resilience. Without frameworks tailored to specific populations, levels of analysis, phases of disaster, and even the unique disaster context, our ability to advance the science of disaster response toward more resilient communities is limited.

Subsequently, in this chapter, we focus specifically on resilience as it relates to one phase of a disaster cycle—the response phase—and is bounded to the actions of one specific collection of actors that we refer to as "the responder network." To accomplish this, we proceed in two ways. First, we advance the definition and conceptualization of a responder network and associated performance goals as a key level of analysis for assessing resilience during the response phase of a disaster. Second, we introduce the concept of *network capacity* and position it within the broader academic discourse on adaptive capacity and community resilience. Drawing from our field research as well as current literature, we discuss five key network capacities for responder networks, focusing our attention on one of the most frequently occurring disasters in the U.S.: wildfire. The clear articulation of key network capacities as well as performance goals is essential to fully understand what we are building capacity for and how it can be measured. Refining concepts like responder network and network capacity are important steps in the effort to empirically claim we are moving toward more resilient community responses as a society.

Following the recommendations of Hilhorst (2003), we proceed inductively, extrapolating broader theory from our field research in this context and integrating it with what is currently known in the literature. In particular, we focus on wildfire that occurs at the wildland-urban interface (WUI) or the area where private homes and other human development meet or intermingle with undeveloped land.

DISASTER MANAGEMENT IN THE WILDFIRE CONTEXT

The field of disaster management has been strongly shaped by the lessons and innovations that have come out of the management of wildfire (Irwin, 1989). This trend is likely to continue as wildfires are one of the most frequently occurring disasters in the U.S. (EM-DAT, 2012) and are increasing in both scale and intensity, and the effects they have on human populations. Many factors contribute to the current wildfire problem including climate change, declining forest health, drought, invasive species, and a history of fire management that has favored suppression (QFR, 2009; Hale et al., 2006; Kitzberger et al., 2001). The problems associated with these factors are magnified by the growing number of people inhabiting places that are increasingly at risk for large wildfire. Wildfire is most likely to occur in these areas that abut forests; consequently, wildfire is largely, but not exclusively, a challenge for more rural areas. Shifts in community and residential patterns during the 1990s and early 2000s have meant that more people have moved into the WUI. The encroachment of human settlements into the areas of the greatest wildfire risk has parallels to other disaster contexts in floodplains and coastal areas. Across the nation more than 10% of land area in the continental United States resides in the WUI, and the population living in the WUI grew by nearly 20% from the 1990s to the 2000s (Hammer et al., 2009). Nearly 50% of all housing units in the West are located in the WUI (11.1 million units), and this percentage is projected to grow in the future as baby boomers retire to amenity rich, rural areas (Hammer, Stewart, & Radeloff, 2009; National Interagency Fire Center, 2009). Consequently, disaster response to wildfire represents a critical context for informing theory development toward the creation of more disaster-resilient communities.

DEFINING A RESPONDER NETWORK

To proceed in developing a framework for conceptualizing and measuring network capacity in responder networks, we must first delineate what is being referred to by a responder network. This task is not as straightforward as it might seem. While it would be simple to imply a responder network is comprised of individuals, organizations, and agencies active in responding to a disaster, upon reflection, it becomes apparent that such a definition is hardly discriminating. This is because during the course of a disaster and its immediate aftermath, everyone who is affected will likely fit this definition. As the incident unfolds, residents will take action to protect their lives and property, seek out information about the status of the situation, and reach out to help their friends and neighbors. Local organizations and agencies with formal responsibilities to minimize and cope with damages will spring into action. As community needs become apparent and go unmet, individuals and organizations may extend, alter, or completely step outside of traditional roles to try and meet those needs (Drabek, 1985; Drabek & McEntire, 2002, 2003).

Sociologists have characterized different types of structural entities that take action during disasters as established, expanding, extending, or emergent (Drabek & McEntire, 2003). None of these terms adequately captures what we mean by the responder network, but elaborating on how these fit together is essential for conceptual clarity. At one end of the spectrum are established structures, which are best characterized by the bureaucratic agencies and organizations that have formal missions, mandates, and charters related to emergency management and are traditionally tasked with emergency response. They often take on a command-and-control structure and have been criticized for their inflexibility (Neal & Phillips, 1995). In contrast, at the other end of the spectrum are emergent structures, which are characterized as those made up of private citizens and public organizations and agencies that develop new tasks and structures organically in response to the crisis (Stallings & Quarantelli, 1985). Expanding and extending structures each embody either more regularized tasks within new structures or non-regular tasks in old structures.

What is missing from these characterizations is the ability to combine forms under one larger networked umbrella. Our definition of responder network is based on a priori understanding of agencies and organizations with formal disaster management responsibilities in the event of a disaster, as well as expectations about organizations, agencies, and individuals who might be expected to take action as part of the emergent structure. It is this collection of established and emergent actors with sustained involvement in coping with the disaster on behalf of the community that researchers look to as the primary participants in the disaster response. For our purposes, we refer to these actors as *responders*.

Collectively, these actors comprise what we refer to as the *responder network*. A basic definition of a network is a collection of actors linked together through some form or relationship or dependency (Brass et al., 2004). We use the concept of a network intentionally to highlight a specific set of attributes that are theoretically important in reflecting on the nature of disaster response. First, the concept of a network infers interdependency among members; however, it is different from a group in that a network has no assumption of an overriding common identity that is shared across members. Whether an organizational leader thinks about their organization as being part of a network is merely an attribute of network membership, not a pre-condition for it. Second, networks are conceptually distinct from other collective enterprises such as organizations, consortia, or collaborative partnerships in that a network by our definition (Provan & Kenis, 2008) does not necessitate an explicitly shared goal except in the broadest sense of minimizing harm from the disaster. Last, the concept of a network does not assume universal connection among all actors. Consequently, it is a concept that is well adapted to characterizing the pluralism and loosely coupled, sometimes underorganized, domains of responders active in a disaster response. Accordingly, we define a responder network as *the collection of individuals, organizations, and agencies that have sustained involvement during the event who aim to serve the community in minimizing and coping with damages brought on by the disaster.*[1]

JUSTIFYING A RESPONDER NETWORK LEVEL OF ANALYSIS

Resiliency can be theorized across different populations and levels of analysis which begs the question of why scholars and practitioners should concern themselves with resiliency as it relates specifically to a responder network. We pose that an emphasis on a responder network level of analysis is a critical addition to scholarship on resiliency for two reasons. The first and most straightforward reason is that during the response phase of a disaster, it is the responder network that actively engages in efforts to help the community cope with and minimize impacts from the disaster. Therefore, they represent a critical domain in which the propositions of resilience theory can be examined.

The second and most important reason is that disaster response is inherently a networked phenomenon that can never be assumed to nest neatly under any one jurisdiction or command structure (Kapucu, 2005, 2006; Magsino, 2009). In light of this absence of hierarchy, network forms of governance have been advocated as the most viable option (Comfort & Kapucu, 2006; Moynihan, 2008, 2009). This latter justification requires some explanation, rooted in an analysis of the Incident Command System and how local, state, and federal governments parcel out responsibility for incident management. We describe this structure as it applies to the wildfire context to provide a more contextually grounded description; however, similar structure and parallel jurisdictional issues exist across disaster contexts.

Like most large disasters, a large wildfire event typically is managed through the Incident Command System (ICS). The ICS was developed in the 1970s out of a need to centralize authority among multiple organizations during wildfire incidents (Irwin, 1989) but has since been adopted by local, state/provincial, and federal/central emergency management organizations to manage disasters. The ICS provides a model for using common language, management concepts, and communication strategy during the chaotic conditions of disasters.

Local and state governments are responsible for responding to a disaster, with assistance from the federal level (Tierney, 2008). However, when a wildfire occurs, the governmental entity upon whose jurisdiction the fire is burning assumes responsibility for managing the incident within the scope of their expertise and resources. We refer to this entity here as the lead agency. If the fire spreads across jurisdictional boundaries (e.g., from federal to county lands) or there are multiple fires (as is often the case in lightning related incidents), the jurisdictions may elect to manage the incident jointly under a structure referred to as unified command. Under the ICS, the incident is generally managed by an organizational structure called an incident management team (IMT). An IMT is part of the ICS. When the costs and complexity of an incident overwhelm the local-level resources and expertise, governments are authorized to request a regional or federal

IMT to assume command of the incident. While these regional or federal IMTs have extensive expertise in incident response, they may have little to no past experience with or knowledge of the affected communities.

The IMT receives its delegation of authority from the lead agency that requested the IMT's assistance. In our research, we have focused on WUI fires either within or adjacent to U.S. Forest Service (USFS) land. The USFS is the largest firefighting organization in the world, responsible for managing 193 million acres of land (NIFC, 2009). There are 155 national forests in the U.S. Each national forest is managed independently and serves as the lead disaster response agency for wildfires started within its borders. Between 2008 to 2010, 59.29% of all wildfires requiring a Type 1 (federal) or Type 2 (regional) IMT occurred on USFS land (NIFC 2011). When a regional or federal IMT is brought in by the USFS Forest (the lead agency on fires on USFS lands), their delegation of authority must lie within the authority of the USFS for managing the incident. The IMT assumes command over all operations and resources operating under the authority of the USFS but generally will not have jurisdictional authority over "the incident" in any holistic sense. For example, an IMT that obtains its delegation of authority from a USFS forest does not generally have direct jurisdictional authority or responsibility for road closures, evacuation of county residents, mass care (e.g., sheltering) of evacuees, or the management of utilities. These functions lie under an often confusing and incomplete patchwork of responders that can span different agencies, levels of government, and sectors.

This ICS structure has two important implications. First, it means that although hierarchical relations are embedded throughout the network, there is no single superordinate entity that has the means, expertise, or authority to assume command of all aspects of the incident. Second, it means the responder network is an emergent entity comprised of both responders from the local area with limited experience with disasters and external responders who may have little or no knowledge of the area. Yet, it is the collective and combined efforts of all of the responders that constitute the disaster response aimed at minimizing harm from the event. As a result, the concept of a responder network and network capacity emerges center stage to our understanding of resilience in the response stage of a disaster.

COMMUNITY RESILIENCY AND CAPACITY

The notion of community resilience has been widely embraced and legitimated by the disaster community in the U.S. and worldwide as a hallmark of effective disaster preparedness and response (Birkman, 2006; Cutter et al., 2010; Manyena, 2006). Recent reviews of the disaster resilience literature have highlighted some of the major themes in the definitions and uses of the term resilience (Birkman, 2006; Cutter et al., 2008; Manyena, 2006;

Norris et al., 2008). Central within these debates is the question of whether resilience is simply the ability to absorb shocks or disturbance and readily return to a pre-incident status or whether resilience also incorporates adaptation and learning toward an improved state (Birkman, 2006; Cutter et al., 2008). As a return to the status quo is problematic in communities identified as systemically vulnerable to disasters, the latter conceptualization has been particularly appealing to those scholars seeking to characterize resilient communities (Manyena, 2006). Consistent with this, Norris et al. (2008) describe community resilience as both a set of adaptive capacities as well as a process through which these capacities are engaged to enable adaptive action in response to a disturbance.

Cutter et al. (2008) offer an integrated definition of resilience as the ability to survive and cope with a disaster with minimum impact and damage, including the capacity to reduce or avoid losses, contain the effects of disasters, and recover with minimal social disruptions. In the Cutter et al. (2008) Disaster Resilience of Place (DROP) model, resilience is viewed as embedded in a set of pre-conditions that influence the coping response, absorptive capacity, and adaptive learning that occurs during a disaster. These, in interaction with the inherent vulnerability of the community and characteristics of the event itself, define the impact of the hazard and degree of recovery. Accordingly, while resilient communities are viewed as less vulnerable (Cutter et al., 2008), vulnerability is also recognized as rooted in larger ecological and societal factors that cannot be fully mitigated through capacity-building endeavors (see Chapter 15). Disaster resiliency is therefore understood as a set of capacities as well as the process through which these capacities are leveraged to minimize a negative impact on the community and strengthen the community's ability to deal with future disasters (Norris et al., 2008). This suggests the need to focus on process as well as outcome elements that can be measured. These propositions lead to the question: What are the capacities within the responder network that facilitate an adaptive response during a disaster?

A PERFORMANCE FRAMEWORK FOR ADAPTIVE RESPONSE WITHIN RESPONDER NETWORKS

To answer the above question, one must first address the question of what an adaptive response to a disaster looks like. As Cutter et al. (2008) aptly point out, an analysis of resiliency necessitates an articulation of resiliency of whom, for what, as the conditions defining resilience differ across domains and levels of analysis. Capacity is defined by its intended goal given that any attribute of a community could be theorized as either a capacity or liability depending on the focal referent of interest. For example, bureaucratic structures that are lauded by some for promoting accountability and stability are criticized by others as fostering structural inertia and limiting adaptability (Morgan, 2006).

It then follows that if we are to think about building capacity within a responder network to promote community resilience, then we must first clarify the goals we seek to advance through these capacities. This requires normatively ascribing to the responder network a set of responsibilities that are believed to advance the overarching goal of promoting resilience. Consequently, the framework of performance goals that follows is essential if we are to understand what comprises a resilient response. Based on our field research and review of the literature, we identify two outcome and three process goals that are generally ascribed to wildfire response networks. These goals can serve as the basis for evaluating performance and for ascertaining which attributes of network capacity are important for achieving these goals, as detailed in Figure 12.1. We focus our attention specifically on the response phase of the disaster. While we illustrate our points with wildfire responder networks here, we anticipate these goals will generalize to a variety of technological and natural disasters.

Goal #1 (outcome): *Minimize injury, loss of life, and damage to values at risk in a manner that advances broader community goals as opportunity allows*. This goal emerges from the National Cohesive Wildland Fire Management Strategy ("Cohesive Strategy") which sought to identify concise, mutually accepted goals and guiding principles that encompassed federal, state, tribal, local, and non-governmental organizations related to land management and wildfirerelated risk (Cohesive Strategy, 2012). The Cohesive Strategy establishes three national goals, including improving wildfire response by diminishing injuries and loss of life to the public and firefighters (Cohesive Strategy, 2012). Goal #1 is essentially an extension of this first performance measure presented by the Cohesive Strategy in two ways. First, it recognizes that the focus on safety embedded within the Cohesive Strategy is necessitated by the reality that protection of other values at risk (e.g., structures, forests, wetlands, watersheds, livestock, biological species

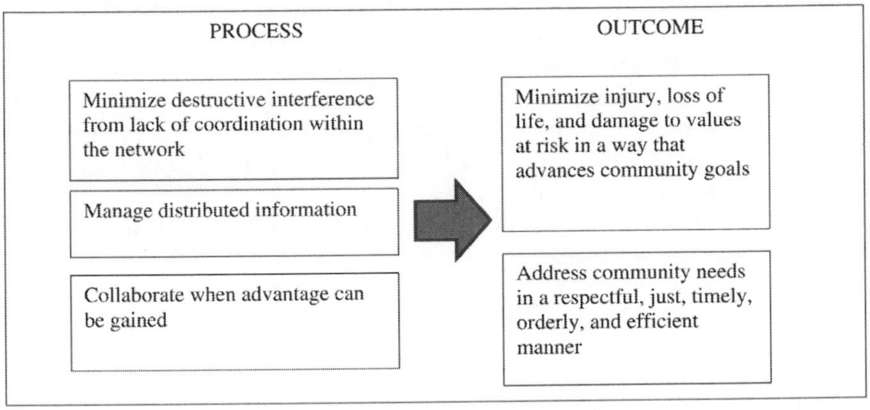

Figure 12.1 Process and outcome performance goals in responder networks.

of concern, sites of cultural significance, etc.) remain a prominent concern during a wildfire event, historically leading to employment of aggressive and sometimes dangerous fire suppression measures being used. Second, we use the phrase "advance broader community goals as opportunity allows" to reflect a dialogue occurring in the wildland fire community about the role of fire in fire-dependent ecosystems and the effects of full suppression fire management policies on increasing the severity and risk of future fires (Dombeck, Williams, & Wood, 2004). The essential recommendation resulting from this dialogue is that fire be viewed as a natural part of fire-dependent ecosystems and that it be managed in accordance with community goals and priorities for protection of lives, structures, and other values at risk. In some cases, this suggests possibilities for more flexible fire management strategies that reduce vulnerability for future fires and pose less risk to firefighters (Steelman & McCaffrey, 2011).

The clear challenge in the actualization of Goal #1 lies in the assumption of community-level goals, values, and priorities. Such terms imply a consensus of thought which may or may not exist. The passage of the Healthy Forests Restoration Act in 2003 stipulated that communities wishing to receive funding for hazardous fuel reduction to mitigate risk associated with wildfires needed to create a Community Wildfire Protection Plan (CWPP). CWPP planning efforts bring together residents, property owners, local, state and federal agencies, and others to create and prioritize a vision for addressing hazardous fuel treatments and wildfire management in the WUI (Steelman & Burke, 2007). However, if there have been no community-level forums for discussing short- and long-term risks, identifying community concerns and goals, and establishing priorities, the assumption that a set of values, understanding of risks, and collection of prioritized goals exists to guide the responder network during the response is tenuous.

Goal #2 (outcome): *Address community needs in a respectful, just, timely, orderly, and efficient manner.* This second goal focuses on the responder network's performance in meeting the needs of the community during the response phase of the wildfire. This would encompass actions that address the physical, economic, social, psychological, and spiritual well-being of those affected by the event. Such needs are widespread and may include needs for information, mass care for the evacuated and displaced including pets and livestock, a meaningful role for volunteers to play in responding to the event, and social support.

One clear challenge in the operationalization and actualization of this goal is the fact that community needs are not monolithic. They vary across groups and are dynamic as the event unfolds. Evacuated populations have different needs than non-evacuated (McCaffrey, Velez, & Briefel, under review), business owners and major landowners may have different needs than community residents for access to their property, individuals who are elderly or have disabilities may have different needs for assistance with evacuation than those who are not (Bloodworth et al., 2007). Another clear challenge is the ever-present reality of limited time and resources. For the responder network to be able to address community needs, they must first be aware of what the perceived needs in the community are and the risks associated with failures to address these needs for the physical, economic, social, psychological, and spiritual well-being of the community. They must then prioritize needs and expectations given scarce resources.

Goal #3 (process): *Minimize destructive interference (from lack of coordination) within the network*. Whereas the first two goals speak to *what* we want from the responder network, the remaining three goals speak to the question of the *manner in which* they should attempt to accomplish these goals to maximize a resilient response. We refer to these as process goals. The first of these speaks to appeals heard throughout the fire management community as well as the hazards community for greater coordination among responders within the responder network. The root of this need stems from functional interdependencies within a responder network. These interdependencies create conditions such that individual actions during the disaster can have significant, and often negative, implications for other actors in the network (Bouwen & Taillieu, 2004). In the absence of coordination, this means that actors making up the responder network can undermine each other's ability to achieve their goals, thereby detracting from the effectiveness of the responder network as a whole in minimizing harm from the event.

Take for example the interdependent actions of an electricity company and the Operations section of the IMT. Each of these actors is bureaucratically autonomous as there is no superordinate authority coordinating their actions. Each has an important function to carry out. The electric company has to restore power to community residences and businesses. IMT Operations has to move large numbers of trucks, equipment, and personnel to fight the fire. For the electric company to accomplish its goal of meeting community access to power, it needs to get access to power lines that lay beyond the fire line. For the IMT Operations to accomplish its goal of minimizing damages to values at risk, it needs to restrict access to the roads behind the fire line to ensure fire equipment can move swiftly and safely. In the absence of coordination among these two parties, either the IMT Operations will interfere with the goal accomplishment of the electric company by restricting their access to needed sites, and/or the electric company will interfere with the goal accomplishment of the IMT Operations by moving electric trucks onto a service road which blocks or slows access by firefighting personnel. Accordingly, a key performance outcome of the responder network relates to the degree to which they minimize destructive interference by coordinating effectively with other members in the network.

Goal #4 (process): *Manage distributed information*. During a disaster, a fundamental management challenge is communication and information exchange in what is usually a highly complex and uncertain environment (Kapucu, 2006). In these conditions, communication and information needs are diverse, unpredictable, and continuously changing with regard to scope, urgency, and information type (Comfort & Kapucu, 2006). Consequently, effectively coordinating relevant information among and across actors within the responder network can be a significant challenge. In our work (e.g., Steelman et al., under review), we have explored this aspect of the responder network as an information market. The metaphor of an information market frames the challenge of effective information exchange during a disaster as a problem of information asymmetry among those who possess information (suppliers) and those who need it (consumers).

Information asymmetries arise in situations when one individual or organization has knowledge that another does not. As a disaster situation evolves, so does the market as new suppliers and consumers enter the marketplace and the types of information in demand shift in response to changing conditions. As with most economic interactions, the patterns of who exchanges information with whom are not explained purely by relations of supply and demand. Rather, they are embedded within broader social networks governed by preferences for similarity, familiarity, and trust (Nowell & Steelman, under review; Frank, 2009; Granovetter, 1985).

Similar to other disaster contexts (e.g., Comfort & Kapucu, 2006), information asymmetry situations are acute in dynamic wildfire contexts. One dynamic that is of particular concern is the asymmetry that exists between responders such as a federal IMT that comes from outside the local area and the local responders. External responders from outside the area who come to assist the community may have broad expertise and experience in disaster response functions but likely have limited knowledge of the local context. Local responders generally have extensive knowledge of the local context but may have more limited knowledge of disaster response functions. However, situational awareness necessary to guide adaptive action requires both broadly grounded theoretical knowledge of wildfire and incident response as well as an understanding of local context. It is through possessing both of these forms of knowledge that general principles can be productively applied to make sense of actual events as they unfold (Weick, Sutcliffe & Obstfeld, 2005). Consequently, the ability of responder networks to effectively address this information asymmetry is proposed as a key performance metric in the process of minimizing harm from the incident.

Goal #5 (Process): *Collaborate when advantage can be gained*. The last performance goal echoes calls heard throughout the broader literature on disaster response for greater active collaboration and collective effort across jurisdictions, sectors, and response functions (Drabek & McEntire, 2002; Auf der Heide, 1989). This goal transcends the ambitions of Goal #3 in simply avoiding destructive interference among responders or Goal #4's aims to get information to those who need it to inform their independent operations. This goal calls for responders to work together and share resources to achieve what would be difficult for any responder working alone to achieve (Bryson, Crosby, & Stone, 2006). Here we highlight two challenges that have emerged from our fieldwork as particularly prominent for understanding collaboration within the responder network during a wildfire.

The first challenge concerns the issue of cost share. Wildfire incident response is big business, costing millions of dollars in suppression and response (Calkin et al., 2005). There is opportunity for some of this money to be spent locally to offset some of the negative economic impact the event generally has on the local economy (Moseley & Toth, 2004). However, concerns and confusion over who is responsible for paying for what becomes a key barrier to effective and timely collaboration. Collaboration in the context of wildfire response often entails different jurisdictions or agencies pooling or contributing resources as needed. However, these resources (e.g., personnel, equipment, vehicles) are part of the costs associated with the fire. Therefore, their utilization requires some agreement on who is going to pay for their use. For example, if the county fire department or local volunteer fire department contributes engines and personnel to the U.S. Forest Service to aid in the fire response, it is expected that the USFS will reimburse these local agencies for the use of these resources just as they would reimburse for any resource dispatched from another forest. However, without cost share agreements assembled ahead of time, valuable time can be wasted waiting for negotiations of reimbursement before local resources are released. Similar challenges have been recorded across other disaster contexts (Urban Institute, 2006; Pipa, 2006; GAO, 2008).

The second challenge that hinders collaboration comes from the need to clearly communicate expectations related to shared resources and actively manage the collaborative relationships over the duration of the event. Another example is instructive here. On the Schultz fire, the local law enforcement collaborated with the USFS and IMT to man road closures on USFS roads to prevent tourists from entering dangerous areas of the forest. While all parties described this collaboration as exemplary of interagency cooperation, there was frustration expressed on the part of the local law enforcement that the USFS and IMT continued to utilize those resources beyond the initial crisis. Local law enforcement saw staffing USFS roads as under the jurisdiction of the USFS and felt they should have prioritized finding USFS resources to man the road closures as soon as possible to release county law enforcement personnel to be reallocated back to their primary law enforcement responsibilities.

Perceived failures on the part of the IMT and USFS to do this or communicate to the local law enforcement when their personnel could be released were viewed by the local law enforcement as problematic. Reports of this same incident by the IMT and the local forest revealed no particular conflict or concern suggesting that they were unaware of the challenge this created for the local law enforcement. The collaboration literature emphasizes the importance of collaborating partners becoming familiar with the strengths, constraints, and limitations of each partner agency so that partners can anticipate how their actions will affect their partners (Nowell & Foster-Fishman, 2011). However, in the disaster context, the responder network is activated by a relatively uncommon event, making the kind of long-term relationship building that occurs in other collaborative contexts more difficult. As a result, partners must be particularly conscientious to communicate expectations clearly concerning resource sharing, focusing attention on cost sharing, and when resources can be expected to be released back to their home units.

A NETWORK CAPACITY FRAMEWORK

The articulation of performance goals and associated challenges is essential to understand what kind of capacity characteristics are needed to facilitate the achievement of these goals. Next, we focus on those collective-level capacities that have emerged from our field research as well as in the current literature as critical in helping responders to overcome the inherent challenges associated with these goals. We refer to these capacities as *network capacities*.

Network capacities are a specific classification of capacity that can be positioned within existing global resiliency frameworks (Cutter et al., 2008, 2010; Norris et al., 2008) almost entirely under the domain of social capital. Social capital has been widely embraced as a critical dimension of disaster resilience (Dynes, 2002; Norris et al., 2008; Cutter et al., 2008). Part of its appeal lies in its very flexibility as a construct. Scholars have considered social capital as both 1) the presence of structures (e.g., neighborhood associations), patterns (bridging and bonding network ties, structural holes), and relational characteristics (trust, friendship, affiliation) of social connectivity that facilitate collective action, as well as 2) the potential resources that are available through a network of relationships (Adler & Kwon, 2002; Kadushin, 2004; Nahapiet & Ghoshal, 1998). Social capital, like community resilience, is defined primarily by its function (Coleman, 1988). Consequently, most of the network capacities discussed here can be conceptualized under a rubric of social capital as they all are network attributes that facilitate cooperation and collective action. However, to advance both scholarship and community assessment, it is critical to unpack specific dimensions of social capital that relate to our targeted performance goals.

Accordingly, our goal in this section is to leverage insights from the wildfire context toward the development of a network capacity assessment framework that could be applied and tested across disaster contexts. In Table 12.1, we pose capacity assessment questions that transcend disaster contexts to provide insight into how our conceptual metrics of network capacity can be operationalized in different settings. These key questions are the first stepping stone toward a standardized set of attributes that can guide network capacity assessment. We envision the value of such a framework to be twofold. First, it will provide a starting point for community dialogue and discussion on current capacity of the local responder network. Second, it will help define a research space aimed at understanding the attributes of responder networks across disaster contexts and their relationship to network performance in coordinating, communicating, and collaborating during disasters. We pose five dimensions of network capacity in the following sections: trust and familiarity, local knowledge, leadership for integrating newcomers, system perspective, and communication infrastructure.

Familiarity and Trust

A generally accepted proposition is that actors who work together prior to a disaster event are likely to work better during the event because they have built up a level of familiarity and trust among each other (Kapucu, 2006). Familiarity is posited to aid in coordination, facilitate the flow of information, and reduce the risks associated with collaboration because it makes responders more predictable to one another (Okhuysen & Bechky, 2009). The basic narrative of this literature argues that when responders know what they can expect from one another, they are better able to adjust their actions accordingly to avoid destructive interference. If they can anticipate the information needs present within the network, then they are in a better position to judge the risks associated with collaboration (Nowell & Foster-Fishman, 2011). Trust is defined by an emphasis on norms and values, which is prevalent in the work of scholars such as Coleman (1988) and Putnam (2001). It entails a belief that other responders will follow through on their commitments and will not act opportunistically (Dirks & Ferrin, 2001). Based on a review of the literature examining trust in organizational settings conducted over the past forty years, Dirks and Ferrin (2001) found evidence that trust can impact the likelihood of cooperative action by both reducing the degree of perceived risk involved as well as through pre-disposing actors to evaluate the actions of those they trust as benevolent in their intention, thereby reducing conflict.

Responders in the wildfire context highlighted familiarity and trust as rooted in the network at three levels: the interpersonal, the interprofessional, and the interorganizational. At the interpersonal level, familiarity and trust are rooted in personal relationships between the leaders within the responder network. Consistent with Granovetter's concept of embeddedness (1985), these interpersonal relationships among local responders often transect professional, personal, and community domains as individuals have worked together, gone to school together, and lived in the same community over time. Consistent with predictions, our research found the level of familiarity between two responders to be significantly positively related to increased frequency in communication and a decrease in the risk for problematic communication (Nowell & Steelman, under review). At the interprofessional level, responders may not know one another personally but they describe familiarity with the professional role another responder plays. This familiarity was identified as valuable in facilitating coordination and information exchange because it gave responders insight into the needs, capacities, and constraints of that position. For example, a USFS archeologist noted the advantage of having worked with an IMT in the past because he knew what their job looked like during different phases of the incident so he was able to gauge the most opportune time to approach them about the protection of archeological values at risk. Finally, interorganizational familiarity and trust are embedded within cultures of cooperation between responder agencies regardless of interpersonal familiarity, as well as formal planning and cost share agreements to facilitate collective action. Organizations and agencies develop identities and reputations that make them more predictable just like people. Individuals active in the responder network are generally acting as agents of these organizations and agencies. Subsequently, familiarity with and trust in an organization or agency can facilitate coordination, communication, and collaboration by the same mechanism as interpersonal trust and familiarity. Conversely, historical tensions among responding organizations and agencies can likewise cast a shadow of suspicion and mistrust that can ultimately undermine network goal achievement. For example, in the wildfire context, responders commonly reference historical tensions between local landowners and the USFS (Carroll et al., 2006) as a challenge to effective communication and cooperation.

In assessing network capacity (Table 12.1), central questions focus on the level and distribution of familiarity and trust across interpersonal, interprofessional, and interorganizational domains within the network. Much scholarly research related to this domain is needed. For example, in settings where there is a lesser degree of social embeddedness, it is reasonable to suspect that interprofessional and interorganizational aspects of social capital will be more important. Further, based on our research, we anticipate that disaster contexts will differ in the degree to which personal relationships are a prominent form of network capacity within the responder network. For example, they may be disproportionately prominent in disasters occurring within more rural settings where personal networks are more likely to overlap. There may also be differences in the density of interpersonal ties within the responder network stemming from professional affiliations. In the wildfire context, a significant proportion of the responder network is comprised of responders who have professional backgrounds in fire or forest management. This profession was described by some responders as a tight-knit group, where key leaders knew one another in various professional capacities over the course of their careers. Understanding the social and demographic characteristics that correlate with different network capacity profiles is an important area for future research.

Local Knowledge

A second dimension of network capacity concerns the bank of local knowledge that is widely accessible by the responder network and relates to what Nahapiet & Ghoshal (1997) referred to as a cognitive dimension of social capital. Cognitive social capital is based on the notion that a shared sense of understanding can facilitate better outcomes because diverse individuals in a system with a similar understanding can benefit the entire collective if there is mutual understanding about desired actions or goals (Portes & Sensenbrenner, 1993). In practice, this dimension concerns the degree to which there exists a shared community consciousness within the responder network regarding key attributes of local knowledge related to the disaster, including the geophysical, cultural, social, political, environmental characteristics of the area. It also concerns the degree to which the community has developed (through some form of community-level dialogue) a set of prioritized goals and understanding about land management, values at risk, and vulnerable populations and ecologies which may have unique needs during a wildfire event. Because our focus here is to define network capacity for the response phase of the disaster, we do not address other key aspects that could facilitate a more comprehensive collective understanding such as priorities in hazard mitigation. However, these would also be aspects of local knowledge.

Local knowledge has broad capacity implications across the five goals of network performance discussed above. Most directly, local knowledge is an essential prerequisite for the responder network to perform well, overcome the challenges associated with minimizing harm to values at risk, meet community needs, and advance broader community goals. Perhaps most importantly in the wildfire context, local knowledge is an essential capacity to enable a local community to interface effectively with external responders such as a federal IMT. Remarks from federal-level Incident Commanders indicate that communities vary greatly in the degree to which they are able to provide concise and well-organized information about local conditions, geography, values at risk, and fire management priorities. In the absence of some collective sense of community goals and values, the local responder network cannot effectively communicate priorities to external responders. Specific needs of vulnerable populations may go unmet if knowledge of these needs is distributed among a few responders while knowledge of potential resources to meet those needs are distributed among a different set of responders. However, if community vulnerabilities and needs are more broadly understood, knowledge of potential resources to meet those needs can be better channeled through the responder network.

Network capacity assessment questions related to this domain focus on the identification and documentation of community needs, vulnerable populations, values at risk, goals and priorities, and vital aspects of community context during a disaster (Table 12.1). Developing a shared vision and

Table 12.1 A Framework for Assessing Network Capacity Characteristics

	PROCESS GOALS		
	Minimize destructive interference from lack of coordination within the network	*Manage distributed information*	*Collaborate when advantage can be gained*
Familiarity and trust—relational characteristics that exist before the incident that can facilitate response (e.g., trust, goodwill, familiarity)	To what extent are responders sufficiently familiar with one another to anticipate how their actions will affect the actions of other responders? Where is this familiarity the strongest? the weakest? Do responders have personal contacts that they can contact to resolve destructive interference as quickly as possible?	To what extent are responders familiar enough with one another other to anticipate information needs of other responders? To what extent are responders willing to go out of their way to supply information to known consumers?	To what extent do responders have personal trusting relationships with other leaders in the responder network that they can call upon if needed? To what extent do responders trust one another to not act opportunistically? How willing are responders to commit resources to collective efforts as need arises? To what extent are cost-share and other agreements in place prior to the event to facilitate collective action between actors during the event?
Local knowledge—cognitive elements of social capital, shared sensemaking, collective vision and understanding	Is there collective agreement and awareness of local land management goals and values at risk that are in potential conflict with broader disaster management goals?	Are local values at risk, including social, economic, archeological, cultural, and ecological, resources identified and documented? Are local priorities in the protection of these values at risk identified through community-level forums? Are the unique vulnerabilities and needs of different populations within the community widely understood? Are pertinent local characteristics documented in a format and location that is known to and accessible by local responders? Can this information be easily communicated to external responders?	Is there shared understanding of community needs and land management/resource protection goals and priorities to facilitate collective action?

Leadership for integrating newcomers	To what extent is there a local responder who can help external responders understand how their actions may influence the work of other responders? Does this local responder have sufficient legitimacy among both external and local responders to serve as a broker?	To what extent is there a local responder with sufficient familiarity and legitimacy who can translate local information and its significance to external responders?	Is there a local responder with sufficient reputation and legitimacy to serve as a convener, bringing responders together as needed, and helping external responders understand who they need to be networked with?
Community-level perspectives	To what extent do local and external responders view themselves as part of a larger response during a disaster?	Do external responders (e.g., IMTs) recognize the importance of local context information, local goals and priorities? Do they have strategies for how to seek it out and incorporate it into their operations?	To what extent is there a strong sense of community and a culture of cooperation among responders?
Communication infrastructure—forums and technology	Are there sufficient fora and communication technology to facilitate coordination?	To what extent are there plans, technology, and forums for facilitating communication in the event of a disaster? Is there sufficient redundancy in these systems in anticipation of damaged infrastructure during the event?	Are there sufficient forums and technology to enable collective action?

making collective sense about local contextual features is crucial to avoid destructive interference, manage distributed information, and collaborate proactively across the network to avoid undesired performance outcomes.

Leadership for Bridging between External and Local Responders

Another aspect of network capacity that has received less attention in discussions of disaster resilience has to do with leadership for bridging between external and local responders. The arrival of external responders is a reality in large-scale wildfires, just as they are in other disaster contexts.When the scale and scope of an event overwhelms local-level capacity, state and federal resources are called upon to assist. Frequently, this assistance includes not only resource and operational support but assistance in the management of the event itself. More seasoned IMTs are brought in to take command of the operations. Consequently, while all disasters are local, they are frequently not managed by locals. This means that a key area of capacity for both the local and external responders concerns how they will integrate together into a coordinated responder network. As discussed previously, pre-existing trust and familiarity can aid coordination among local responders. However, these attributes are frequently lacking in network ties between external responders and local responders, necessitating a different form of capacity for promoting effective communication, coordination, and collaboration.

During our field research, we have discovered that the lack of relational social capital can be mitigated by the presence of a local agency who can serve as a bridge between the external IMT units and the rest of the local responder network. In the wildfire context, the agency best equipped to fill this role tends to be the local branch of the USFS. The role of this agency is threefold: First, they serve initially as an information broker, moving information back and forth between the local responders and IMT. They channel local context information to the IMT and facilitate the distribution of information from the IMT to the local responder network and broader community. Second, they serve as a relationship broker, facilitating introductions between IMT unit leaders to key local counterparts with whom the IMT will need to interact with directly, helping to establish the legitimacy of both parties to build trust more quickly. Last, they serve as a network manager throughout the incident, monitoring the network to ensure information asymmetries are being addressed, efforts are coordinated, and opportunities for collaborative advantage are recognized and acted upon. The ideal agency for this position is one that has significant reputational legitimacy with both the external and local responders. The more both external and internal responders trust and respect this agency, the better positioned they will be to perform this leadership function within the network. It is also important that this intermediary agency has sufficient familiarity of the local context, local responder network, and external responders to navigate interactions with this diverse group, helping to translate across professional boundaries.

Network capacity assessment questions related to this domain focus on key elements associated with bridging roles between internal and external responders, as indicated in Table 12.1. These questions help identify whether and how leaders play a role in minimizing destructive interference, manage distributed information, and capitalize on opportunities where advantage can be gained. Additional scholarly work is needed to better understand this capacity domain across different disaster contexts. Key research questions include: 1) Is this form of brokerage more or less important in different disaster contexts and if so, how? 2) Who are the key brokers in different contexts? To what extent do they share similar characteristics? 3) What are the organizational capacities needed to best fulfill this brokerage role?

Community-Level Perspectives

This domain speaks to whether there is the awareness and motivation needed for responders to act in a coordinated manner. A coordinated, multi-organizational response to a disaster frequently requires organizational representatives to view themselves as part of a larger collective response and prioritize cooperation over organization-centric interests. Psychological Sense of Community (PSOC) is defined as "the perception of similarity to others, an acknowledged interdependence with others, a willingness to maintain this interdependence by giving to or doing for others what one expects from them, and the feeling that one is part of a larger dependable and stable structure" (Sarason, 1974: 157).

The theory of PSOC posits that individuals and collectives with a greater sense of community have a high concern for community and a cognitive disposition toward advancing community interests (McMillan & Chavis, 1986). As a result, the theory states individuals with higher PSOC are more likely to transcend individualistic or opportunistic behaviors and engage in collective efforts to promote community interests and well-being (Nowell & Boyd, 2010). This theory has received substantial empirical support in non-disaster contexts, with PSOC predicting a range of community-oriented types of behaviors such as civic engagement, political participation, and volunteerism within the general population (Chavis & Wandersman, 1990; Hughey, Speer, & Peterson, 1999). PSOC has also been theorized to lead to better performing and more socially responsible organizations (Block, 2008; Mintzberg, 2009; Nowell & Boyd, 2010). Subsequently, there is reason to suspect that local responder networks characterized by higher sense of community and responders who view their agency as one part of a larger system of response will be more active coordinating and sharing information, and have a greater willingness to collaborate.

Accordingly, network capacity assessment questions related to community-level perspectives focus on the extent to which responders in the network share a strong sense of community and view themselves as part of a larger response system, as detailed in Table 12.1. For external responders, assessment questions relate to their understanding of how they fit within the local responder network and are motivated to seek local knowledge to inform their operations. While scholars have argued for the importance of sense of community as an aspect of community capacity for disaster resilience (e.g., Norris et al., 2008), there is very limited empirical work conducted within this domain. Therefore, there is much to be understood about how responders' connection to place influences their interactions within a responder network before and during a disaster. Further, we know little about how responders view themselves in relation to the responder network and the influence of these cognitions in aggregate on the performance of the network to act in a coordinated and coherent fashion. These questions will be critical to address to gain insight into how destructive interference within the network can be avoided, diffuse information can be better managed, and responders can identify areas for more profitable collaboration.

Communication Infrastructure

Disaster scholars have emphasized the necessity of adequate infrastructures for supporting communication, coordination, and collaboration during the response (Carley & Harrald, 1997; Comfort & Haase, 2006; Green, Kliener, & Montgomery, 2007), particularly those that facilitate communication between local and external (e.g., state and federal) responders (Eikenberry, Arroyave, & Cooper, 2007; Kapucu, et al., 2010). Network capacity related to this includes communication technology that is robust, scalable, and interoperable to allow for integration of an increasingly complex number of responders and array of information. As disasters frequently damage communication infrastructure, taking out phone lines, electricity, and cell phone towers, the presence of backup systems in anticipation of such disruption is a valuable aspect of network capacity (Comfort & Haase, 2006).

However, technology such as radios, cell phones, satellite phones, and computers are only one dimension of a community's infrastructure capacity for supporting communication, coordination, and collaboration within the responder network. Of equal concern are spaces, fora, and boundary objects for facilitating information sharing and coordination. We found that key meeting spaces among those in the responder network, including structured as well as ad hoc meetings, were essential in light of disrupted communication technologies. For instance, an established meeting at 8 am for all cooperating agencies as well as the IMT was a reliable mechanism for giving and receiving information in light of uncertainties associated with cell phone and electrical disruption. Additionally, our research found that people most frequently tap into the information sources during a disaster that were most familiar to them before the disaster, even if they recognized these sources as less trustworthy and less adequate for their needs than other sources available to them (Steelman, McCaffrey, Velez, & Briefel, in prep.). This suggests that existing community infrastructure such as radio stations, television stations, newspapers, friends, family, listservs, websites, contact lists, and community meeting spaces that facilitate communication among local responders before a disaster will also likely serve as resources for supporting communication and coordination during an incident (Majchrzak, Jarvenpaa, & Hollingshead, 2007).

Network capacity assessment questions related to this domain focus on communication technology as well as lower tech meeting space and opportunity to operationalize this final aspect of network capacity for our framework (Table 12.1). Additional questions that remain to be addressed in this domain include more fully understanding the comparative advantage of different communication modalities in different contexts, what factors contribute to perceptions of trustworthiness as a communication option, and how various high-tech and low-tech communication infrastructure modalities fit together for optimal mixes under the unpredictable circumstances of disaster.

CONCLUSION

Measuring community resilience to disaster is a key challenge for disaster scholars at this time. Across the fields of disaster research there is interest in identifying local-level capacities that can improve a community's ability to prepare for and respond to disasters. In this chapter, we have attempted to increase understanding of how local-level capacities are related to greater community resilience. Our contribution has been to focus on more narrowly identified capacity characteristics that exist among a specific subset of actors that we define as the responder network. We distinguish network capacities in their relationship to network-level performance goals as well as their level of analysis. While community competence may reside at the resident or organizational level, network capacity is conceptualized as an attribute of the responder network. We focus specifically on network capacity as it relates to performance during the response phase of the disaster.

To understand how one measures network capacity in this context, it was essential to specify performance goals for which network capacity characteristics must be targeted. The specification of process and outcome performance goals in conjunction with network capacity characteristics provides a framework for understanding whether responder networks are indeed moving toward or away from more resilient outcomes. We feel that these concepts are an essential stepping stone in the broader collective effort to make empirical claims about whether we are moving toward more resilient responses as a society.

Some limitations associated with our proposed framework are important to note. We investigated these issues within the context of wildfire. It is not clear how this framework would apply to other disaster contexts. We believe that the concepts are applicable, but that is an empirical question worthy of further investigation. Wildfire occurs in a predominantly, but not exclusively, rural setting. We have reason to believe that the number and density of responder relationships may increase in more urban settings relative to rural settings.The concepts suggested here should still hold, but may need to be implemented among a larger number of responders who presumably would be needed in a more populous urban area.

Much research remains to be done on disaster network capacity. In addition to the specific areas suggested in the previous sections related to network capacities, we propose a broader agenda for more fully developing network capacity as a robust measurement framework for community resiliency. A tentative research agenda for moving forward would: 1) confirm and expand upon these network capacity characteristics, examining their generalizability and translation across disaster contexts; 2) test and refine assessment questions to see if there is broader agreement on them within the disaster community; 3) once there is agreement on the network capacity characteristics and questions, develop a set of standardized measurements to provide more empirically precise insight into how the capacities and process goals relate to the performance goals. The ultimate vision with these steps is to move us in the direction to better understand what we mean by resilience at different levels of analysis and across different contexts as presented by several examples in this book.

NOTES

1. In operationalizing the responder network on large-scale wildfires, we have found it useful to distinguish three key stakeholder groups: 1) the federal Incident Management Team (IMT) that assumes command in managing a large-scale disaster response; 2) the local lead disaster response agency that requests the IMT and may coordinate local disaster planning and preparedness efforts; and 3) local cooperators which consist of the array of local organizations and agencies that have specialized roles and responsibilities related to the disaster response and preparedness (e.g., public safety, information evacuation, sheltering, infrastructure protection and repair) not under the jurisdiction of the lead agency.

REFERENCES

Adger, W., Brooks, N., Bentham, G., Agnew, M., & Eriksen, S. (2004). *New indicators of vulnerability and adaptive capacity.* Technical report no. 7. Tyndall Centre for Climate Change Research. Norwich: East Anglia University.

Adler, P.S., & Kwon, S. (2002). Social capital: Prospects for a new concept. *Academy of Management Review*, 27(1), 17–40.

Auf der Heide, E. (1989). *Disaster response: Principles and preparation and coordination.* St. Louis, MO: The C.V. Mosby Company.

Birkmann, J. (2006). Measuring vulnerability to promote disaster-resilient societies: Conceptual frameworks and definitions. In United Nations, *Measuring vulnerability to natural hazards: Towards disaster resilient societies*. Tokyo: United Nations University.

Block, P. (2008). *Community: The structure of belonging*. San Francisco: Berrett-Koehler.

Bloodworth, D., Kevorkian, C., Rumbaut, E., & Chiou-Tan, F. (2007). Impairment and disability in the astrodome after Hurricane Katrina: Lessons learned about the needs of the disabled after large population movements. *American Journal of Physical Medicine and Rehabilitation*, 86(9), 770–775.

Bouwen, R., & Taillieu, T. (2004). Multi-party collaboration as social learning for interdependence: Developing relational knowing for sustainable natural resource management. *Journal of Community & Applied Social Psychology*, 14(3), 137.

Brass, D., Galaskiewicz, J., Greve, H., & Tsai, W. (2004). Taking stock of networks and organizations: A multilevel perspective. *Academy of Management Journal*, 47(6), 795–817.

Brower, R., Choi, S., Jeong, H., & Dilling, J. (2009). Forms of inter-organizational learning in emergency management networks. *Journal of Homeland Security and Emergency Management*, 6(1).

Bryson, J., Crosby, B., & Stone, M. (2006). The design and implementation of cross-sector collaboration: Propositions from the literature. Special issue, *Public AdministrationReview*, 66, 44–55.

Calkin, D., Gebert, K., Jones, J., & Neilson, R. (2005). Forest service large fire area burned and suppression expenditure trends, 1970–2002. *Journal of Forestry*, 103(4), 179–183.

Carley, K. & Harrald, J. (1997). Organizational Learning Under Fire: Theory and Practice. *American Behavioral Scientist*, 40(3), 310–332.

Carroll, M.S., Higgins, L., Cohn, P., & Burchfield, J. (2006). Community wildfire events as a source of social conflict. *Rural Sociology*, 71(2), 261–280.

Chavis, D., & Wandersman, A. (1990). Sense of community in the urban environment: A catalyst for participation and community development. *American Journal of Community Psychology*, 18(1), 55–81.

Cohesive Strategy (National Cohesive Wildland Fire Management Strategy). (2012). National Cohesive Wildland Fire Management Strategy. Retrieved July 26, 2012 http://www.forestsandrangelands.gov/strategy/index.shtml

Coleman, J.S. (1988). Social capital in the creation of human capital. *American Journal of Sociology*, 94, S95–S120.

Comfort, L., & Haase, T. (2006). Communication, coherence, and collective action: The impact of Hurricane Katrina on communications infrastructure. *Public Works Management Policy*, 10(4), 328–343.

Comfort, L., & Kapucu, N. (2006). Inter-organizational coordination in extreme events: The World Trade Center attacks, September 11, 2001. *Natural Hazards*, 39(2), 309–327.

Cutter, S.L., Barnes, L., Berry, M., Burton, C., Evans, E., Tate, E., & Webb, J. (2008). A place-based model for understanding community resilience to natural disasters. *Global Environmental Change-Human and Policy Dimensions*, 18(4), 598–606.

Cutter, S.L., Burton, C.G., & Emrich, C.T. (2010). Disaster resilience indicators for benchmarking baseline conditions. *Journal of Homeland Security and Emergency Management*, 7(1), 1–22.

Dirks, K.T., & Ferrin, D.L. (2001). The role of trust in organizational settings. *Organization Science*, 12(4), 450–467.

Dombeck, M.P., Williams, J., & Wood, C. (2004). Wildfire policy and public lands: Integrating scientific understanding with social concerns across landscapes. *Conservation Biology*, 18(4), 883–889.

Drabek, T. (1985). Managing the emgercy response. *Public Administration Review*. 45, 85–92.

Drabek, T.E., & McEntire, D.A. (2002). Emergence phenomena and multiorganizational coordination in disasters: Lessons from the research literature. *International Journal of Mass Emergencies and Disasters*, 20(2), 197–224.

Drabek, T.E., & McEntire, D.A. (2003). Emergent phenomena and the sociology of disasters: Lessons, trends and opportunities from the research literature. *Disaster Prevention and Management*. 12(2), 97–112.

Dynes, R. (2002). *The importance of social capital in disaster response*. Newark, DE: Disaster Research Center.

Eikenberry, A., Arroyave, V., & Cooper, T. (2007). Administrative failure and the international NGO response to Hurricane Katrina. *Public Administration Review*, 67, 160–170. doi:10.1111/j.1540–6210.2007.00825.x.

Frank, K.A. (2009). Quasi-ties: Directing resources to members of a collective. *American Behavioral Scientist*, 52(12), 1613–1645.

GAO (Government Accountability Office). (2008). *National disaster response: FEMA should take action to improve capacity and coordination between government and voluntary sectors*. Washington, DC: GAO.

Granovetter, M. (1985). Economic action and social structure: The problem of embeddedness. *American Journal of Sociology*, 91(3), 481–510.

Green, J.P., Kleiner, A.M., Montgomery, J.P. (2007). The Texture of Local Disaster Response: Service Providers: Views Following Hurricane Katrina. *Southern Rural Sociology*, 22(2), 28–44.

Hammer, R.B., Stewart, S.I., & Radeloff, V.C. (2009) Demographic trends, the wildland-urban interface, and wildfire management. *Society & Natural Resources*, 22(8), 777–782.

Hilhorst, D. (2003). Responding to disasters: Diversity of bureaucrats, technocrats and local people. *International Journal of Mass Emergencies and Disasters*, 21(1), 37–55.

Hughey, J., Speer, P.W., & Peterson, A. (1999). Sense of community in community organizations: Structure and evidence of validity. *Journal of Community Psychology*, 27(1), 97–113.

Jensen, J.K. (2009). Climate change and rural communities in the US. Briefing paper. Rural Policy Research Institute. Retrieved from http://www.rupri.org/Forms/Climate_Change_Brief.pdf.

Kadushin, C. (2004). Too much investment in social capital? *Social Networks*, 26, 75–90.

Kapucu, N. (2005). Interorganizational coordination in dynamic context: Networks in emergency response. *Connections*, 26(2), 33–48.

Kapucu, N. (2006). Interagency communication networks during emergencies: Boundary spanners in multi-agency coordination. *The American Review of Public Administration*, 36(2), 207–225.

Kapucu, N., Arslan, T., & Collins, M. (2010). Examining intergovernmental and interorganizational response to catastrophic disasters. *Administration & Society*, 42(2), 222–247.

Majchrzak, A., Jarvenpaa, S., & Hollingshead, A. (2007). Coordinating expertise among emergent groups responding to disasters. *Organization Science*, 18(1), 147–161.

Magsino, S.L. (2009). *Applications of social network analysis for building community disaster resilience*. Washington, DC: The National Academies Press.

McCaffrey, S., Velez, A.L., & Briefel, J. (under review). Difference in information needs for wildfire evacuees and non-evacuees. Submitted to *International Journal of Mass Emergencies and Disasters* on Dec 31, 2011.

McMillan, B., & Chavis, D. (1986). Sense of community: A definition and theory. *Journal of Community Psychology*, 14(1), 6–23.

Mintzberg, H. (2009). Rebuilding companies as communities. *Harvard Business Review*, July–August.
Morgan, G. (2006). *Images of organization.* 3rd edition. Thousand Oaks, CA: Sage.
Moseley, C., & Toth, N. (2004). Fire hazard reduction and economic opportunity: How are the benefits of the national fire plan distributed? *Society and Natural Resources*, 17(8), 701–716.
Moynihan, D. (2008). Learning under uncertainty: Networks in crisis management. *Public Administration Review*, 68(2), 350–365.
Moynihan, D. (2009). The network governance of crisis response: Case studies of incident command systems. *Journal of Public Administration Research and Theory*, 19(4), 895–915.
Nahapiet, J., & Ghoshal, S. (1998). Social capital, intellectual capital, and the organizational advantage. *The Academy of Management Review*, 23(2), 242–266.
Neal, D., & Phillips, B. (1995). Effective emergency management: Reconsidering the bureaucratic approach. *Disasters*, 19(4), 327–337.
NIFC (National Interagency Fire Center). (2009). *Quadrennial Fire Review 2009.* Washington DC: National Association of Stte Forests. Retrieved from http://www.nifc.gov/PUBLICATIONS/QFR/QFR2009Final.pdf.
Norris, F., Stevens, S., Pfefferbaum, B., Wyche, K., & Pfefferbaum, R. (2008). Community resilience as a metaphor, theory, set of capacities, and strategy for disaster readiness. *American Journal of Community Psychology*, 41(1), 127–150.
Nowell, B., & Boyd, N. (2010). Viewing community as responsibility as well as resource: Deconstructing the theoretical roots of psychological sense of community. *Journal of Community Psychology*, 38(7), 828–841.
Nowell, B., & Foster-Fishman, P. (2011). Examining multi-sector community collaboratives as vehicles for building organizational capacity. *American Journal of Community Psychology*, 48(3–4), 193–207.
Nowell, B., & Steelman, T. (under review). Familiarity, similarity, and performance effectiveness in disaster response. 2012 Academy of Management annual meeting, Boston, MA.
Okhuysen, G.A., & Bechky, B.A. (2009). Coordination in organizations: An integrative perspective. *The Academy of Management Annals*, 3(1), 463–502.
Pipa, T. (2006). *Weathering the storm: The role of nonprofits in the Hurricane Katrina relief effort.* Washington, DC: The Aspen Institute.
Portes, A., & Sensenbrenner, J. (1993). Embeddedness and immigration: Notes on the social determinants of economic action. *American Journal of Sociology*, 98(6), 1320–1350.
Provan, K., & Kenis, P. (2008). Modes of network governance: Structure, management, and effectiveness. *Journal of Public Administration Research and Theory*, 18(2), 229–252.
Sarason, S. (1974). *Psychological sense of community: Prospects for a community psychology*. San Francisco: Jossey Bass.
Stallings, R., & Quarantelli, E. (1985). Emergent citizen groups and emergency management. *Public Administration Review*, 45, 93–100.
Steelman, T., & Burke, C. (2007). Is wildfire policy in the united states sustainable? *Journal of Forestry*, 105(2), 67–72.
Steelman, T., & McCaffrey, S. (2011). What is limiting more flexible fire management—public or agency pressure? *Journal of Forestry*, 109(8), 454–461.
Steelman, T., Nowell, B., Bayoumi, D., & McCaffrey, S. (under review). Understanding information exchange during disasters: Methodological insights from info-centric social network analysis.
Weick, K., Sutcliffe, K., & Obstfeld, D. (2005). Organizing and the process of sensemaking. *Organization Science*, 16(4), 409–421.

13 Coping with Catastrophe

"Black Swan" in Northeastern Japan, March 11, 2011

Louise Comfort and Aya Okada

CASCADING FAILURE IN INTERACTING SYSTEMS

What do you do when you are facing an event that you have never seen before, you don't know what to do, but you have legal responsibility for protecting the lives, property, and continuity of operations for your community, so you must do something? This dilemma confronted public managers at municipal, prefectural,[1] and national levels of operation when a set of triple disasters struck Northeastern Japan on March 11, 2011. This classic "Black Swan"[2] set of events was beyond anyone's imagination: a massive 9.0 Richter scale undersea earthquake generated an immense tsunami wave reaching heights of more than thirty meters in certain locations along a 290 km coastline (Cabinet Office, 2011). The main shock at 2:46 pm local time was followed by a rapid series of aftershocks, including two over magnitude 7.0 within the next twenty-nine minutes (JMA, 2012), shattering infrastructure, shutting down electrical power, transportation, and communications throughout the region. The powerful tsunami wave, reaching heights of over fifteen meters on the Fukushima coast, flooded the Fukushima Daiichi Nuclear Power Plant, disabling the backup generators and causing two reactors to malfunction, overheat, and release radioactive pollution into the air, land, and sea of Fukushima Prefecture. This cascade of events overwhelmed the capacity of local municipal and prefectural governments to mobilize response for the damaged communities and region, and tested the national agencies severely in the first days, weeks, and months of the disaster.

Most difficult for public managers in positions of responsibility in this rapidly evolving set of disasters was recognizing the scope, scale, and impact of this interacting set of events, each one separately a major disaster, on the affected communities and the entire region of Northeastern Japan. It was the interaction of the physical systems of the earth and sea upon the engineered systems of buildings, roads, and critical infrastructure that generated the severe impact on, and consequent losses in, the social systems of communities, organizations, businesses, schools, hospitals, households, and individuals. Each separate hazard had, in fact, been considered in disaster planning processes at local, prefectural, and national levels of operation, but no planning process had considered the consequent escalation of damage and continuing risk from interacting events.

As a result, the affected communities were ill-prepared to cope with the loss of the most critical functions of communication, coordination, and control, or stabilization, in coping with the continuing processes of disruption and dislocation that were triggered by this extraordinary set of events.

Managers at each level of operations needed to take action, but in virtually all cases, their capacity to do so was limited by lack of an overall profile of damage to the region and lack of access to their normal means of communication and immediate sources of support, as neighboring communities had suffered equivalent or worse damage. The immediate challenge for public managers was mobilizing response operations in a devastated environment with limited personnel, non-functioning equipment, and traumatized communities. The longer term challenges include comprehending the scope and scale of this set of events and rethinking the fundamental process of recovery from, and planning and preparedness for, catastrophic events.

THE TOHOKU CATASTROPHE VIEWED FROM THE UNITED STATES

Emergency managers, researchers, public managers, officials, and the general public in the U.S. watched the events unfolding in Japan on March 11, 2011 with great concern, empathy, and dread. The sequence of events seemed eerily familiar. With painful memories of the collapse of interorganizational and interjurisdictional response to Hurricane Katrina in 2005 still fresh, emergency managers in the U.S. understood instantly the monumental burden and sense of loss that their counterparts in Japan were facing in an even larger set of disasters. Nuclear engineers, many familiar with the design of the Fukushima Daiichi reactors that failed in this event, were eager to offer assistance, if needed. Researchers tracked the sequence of events and sought to estimate possible strategies of action, using computational modeling and geospatial analysis. Ordinary American citizens sent personal donations for disaster relief to the damaged communities, and many volunteered services. The common bond that American citizens shared with the Japanese society was the very human realization that if a catastrophe of this magnitude could occur in advanced, industrialized Japan that has made a significant investment in disaster preparedness, such an event could also happen in the U.S., with its Western coastal cities exposed to seismic and tsunami risk and 140 nuclear reactors in operation across 49 states.

Importantly, the triple disasters in Japan initiated a wave of reflection and reconsideration of disaster planning and preparedness in the United States. Governor Andrew Cuomo of New York recommended closing the

Indian Point nuclear reactors, located thirty-five miles north of New York City, with twenty million people within a fifty-mile radius of the plant (Hakim, 2011). Officials at FEMA reviewed the National Response Framework (NRF) and its training programs for linking federal, state, county, and municipal agencies with responsibilities in disaster response, revised since Hurricane Katrina in 2005, and questioned whether the planned interjurisdictional framework would effectively structure response operations to such a massive disaster (DHS, 2011). Local officials in the coastal cities of Los Angeles in California and Seattle in Washington State reviewed their plans for tsunami warning and evacuation of complex urban regions (City of Seattle, 2011; National Weather Service Program Office, 2011). Citizens in communities exposed to seismic, tsunami, and nuclear hazards sought information regarding preparedness plans for their communities. The events in Japan had a clear impact on raising awareness in the U.S. of potential long-term risk from seismic and tsunami events, as well as continuing concern for managing the risk from large-scale technical systems, such as nuclear plants that affect the safety of millions of people in given geographic regions. The question is whether this newly honed awareness of cascading risk would trigger a different form of planning and preparedness for long-term, catastrophic risk reduction in the U.S., Japan, or globally.

INTERGOVERNMENTAL RESPONSE TO CATASTROPHE

A key observation drawn from the Tohoku disasters is that the size, scale, and scope of destruction from such a catastrophic event requires a different mode of planning and preparedness from single-hazard events that impact a specific location or a particular community. Catastrophic events generate a degree of complexity that cannot be managed by a single agency at any jurisdictional level of operation, but require the collaborative resources of multiple agencies operating across jurisdictions simultaneously to mobilize a timely, effective response to the demands generated by the event. This observation reinforced the judgment of Michael Chertoff, Secretary of the U.S. Department of Homeland Security, during Hurricane Katrina in 2005, that catastrophic events require a higher level of planning, preparedness, and mobilization of resources to engage the full response capacity of the nation (Goodnough, 2006). A critical effect of such a change would be to shift the burden of first response to a major catastrophic event from the local governments that are the first to act, but have the least resources and training, to the national government, with deeper pockets but a broader range of responsibilities. Such a change in responsibility also acknowledges that informed, timely decision making in the first minutes, hours, and days of a major damaging event can alter the trajectory of interdependent consequences that follow from the initial incident. Bringing the full force of informed, trained personnel to reinforce local efforts initiated by local personnel, while preserving the advantages of local knowledge and local contacts creates a more powerful, unified approach to managing a catastrophic event.

Mobilizing an interorganizational, interjurisdictional response to a catastrophic event, however, requires a different mode of organization and communication to ensure a coherent, responsible workflow among participating organizations. Disaster planning creates the legal and organizational structure through which response operations can be mobilized quickly and effectively. Planning for catastrophic events is not easy, and often runs counter to the standard concept of planning for small- to moderate-scale incidents that is based on a clear assignment of roles, specific responsibilities attached to each role, and a hierarchy of command and control. In situations of high uncertainty and rapidly changing conditions characteristic of catastrophic events, pre-assigned roles may not fit the demands for action; personnel to fill those roles may not be available, and the hierarchy may not be able to exercise effective control. In such situations, planning is still critical, but these plans are developed in real time and communicated quickly and effectively to relevant responders and community residents. Some coastal cities, such as Kesennuma and Kamaishi, had initiated planning for earthquake and tsunami events, but had grossly underestimated the size, scale, and scope of the March 11, 2011 catastrophe. The intent is not to seek control of the event which may not be possible, but to inform, enhance, and enable responsible local action within the structure of a larger system of interacting response operations.

DISASTER PLANNING FOR EVENTS AT DIFFERENT SCALES OF OPERATION

The Tohoku Triple Disasters demonstrated the critical need for different types of planning appropriate to different scales of damage, destruction, and disaster operations. While disaster plans were in place on March 11, 2011, i.e., Disaster Countermeasures Basic Act of 1961 and Basic Disaster Management Plan (last revised in 2008 prior to the Tohoku disasters), there were three different plans—one for earthquake hazards, one for tsunami hazards, and one for nuclear accidents (Government of Japan, 2012). Further, the agencies responsible for managing disaster operations differed by type of hazard. The respective plans were basically designed for smaller incidents and organized by separate jurisdictions through regional disaster management plans, with little, if any, coordination among the jurisdictions. In the most severely damaged coastal communities of Fukushima, Miyagi, and Iwate prefectures, the local municipalities were seriously damaged, if not totally destroyed. Most of these municipalities lost not only their buildings, equipment, and communications capacity, but also key operations personnel. Consequently, they were severely limited in their capacity to respond to the overwhelming need. The severe damage to local municipalities meant that the prefectural governments in the region had limited information with which to plan, and limited means to collect timely, accurate information as a basis for action. Under these damaging conditions that did not fit existing disaster plans, action stalled at the prefectural level as managers with limited experience with large-scale socio-technical-ecological disaster and less available equipment sought to comprehend the scope of the events and determine viable courses of action within the limits of their resources and capability. The existing plans offered little guidance for informed action under such catastrophic conditions. Without a legal structure within which to organize cross-jurisdictional operations, response actions formed on an ad hoc basis, with nonprofit and private organizations operating in response to needs as they saw them, rather than interacting with public organizations, to provide a systematic approach for timely, comprehensive assistance to the thousands of people affected by these severe events.

KEY FUNCTIONS IN DISASTER RESPONSE

Four key functions—cognition, communication, coordination, and control—are central to effective disaster operations, but each requires investment of time, attention, and resources before a disaster occurs (Comfort, 2007). Further, these four functions need to be calibrated in time and scale to one another interactively for the response system to operate most productively. Cognition means the capacity to recognize risk quickly and accurately as a basis for action. Communication means the ability to inform others in order to mobilize common action to avert or reduce the risk. Coordination represents the capacity to organize timely, efficient action among multiple actors simultaneously to achieve the desired goal of averting risk and protecting the community. Control signifies the capacity to stabilize operations in damaged communities to avoid escalating danger and to bring the incident to a manageable level of performance in meeting the needs of the community. Each of these functions was present to some degree in managing the response to this massive disaster, but absent in others. Actual incidents, drawn from interviews with actors from the region, illustrate each type of function and the degree of interdependence among the four types. Overall, the performance of the whole disaster response system depends on the degree to which the human actors understood each function and acted on the information available to them to take the next logical step, operating both separately and interactively with the other three functions to achieve a reasoned response.

Meeting with a family living in a small, isolated community in the hills above heavily damaged Minami Sanriku, the mother of the family described her instant recognition of danger from the tsunami when she looked out to sea from her hillside position and saw the huge tsunami wave advancing toward the town. Her husband and son were in the family's fishing boat, bringing the boat to shore. She called to her husband to leave the boat and run to the hills in advance of the wave. Her husband and son hesitated, reluctant to leave their highly valued boat behind. She insisted, and they finally left the boat, fleeing to their house on the hill, just as the wave crashed only a few meters from their front door. The fishing boat was totally destroyed by the advancing tsunami wave, but her husband and son were safe. Had she not recognized the risk and communicated the urgency of immediate action, they too would have been lost to sea.[3]

The tsunami wave destroyed the utilities that provided electricity and water to this small community of only a dozen families, as well as the only road that connected them to outside sources of assistance. Cut off from external assistance and with no working means of communication for two full weeks following the disaster, the families in this tiny community coordinated their scarce resources. By sharing their limited food and water with one another, they managed to survive under extremely limited conditions. In this respect, they stabilized the conditions under which they were living, limited to be sure, but they had established control over their environment in a substantial way, and were no longer at risk for their safety. A strong sense of solidarity as well as self-restraint enhanced community resilience, as was seen in most cities and towns in disaster-affected regions.

Are there ways in which these four functions of cognition, communication, coordination, and control could be enhanced in practice to serve the broader regional community of Minami Sanriku, or other communities along the severely damaged Miyagi and Iwate coasts? To do so would require a different mental model of the roles involved in disaster preparedness and mitigation/prevention, for public agencies, private companies, nonprofit organizations such as schools and hospitals, and most importantly, for individual families and households. The key change is acceptance of responsibility for the safety of one's household, school, hospital, business, or public agency, and recognition that the initial actions taken at the first indication of danger are likely to shape the trajectory of response. It is based on a different conception of risk, that is, risk shared by the community is public risk. It also implies shared responsibility for reducing that risk, and the actions of each member of the community either contribute to, or reduce, that risk. Discovering the appropriate balance between shared risk and shared responsibility is a continual process of matching knowledge against resources, and designing a process of iterative review, reflection, and redesign that updates the actual risk against the growing awareness and capacity of a community to manage that risk.

Decision making by emergency managers under conditions of extraordinary risk and danger entails the voluntary acceptance of responsibility for the safety of others—community residents, organizational personnel, and public agencies—as well as continuity of operations for the community. It requires the capacity to recognize the shift in scale of operations and to adjust one's actions accordingly (Kahneman & Klein, 2009), tasks that, taken collectively, involve cognition, communication, coordination, and control.

NETWORKS OF ACTION IN DISASTER RESPONSE

In response to these devastating events, networks of action emerged to provide assistance to damaged communities and shattered families. These networks were identified and documented through a content analysis of news reports collected from the *Yomiuri* newspaper, published in Tokyo, for the first three weeks following the disaster, March 11–April 2, 2011. All articles reporting engagement in response operations during this period were included in the analysis. The daily record of organizational involvement reported the activities of public, private, and nonprofit organizations at village, municipal, prefectural, national, and international levels of operation. These reports were coded by name of organization, jurisdiction, source of funding, type of transaction, and name of interacting organizations, and entered into the ORA software program, developed at the Center for Computational Analysis of Social and Organizational Systems (CASOS), Carnegie Mellon University. This set of organizations, all engaged in actions directed toward meeting demands generated by the set of triple disasters, was identified as a network of organizations responding to the needs generated by this catastrophic event. Not all organizational transactions involved interactions with other organizations. Those that did were identified and analyzed separately as a distinct subnetwork within the larger response system that exerted more influence than other organizations on the outcomes of actions taken (or not taken) to meet the needs of the affected communities.

While newspaper accounts of disaster operations do not provide a complete record of disaster operations, nonetheless they provide consistent reporting of major activities by key organizations, and include reports of actions by private and nonprofit organizations which may not be included in the records maintained by public organizations. Consequently, analyses of news reports provide a basic profile of the actions taken in disaster operations. These reports were complemented by review of reports from investigations requested by public agencies, reconnaissance reports by national and international professional associations, and on-site interviews with key decision makers at municipal, prefectural, and national agencies.

The actual networks of response organizations documented in news reports published in the *Yomiuri* newspaper for the first three weeks following the events of March 11, 2011 reveal distinctive patterns that show the organizational response strategies of the different organizations that have legal obligations to provide assistance in disaster. Figure 13.1 below shows key entities that responded to the earthquake threat.[4] Figure 13.2

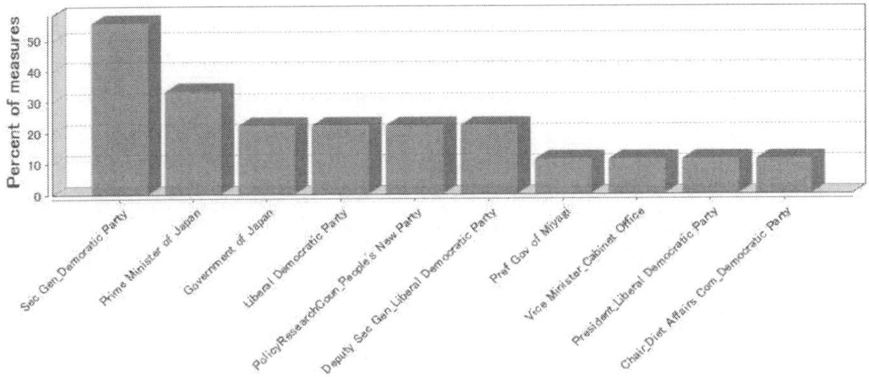

Figure 13.1 Earthquake response network.
Source: Yomiuri newspaper, Tokyo, Japan, March 12–April 2, 2011

shows the key entities that responded to the tsunami threat, and Figure 13.3 shows the smaller set of key entities that responded to the nuclear reactor crisis. These figures document the distinctive stove-piped pattern of Japanese public agencies, with very little cross-communication either among organizations within a jurisdiction or between jurisdictions.

As shown in Figure 13.1, the most frequently mentioned organization in news reports of actions taken in reference to the earthquake response was the Office of the Secretary General of the Democratic Party of Japan, not the local governments of the damaged communities nor even the Office of the Prime Minister of Japan. Of the top ten organizations identified in news reports in reference to the earthquake disaster, only one prefectural government, Miyagi, was named. The governments of Iwate and Fukushima prefectures were not included, nor were the councils of the heavily damaged coastal towns and cities. This is not surprising, as these entities had little capacity for communication and therefore less capacity for coordinating response actions in their communities, but the evidence shows the significant gap in practice from the responsibilities assumed in the pre-disaster planning process.

Figure 13.2 reports the most frequently named organizations in reference to actions taken in response to the tsunami. The list of top ten organizations is virtually the same as those named in reference to the earthquake response network. This finding is likely the result of the planning process which essentially considered tsunami response a subordinate set of actions to be taken in conjunction with the earthquake response. The consequences of March 11, 2011 vividly demonstrated, however, that the tsunami generated by the earthquake was by far the more damaging event, and required different types of response in terms of shelter, medical care, restoration of utilities, removal of debris, and reconstruction of the severely damaged communities. The figure shows that seven of the top ten organizations reported in this network are organizational units of political parties, not

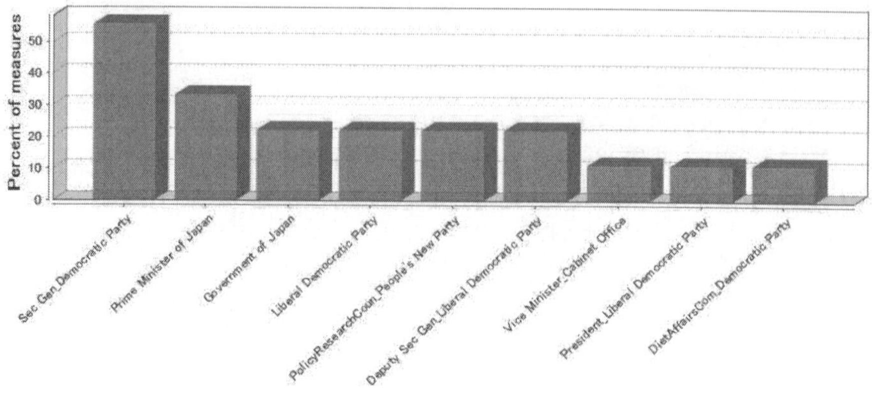

Figure 13.2 Tsunami response network.
Source: Yomiuri newspaper, Tokyo, Japan, March 12–April 2, 2011.

administrative units of governing agencies. This finding indicates the politicized nature of efforts to mobilize response actions.

Figure 13.3 shows a quite different response network to the nuclear power plant failure. In reference to this event, the Office of the Prime Minister of Japan is identified in the highest number of rankings, while the private company, Tokyo Electric Power Company, is named second and the Office of the Secretary General of Democratic Party of Japan is virtually tied in third place on measures of centrality to the nuclear response network. While three of the actors identified in this network are political parties, three other actors represent the Government of Japan, the key Ministry of Health, Labour, and Welfare, and the Office of the Chief Cabinet Secretary of Japan. These findings demonstrate the active engagement of

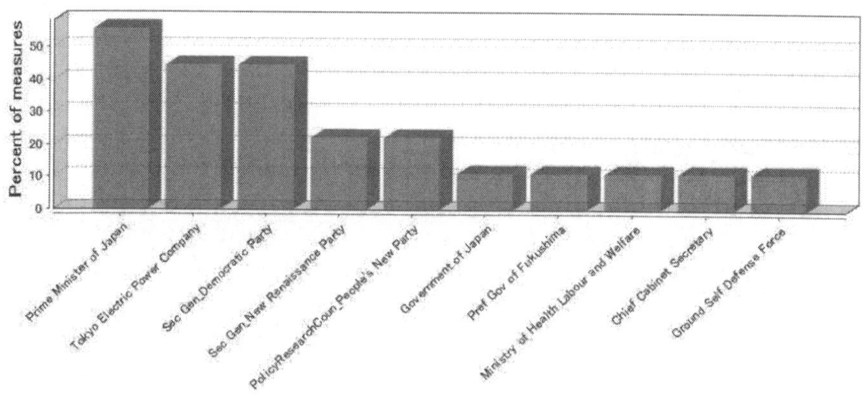

Figure 13.3 Nuclear response network.
Source: Yomiuri newspaper, Tokyo, Japan, March 12–April 2, 2011.

the highest offices of the Government of Japan in responding to this disaster. Importantly, the Prefectural Government of Fukushima is also identified in approximately 10% of the rankings based on news reports, as well as the Japan Self-Defense Force.

Interestingly, response actions to the nuclear reactor failures reflected the greatest uncertainty, and therefore drew the attention of the highest levels of government in Japan. Yet, the density of this network in terms of actions reported was considerably less than for the earthquake and tsunami response networks, reflecting the relatively small number of actors engaged in managing this complex, high-risk, technical system.

CONCLUSION: DESIGNING RESILIENT COMMUNITIES

Events of this magnitude are likely to occur again—at a different time, in a different place, and generated by a different type of hazard. The first challenge for those responsible for protecting their communities against disaster is to anticipate the likely conditions that would trigger a cascading set of events that can only be termed catastrophe. The second challenge for disaster managers is the continuing effort to reduce the vulnerability of their communities to natural and technological hazards. To do so, however, requires substantive change in the design and development of strategies for disaster management in complex, socio-technical societies. The third challenge is to increase the resilience of communities to respond to these events when they do occur, particularly by identifying key points of intervention within interdependent systems. For example, in response operations following the 2011 Tohoku disasters, the Ministry of Land, Infrastructure, Transport, and Tourism made the critical decision to conduct emergency repair of roads and to place restriction on use of main highways that run from Tokyo to Northeastern Japan. This timely action facilitated the flow of aid to the disaster-damaged region.

At least five major changes are necessary to alter the current pattern of performance in assessing, responding to, and managing disaster risk. First, managing risk of catastrophic disaster requires a different mental model of collaborative action under conditions of uncertainty. Such a model differs from standard management models by identifying, first, the core elements in the operating system of the community at risk that are essential to maintain continuity of operations, and second, the alternative strategies that will enable the system to function, given a severe threat to any one of the core elements. This approach assumes that the community will be subject to threat, and that the principal task of preparedness is to maintain the capacity of the community to function under threat, whatever that threat is and whenever it occurs.

Second, because a major function of disaster preparedness is to reduce the uncertainty regarding the interaction among the physical, engineered, and social systems that produces catastrophic disaster, it is essential to integrate science into the planning process to understand risk and monitor more accurately the sources of threat. This knowledge can then be used to inform key decisions regarding the allocation of resources, time, and attention to disaster preparedness and planning, reducing uncertainty and enabling action more broadly throughout the system. The limits of current knowledge regarding near-field tsunami risk was demonstrated when the Japan Meteorological Agency initially made an inaccurate calculation of the height of the tsunami wave generated by the near-field earthquake (*Yomiuri*, 2011). By the time a more accurate calculation was transmitted to local agencies, people either did not hear the rectified warning or had too little time to evacuate (Cabinet Office et al., 2011). Changing not only the mental model of managing risk, but also the allocation of scarce resources and attention to explore alternative strategies of risk reduction, will facilitate a more flexible, adaptive set of organizations that have the capacity to recognize threats when they occur and adjust their performance accordingly.

Third, integrating science into the measurement and monitoring of risk conditions will produce a continuing flow of information regarding risk that allows updating of changing conditions and alerts managers to potential failure. Managing this flow of data to relevant personnel with responsibilities for action in event of serious threat requires the design and maintenance of an information infrastructure that facilitates cross-organizational and cross-jurisdictional communication, information sharing, feedback, and timely updates of valid information.

Fourth, while organizational structure remains central to facilitating rapid, effective action in disaster events, it is essential that this structure be adaptive to the needs and obstacles that may hinder or help performance. Hierarchical organizations tend to fracture under the stress of disaster, while networked organizations develop strength from distant sources that may enable fresh perspectives in managing known risk. Finally, to remain viable, networked organizations need to engage public, private, and nonprofit organizations in a region-wide assessment of risk, resources, and skills available for collaborative action. This intersectoral approach is essential in a catastrophic event when the full resources of the society are needed to staunch the demand for assistance to the affected population. Importantly, such changes are essential to increasing resilience to disaster not only for Japan, but to all nations exposed to risks that derive from living on a restless planet in an increasingly global community.

ACKNOWLEDGEMENTS

An earlier version of this paper appeared in a Japanese academic journal *Jichitai Kiki Kanri Kenkyu* (Emergency management studies) published by the Japan Emergency Management Association in 2012. We gratefully acknowledge the University Center for International Studies and Graduate School of Public and International Affairs, University of Pittsburgh, for support of a reconnaissance study to the disaster-affected region of Northeastern Japan, as well as Kobe University, Kobe, Japan; Professor Yoshio Murayama, Hokkaido; and Iwate Prefecture staff. We thank Professor Akira Nakamura, Meiji University, for his kind invitation to the symposium "Rethinking Crisis Management after 3.11" at Meiji University, Tokyo, Japan on February 25, 2012.

NOTES

1. Prefecture is the subnational administrative unit of government in Japan, equivalent to state or province levels in other countries.
2. The term "black swan" refers to an event that no one believed was possible until it was directly observed (Taleb, 2010).
3. Personal communication, Minami Sanriku, Miyagi Prefecture, Japan, October 31, 2011.
4. ORA identifies "key entities" within the network by calculating a set of centrality measures: total degree centrality (controlling agency), Eigenvector centrality (interorganization leader), hub centrality (acts as a hub), authority centrality (acts as an authority), betweenness centrality (information conduit), and high betweenness and low degree centrality (connects agencies). Key entities are thus organizations with the most ability to monitor and direct information flows and to reach out to other organizations to perform tasks and gain resources.

REFERENCES

Cabinet Office of Japan. (2011). White paper on disaster management. In Japanese. Retrieved from http://www.bousai.go.jp/hakusho/H23_zenbun.pdf.

Cabinet Office of Japan, Fire and Disaster Management Agency, & Japan Meteorological Agency. (2011). Analysis results of interview survey on evacuation behaviors for the 2011 Great East Japan Earthquake (residents). Retrieved from http://www.bousai.go.jp/jishin/chubou/higashinihon/7/1.pdf.

City of Seattle. (2011). Seattle City Council emergency preparedness initiative to launch Monday. News advisory, June 9. Retrieved April 30, 2012 from http://www.seattlechannel.org/news/detail.asp?ID=11798&Dept=41.

Comfort, L.K. (2007). Crisis management in hindsight: Cognition, communication, coordination, and control. Special issue, *Public Administration Review*, 67, 188–196.

DHS (Department of Homeland Security). (2011). Presidential policy directive 8 (PPD-8): National preparedness. Retrieved from http://www.dhs.gov/xabout/laws/gc_1215444247124.shtm.

Goodnough, A. (2006). Chertoff pushes for more hurricane readiness. *New York Times*, April 13. Retrieved from http://select.nytimes.com/gst/abstract.html?res=F10A13F934540C708DDDAD0894DE404482.

Government of Japan. (2012). Summary of disaster countermeasures and disaster management plan of 2011. In Japanese. Retrieved from www.bousai.go.jp/1info/pdf/saigaipanf.pdf.

Hakim, D. (2011). Cuomo takes tough stance on nuclear reactors. *New York Times*, June 28. Retrieved from http://www.nytimes.com/2011/06/29/nyregion/cuomo-emphasizes-aim-to-close-indian-point-plant.html.

JMA (Japan Meteorological Agency). (2012). *Observed earthquakes larger than seismic intensity of five: Within aftershocks, after the initial shake.* In Japanese. Retrieved from http://www.seisvol.kishou.go.jp/eq/2011_03_11_tohoku/i5.pdf.

Kahneman, D., & Klein, G. (2009). Conditions for intuitive expertise: A failure to disagree. *American Psychologist*, 64(6), 515–526.

National Weather Service Program Office. (2011). NTHMP semi-annual report. Retrieved from nthmp.tsunami.gov/ . . . /Semi-Annual_Report_8_30_11_FINAL-CA.doc.

Taleb, N.N. (2010). *The black swan: The impact of the highly improbable.* 2nd edition. New York: Random House.

Yomiuri newspaper. Tsunami warning, in need of system improvement. In Japanese. June 19.

14 Enhancing Community Resilience

A Matter of Multi-Level Framework, Mixed Methods, and Multi-Sectoral Tools

Louise Lemyre and Tracey O'Sullivan

INTRODUCTION

Resilience is gaining popularity as a term to rally forces for the betterment of societies in the face of hardship and adversity. Like stress and strain, the word "resilience" has roots in engineering, psychology, and system sciences. The ubiquity of the term flags a call for interdisciplinary and multi-level considerations in studying and fostering resilience factors. Observing disaster management, crisis recovery, and preparedness efforts, the findings presented in this chapter indicate the need for a comprehensive framework, research on individual and collective processes, and interventions that are multi-prong and multi-sectoral. In this chapter, we will review the conceptual underpinnings and scope of resilience, document the multi-layers of evidence from mixed methods approaches, and will present a series of tools that address some of the critical requirements too often overlooked in planned interventions. The effort towards developing resilience that will be described below come from our ongoing research in individual and collective social and organizational psychology, cross-fertilized by interdisciplinary work in social sciences, in risk sciences, in health sciences, and public policy.

THE CONCEPT OF RESILIENCE

Beyond the engineering analogy of a resilient material absorbing shock and bouncing back into shape, in psychology the concept of resilience relates back to early studies of children who experienced trauma and who, against the odds, survived as normally functioning and mentally healthy individuals (Rutter, 1987). In that respect, the work of Garmezy (1974) marked the field. She identified the critical role of a reliable substitute support figure to explain the resilience of deprived children (Werner & Smith, 1982, 1992). The work of pioneers such as Bowlby (1978, 1982), Piaget (1975), Bandura (1979, 1989, 1995), and Bronfenbrenner (1977) also identified the determinant aspects of attachment relations, cognitive accommodation, experience of self-efficacy, and co-constructive interpersonal relationships, respectively, in the building of resilient individuals. More recently, Masten (2006) and Luthar (2003) have systematized these influences in early interventions for children and brought forward the transformational aspects of resilience which go beyond mere resistance and recovery. The interest for the concept of resilience has also revived in the context of the salutogenic movement (Antonovsky, 1987) and of positive psychology. Hence, resilience involves developing a sense of mastery rooted in a history of reliable trustworthy relationships.

Stress and coping. Many of the studies on resilience can be situated in the large body of research on stress and coping. Stress is a term borrowed from engineering to refer to the tension in a material. The term was coined by Hans Selye (1956) to describe the General Adaptation Syndrome of all organisms to external stimuli in three phases of Alarm, Resistance, and Exhaustion. From the early work on the endocrine response of frogs, rats, and humans, the field of experimentally induced stress moved to examine acute stressors such as life-events and chronic hardship in individuals. It related to subsequent physical and mental health, or overall well-being.

A number of studies on stress focused on trauma as an extreme case of a circumscribed acute stressor that may perturb individuals and bring them to a state of clinical psychopathology with Post-Traumatic Stress Disorder (PTSD). In the context of extreme stressors, individuals who survive functionally intact or who recover spontaneously from these devastating exposures are said to be resilient. The field of psychiatric clinical trauma has grown exponentially in the recent years (Bonnano, Westphal, & Mancini, 2010; Norris, Tracy, & Galea, 2009; Norris et al., 2008; Raphael, 1986). Beyond the relevance of clinical application, resilience has also been examined in the context of large groups of people, organizations, and communities, in relationship to maintaining adequate social and economic functions in the aftermath of a disaster or crisis (Chandra et al., 2011; Lemyre et al., 2005; Norris et al., 2008; O'Sullivan et al., 2012).

Resilience or resiliency. Two terms are in use in the literature: resilience and resiliency (Garmezy 1991, 1993; Masten, 2001). Sometimes they are used interchangeably. Other times they refer distinctively to either a pre-event capacity, or to a post-event outcome (similar to the distinction between likeability and actually being liked) (Anault, 2003). This distinction is important as it points to the temporal aspect of the evaluation of the construct. Is resilience dependent on the experience of a hardship? The very reference to an actual disaster, stressor, or crisis constitutes one point of contention in the field. Indeed, some argue there is an absolute requirement of a traumatic event to invoke the concept of resilience, and that the process of resilience is the return to normalcy (Cyrulnik, 2002; Masten, 1999). Others speak of a capacity to bounce back, with resilience being a potential, a pre-disposition, a set of skills or of assets, a reservoir or repertoire of resources, and therefore being amenable to pre-event capacity building and preparedness (Cyrulnik & Duval, 2006; Garmezy & Masten, 1991; Rutter, 1993, 2000).

The debate between capacity and impact is useful to the extent that it forces researchers to clarify their points of reference and to make explicit their assumptions. As people talk of community resilience and want to develop it, it demands clarification also on the levels or units of analysis that become involved (eg. individuals or communities), as well as the timeline, whether it is being examined pre- or post-event (Colussi, 2000). Henceforth, these considerations relate to the appropriateness of specific measurements and indicators. If we do not negate that all applications of the word "resilience" have their relevance, we only emphasize here the necessity to be clearer on the target and context: individuals or communities, pre-event capacity or post-event impact.

Time frame. The debate over capacity or outcome poses the question of time. Is resilience the residual functioning level after adversity, ex post facto, or the preparedness state of a built-in capacity to face crises and survive them? The answer lies in the mapping of the time dimension as per Figure 14.1, portrayed on a timeline from -3 to +3, from pre-event baseline (-3) to threat (-2) to pre-warning (-1) to occurrence of an incident (t0), rescue (t1), recovery (t2), restoration (t3).

Addressing the time frame not only helps to situate the type of resilience that is being sought and the appropriateness of efforts, programs, and measurements, it also calls directly the doctrines of two strategies vis-à-vis the massive investments required by emergency response: 1) the prevention approach, according to which resilience building should be embedded in routine practices and social fabric; and 2) the sustainability approach, by which resilience should be self-maintained, self-regulated, and self-multiplying, via the engagement of collectivities, to meet the necessary surge capacity and the long-term needs.

Measurement. The measurement of resilience also remains a challenge, above and beyond the conceptual fuzziness of the construct. To our knowledge, there is not a single instrument that fully captures the notion of resilience. It is a multi-faceted construct that expresses itself under various modalities according to time phases and units of analysis. Researchers divert towards a collection of indicators, which usually tend to focus more

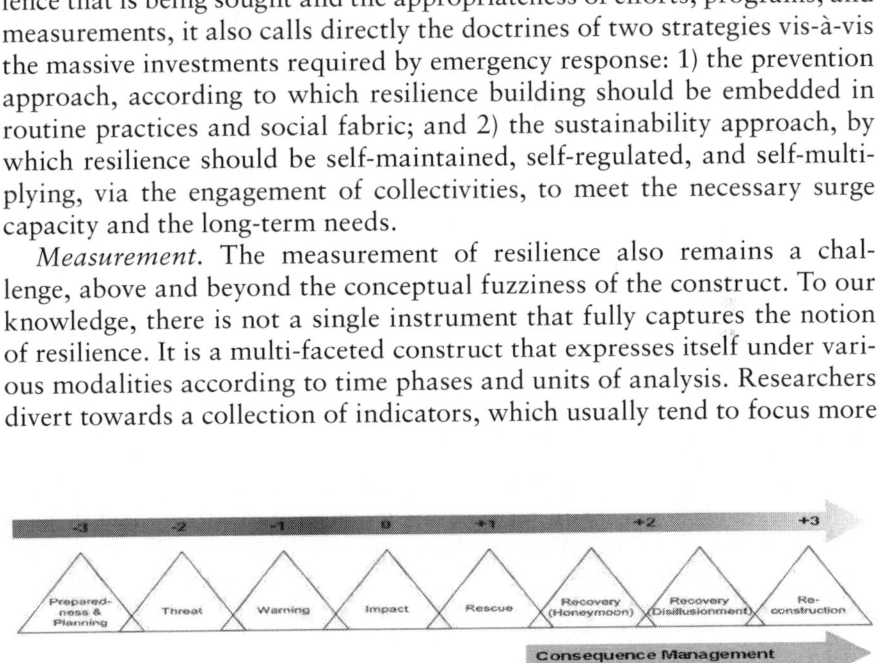

Figure 14.1 Extended timeline of disaster management by time phase -3 to +3.

on either individual mental health measures or more on social network proxies. Some are exclusively psychological (Masten, 2006) while others purely economics (Kretzman & McKight, 1993).We propose in this chapter that a multi-level, multi–time phase, multivariate, and multi-disciplinary approach is needed.

RESILIENCE: A SYSTEMS-APPROACH

Resilience is more than units of willpower, a set of skills, or as proposed by Kobasa (1987), a quantity of hardiness in individuals; instead it is it composed of layers of supporting environments and involves the very processes to leverage this social capital. Therefore a systems-approach implies that resilience includes the property and characterization of the relationships and interconnectedness of the complex system of components and mechanisms.

Social ecology paradigm. As a psychosocial case of systems, the socio-ecological approach, inspired by Bronfenbrenner's seminal work, the Ecological Theory of Human Development (1979), now widely known as the Ecological Systems Theory, describes the social environment as various nested "systems" of our lives from which we derive our norms, rules, and resources. These systems include the *microsystem* (family and close social network); the *mesosystem* (work and neighborhood); and the *macrosystem* (the larger socio-cultural context and institutions). This approach is also reflected in population health research (Lemyre & Orpana, 2002), and as commented by Fleury and Lee (2006): "A consensus is emerging that multilevel perspectives consistent with social ecological models are promising approaches in health behavior research and health promotion efforts" (130). This perspective can also be captured by the term "whole community" which is detailed in the chapter by Edwards.

Gibson and Lemyre (2012) have been conducting multi-level analyses to identify layers of variance associated within the social environment that relate to preparedness and the building of resilience. They characterized the various units of analysis: the individual, the family, the neighborhood, the city (Figure 14.2).

Figure 14.2 illustrates how both individual and collective preparedness are linked to four levels of the social environment: 1) individual characteristics such as income, education, and gender, 2) microsystem of support and of networks, 3) mesosystem of neighborhoods, and 4) macrosystem of juridisctional policies.

Tiers of ripple. In parallel to the layers of social environment, the impacts of major events, disasters, or crises can also be described in systems terms, as cascades of ripple effects from a tier perspective of direct damage on individuals, goods, and infrastructure (Tier 1), to a subsequent wave of disruption in the interconnected web of public services of a community (Tier 2), to an undertow that shakes the fundamentals of societies, institutions,

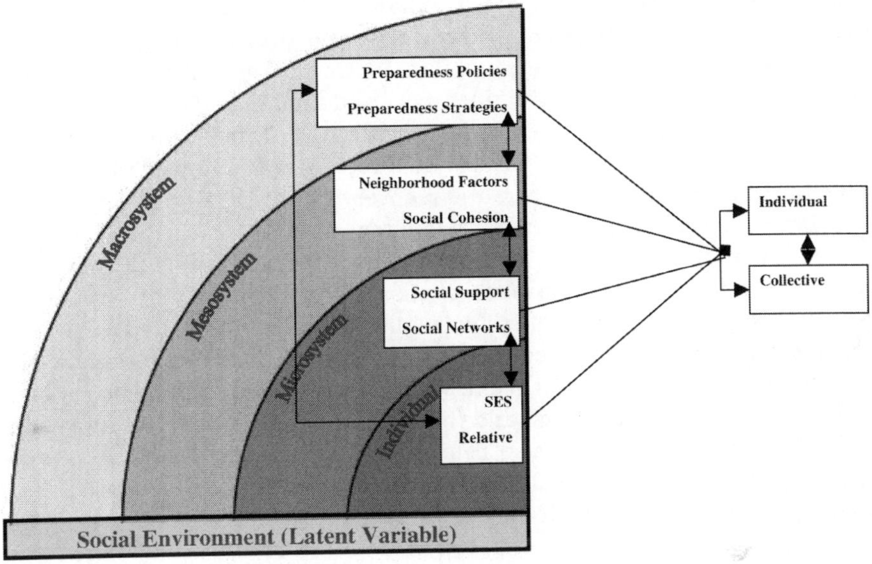

Figure 14.2 Characterization of the socio-ecological system of preparedness and its determinants.

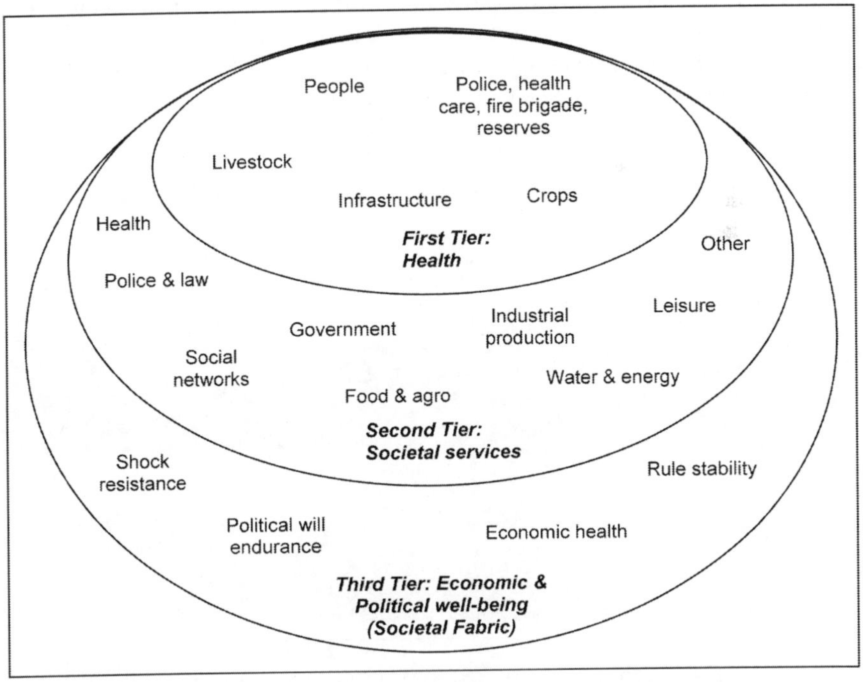

Figure 14.3 Tiers of ripple effects (EU-ASSRBCVUL & Lemyre, 2006).

social order, and values (Tier 3) (EU-ASSRBCVUL & Lemyre, 2007). The magnitude of events, disasters, and crises can be rated like a seismic "Richter" scale, on a logarithmic scale of 0 to 10 according to their ripple effects across tiers, in numbers of casualties, cost of damage, or days without public service. Resilience of a community relies on all of these foundations and manifests itself on all of their associated indicators, as followed through the timeline of preparation, impact, and recovery phases.

Risk management perspective. The integration of the various considerations—a) individuals and communities as units of analysis, b) adversity defined as stressors or life hazards, c) phases of the timeline from pre-event to post-event, and d) levels of systems—has led us to adopt a risk management perspective to resilience. The risk language allows us to situate the adversity in a hypothetical space in which both the capacity to be resilient "in case of a crisis occurring" and the residual resilient integrity and functionality are all relevant. It also helps to change the focus from a hazard-driven paradigm, investing in technological solutions, to a consequence management strategy, which focuses on response and recovery.

A risk management perspective requires stretching the usual reference to risk, as the likelihood of an event times its damage, to encompass psychosocial consequences such as distress and dysfunctionalities of organizations beyond the sole indicators of mortality and material losses. Risk becomes an integrative social and environmental paradigm. The risk perspective comes with tested tools and methodologies. It therefore brings a new set of powerful quantitative and qualitative devices to measure, model, and integrate facets, time, and levels of analysis. It serves as a heuristic to integrate and promote psychosocial resilience at the community level. The psychosocial risk analysis and management (P-RAM) of Lemyre et al. (2005) positions our perspective and study of the resilience dynamics.

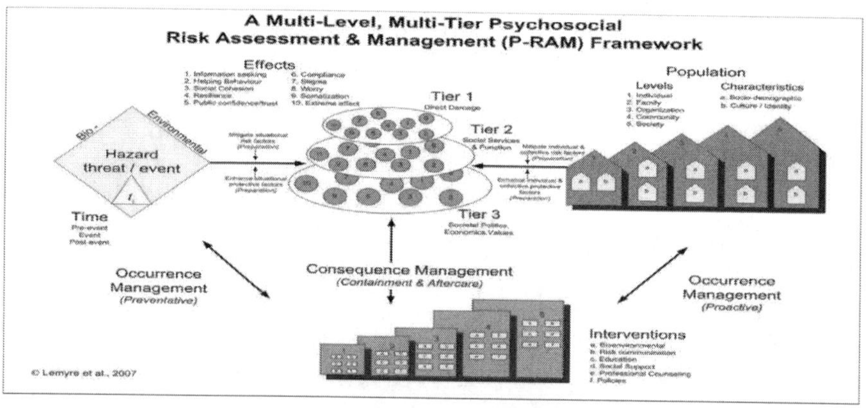

Figure 14.4 The psychosocial risk assessment and management (P-RAM) framework (Lemyre et al., 2005).

System processes. The last consideration that we will address in our framework of study bears on the leveraging of the various layers. It relates to the organizational interdependencies that foster or impair resilience. In the context of a complex situation, three types of interrelations between systems or organizations stand out: coordination, cooperation, and collaboration. These approaches to interactions of systems or organizations shape and color the potential for resilience (see Figure 14.5).

Coordination, as per our synthesis of the literature, corresponds to the process of communication in the planning and sharing of resources, risk, and rewards for the purposes of efficiency and effectiveness in achieving complementary goals (Alberts & Hayes, 1995, 2003). This approach involves an emphasis on efficiency and on preventing overlapping of resources and services. Activities and decision making occur within organizational silos in parallel with other organizations. These parallel activities are represented by the vertical groupings of organizations (blocks) below.

Cooperation is conceptualized as a process where parties with similar interests plan together, negotiate roles, and share resources to achieve joint goals, but maintain separate identities (Taylor-Powell et al., 1998; Paquet, 1999, 2011). With a cooperative approach to problem solving, resource and service gaps are filled by additional organizations. Organizations are more interdependent under cooperation, with joint decision making and joint outcomes being key features of this approach. This interdependency is represented in Figure 14.5 with the horizontal integration of organizations (blocks) and the introduction of new partners to fill resource gaps.

Collaboration is defined as a process whereby organizations see different aspects of the problem, identify a common goal, and explore solutions within their differences; as a result, solutions go beyond their individual limited visions of what is possible (Block, 1993; Paquet, 2005). The collaborative approach places an emphasis on a shared definition of the problem as well as on the generation of creative solutions. Under this approach, decision making can be described as "shared" or "networked"; in addition to information, activities, resources, power, and authority are all shared. While organizations maintain their unique identities when collaborating,

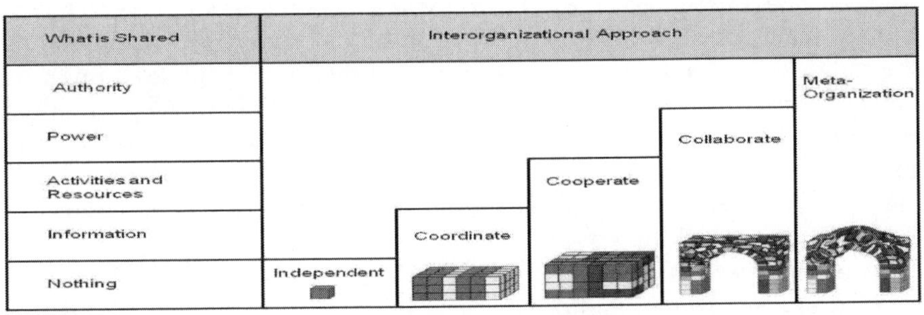

Figure 14.5 Characterization of interorganizational approaches.

the process may require the organizations to accommodate different visions of the problem using flexible interpretive frameworks. Collaboration is represented here as a bridge that is made up of organizations that support one another, fill in service gaps, and adapt their frameworks (shape) to respond to the complexity of the situation. Therefore resilience entails a set of resources that are being mobilized within a timeline via specific types of interactions and relationships.

A RESEARCH PROGRAM

With the above conceptual clarifications, our research program has followed the agenda of addressing individual, collective, and organizational parameters. A mixed methods approach is best suited for this endeavour, and we will present briefly some key features of our evidence-based model. At the individual and collective levels, contextual elements needed to be ascertained and understood in something of an anthropological and ethnographic way. Case studies of actual major events allowed revisiting the configurations of factors that yielded to more or less resilience at the community level. More subtle and intricate individual appraisal of situations, challenges, and resources required a psychometric approach of survey type, assessing perception of capacities to face adversity, preparedness, and beliefs about resourcefulness. Finally, the understanding of the complex interactions of multiple organizations in the wicked problems on the ground benefited from an experimental approach in which tasks are simulated and tested in a controlled setting. These qualitative and quantitative methods brought different insights on the requirements for resilience.

CASE STUDIES

A series of case studies based on Canadian major events are documented below. From a list of sixty-three extreme events that had occurred in Canada in the recent past, six case studies in particular were selected. They were chosen on the basis of the following criteria: i) they were recent enough (past fifteen years), ii) involved all levels of governments including the military (at least at a standby level), iii) were multi-jurisdictional, iv) impacted multiple populations, and v) involved multiple types of responders from all sectors. These included 1) the 1998 Ice Storm in Quebec and southern Ontario which deprived 50% of the population of electricity during winter for weeks, 2) the almost annual Red River floods of Manitoba, 3) the Kelowna wildfires which destroyed dozen of homes, 4) the SARS crisis as experienced in Toronto, with six deaths and hospitals closing, 5) Operation Sleepover in Gander on September 11, which allowed a dozen airplanes from the U.S. to enter no-fly zones, and 6) the Ontario blackout in the summer of 2003 which shut down electricity in most of the provinces of the East for days. Each of these events had success stories, which helped us identify the strengths and weaknesses of resiliency during cases.

The case studies were documented using secondary data sources derived from peer-reviewed literature, the "grey" literature of agency reports, and, in some instances, media reports.

An analytic grid, described in more detail in Lemyre et al. (2011), was designed to form a matrix that outlined the key analytic dimensions, crossed with the timeline of the event. The key elements covered included: a) who made the major decisions (organizations involved); b) what the decision was (content and outcome); c) when the decision was made (in relation to the event timeline); d) where (at which level) the decision was being made; and e) how the decision was made (e.g., unilaterally, coordinated, cooperatively, collaboratively). These decisions were then positioned within the event timeline of pre-event (-3) to reconstruction (+3). Observations were also made on the outcomes and on the complexity of the event at each particular stage.

Analyses suggested that lack of planning, lack of anticipation, and lack of training were the factors that contributed significantly the most to the complexity of a situation (e.g., 1998 Ice Storm; SARS). Communication difficulties encountered by both the private and public sectors involved paths of communication that had not been pre-defined and rehearsed. This, in turn, contributed to increased uncertainty and heightened the complexity of the situation. Many organizations were unable to communicate with each other, some hindered by lack of technology, or sometimes by lack of protocols in spite of the technology, thus limiting the distribution of information and resulting in divergence of the perceived impacts (Lemyre et al., 2011).

On the opposite, nuggets of success could be established as sort of ministories built on pre-established interorganizational networks that diminished the complexity of the situation during the event phase, such as with the informal networks that could mobilize during Gander, Operation Sleepover to find accommodations for travelers, or in SARS to get available staff to work. Interorganizational relationships through extensive preparedness and disaster management planning and training decreased the level of complexity by mitigating uncertainty levels.

Post-event resilience was also facilitated by flexible preparedness and response plans that had been rehearsed as well as training programs in the pre-event phase. Private and public sectors that had developed flexible preparedness and response plans as well as training programs that were adaptable to a wide variety of events, varying in scope and severity, showed better and faster adaptation. Although the level of uncertainty remained relatively high, the flexible guidelines allowed organizations to make efficient decisions and reduce the complexity of the situation.

The absence of coordination and transparency between municipalities inhibited effective decision making in the impact and rescue phases during the Red River Flood. Plans needed to be shared and coordinated between neighboring municipalities, and the lessons learned which were identified in the post-event phase needed to be integrated in the planning of future disasters.

In the Kelowna wildfires, it was the transparent communication with the public that significantly improved public trust and reduced anxiety levels and uncertainty. By engaging the media, the scope and severity of the situation were conveyed more accurately to the public, increasing public confidence, and reducing the level of uncertainty. However, during the Ontario blackout and the SARS outbreak, lack of communication with the public and confusing interorganizational communication remained the biggest factor associated with increased complexity. This hindered decision making and contributed to a lack of trust and poor coordination.

These case studies therefore showed nuggets of strengths and streams of gaps which point to the resilience and weaknesses of systems in times of hardship. Although they all involved different contexts, events, and players, they informed us of some of the key factors. Communication in between organizations was critical, and perception by individuals of their state of efficacy was determining for people to initiate concrete actions. Hence these two variables are the subject of our next studies.

A COMMUNITY SURVEY OF CANADIANS

Following a series of national representative surveys on perception of risks in a population health framework, a public survey on preparedness was designed to benchmark the preparedness level and resilience capacity of Canadians (Lemyre et al., 2005). A nationally representative survey on Canadians' perceptions of security threats and preparedness was conducted via telephone using stratified random sampling digit dialling. Respondents included 1502 individuals over the age of eighteen, weighted to be representative of the Canadian population in terms of age, gender, and distribution across five residential regions (Atlantic region, Quebec, Ontario, Prairies, and British Columbia). The territories were not included.

During the interviews, which lasted approximately thirty-five minutes, respondents rated their willingness on a five-point Likert scale to collaborate with authorities in response to government-issued instructions, such as the behavioral responses of undergoing decontamination treatment; agreeing to remain inside a building for protection (sheltering-in-place); agreeing to strictly isolate oneself from others (quarantine); and going to a public shelter. Actual preparedness of oneself, one's family, and perceived preparedness of authorities (municipal, provincial, and federal) were also assessed. Additional items addressed trust in sources of information and indicators of social capital.

Based on the community survey, Lee, Gibson, Markon, and Lemyre (2009) showed a three-factor structure of preparedness, posed to be predictive of resilience. The first and main factor involved anticipatory behaviors such as obtaining information on shelters, establishing a plan, establishing a meeting area, preparing a supply kit, consulting for advice, learning evacuation plans, seeking social support, and receiving first aid training. The second factor centered on information seeking and turning towards one's network of support, and a third factor clustered around avoidance behaviors.

What best predicted stress, used here as a generic (inverse) marker of sense of resilience, was preparedness, controlling for age, education, and sex (adjusted R^2 of .10, $F(9, 1387)=17.87$, $p<.001$). Specifically, preparedness and avoidance behavior both emerged as respectively significant positive and negative predictors of psychological stress; and actual current formal preparedness (such as having emergency kits) further improved the prediction of psychological stress. Current avoidance behavior (not wanting to write down phone numbers or verify locations of shelters) was a significant positive predictor of psychological stress, and current anticipatory preparedness (seeking information) significantly *negatively* predicted psychological stress. It accounted for 10% of the variance in current psychological stress, used as an indicator of resilience in a normal population (Lemyre & Tessier, 2003).

AN INTERORGANIZATIONAL EXPERIMENT

The other critical aspect of response to disasters was interorganizational. In our case studies, leverage of resources by each organization required interactions of systems between organizations. Hence we examined interorganizational problem solving and decision making in the context of facing a major event with limited resources. The Problem-solving and Organizational Decision-making Simulation (PODS) experiment we designed consisted of in-vivo sessions conducted with professionals in senior emergency management decision-making roles working in small groups ("pods") connected via video teleconferencing equipment to perform two tasks related to responding to an emergency scenario. The experiment was a 2x2 design with two conditions related to the problem-solving approach based on the types of tasks (coordination versus collaboration), and two conditions related to the composition of pods according to organizational type (homogenous versus mixed). The study included both experimental quantitative methods, and qualitative interview methods. Participants consisted of professionals in senior decision-making roles related to emergency management within a variety of municipal, provincial, and federal organizations, including the military and NGOs.

The experimental procedure involved participation in pods responding to a specific emergency scenario. Participants were divided into three separate pods (groups), with two to four people per pod. Some sessions had homogenous pods as a condition. For example, participants withsimilar decision-making structure were grouped together in a command-and-control military pod, an Incident Command System (ICS) POD with first responders, and a non-ICS POD for public sector agencies and NGOs. For the mixed sessions, participants were purposely placed in PODS that combined different types of organizations.

Each POD room was equipped with a multimedia set-up with a computer station to share, audio and video recording, and webcams fed into the NEF-SIS video conferencing software. An overall session "controller" guided the PODS through the various tasks, presented the city of "GAP-ville," and briefed them on the emergency situation. Various slides and scripted voice-overs prompted participants to open certain envelopes according to a set timing and sequence throughout the experiment session. The tasks focused on either coordinating or collaborating with respect to public communications (Task 1) and responder health and safety (Task 2).

The dependent variables included task outputs, communication patterns intra-POD and inter-POD, leadership-stewardship behaviors, as well as questionnaires on prior knowledge and relationships, evaluation of the task, of the performance, and of the process, plus a network assessment. Analyses were conducted to compare problem-solving approaches (coordination versus collaboration) and multi-organizational environments (homogenous versus mixed PODS).

Participants in the collaboration tasks rated their POD's participation in problem solving significantly higher than in the coordination tasks ($t(25)=-2.82$, $p<.01$). They also reported higher individual satisfaction levels with opportunities to provide input to the problem solving among those who participated in collaboration tasks when compared with those who had participated in coordination tasks ($t(25)=-2.82$, $p<.01$). Participants in the coordination sessions actually reported higher levels of frustration working with the people in the other PODS when compared with those participating in the collaboration sessions ($t(25)=-2.44$, $p<.02$). Those participating in collaboration tasks were more likely to indicate higher levels of communication effectiveness when compared with participants in coordinating tasks ($t(25)=-2.30$, $p<.03$); and they were also more likely to indicate that there was greater discussion during the task when compared with those undertaking coordination tasks ($t(25)=-2.21$, $p=.04$).

Perception of leadership was higher in the coordination task than in the collaboration tasks ($t(25)=-2.72$, $p<.01$) but there was more satisfaction with the outcome ($t(25)=-2.82$, $p<.01$) among those who participated in a collaborating task, and they reported more that the decisions were consensus-based ($t(25)=-2.89$, $p<.01$). Then, the qualitative follow-up semi-structured interviews explored different aspects of decision making and problem solving within a multi-organizational context during various stages of an extreme event. The content analysis yielded to the notion of "stewardship" as an alternative style to leadership for serving the group and the mission. Stewardship, as described by Block (1993), involves a high-level sense of the goal, openness to diversity, sharing of information, value of engagement and participation, and a sharing of authority and power in a distributive subsidiary way (Paquet, 2011).

Although this study was preliminary and exploratory, the experiment validated the pattern of sharing information, resources, and power according to the constituency of the working groups. It allowed for evaluation of different types of tasks which involved distinct approaches, in term of processes and of outcomes. For example, in a complex situation, communication benefited from a collaborative approach staged in a mixed environment, albeit requiring more time and involving more frustration. Participants reported appreciating the diversity and alternative solutions and were overall more satisfied with the outcome.

TOOLS FOR ENHANCING COMMUNITY RESILIENCE

Using the knowledge developed from this multi-layer research base, we also simultaneously produced intervention tools for stakeholders. These tools could be employed at the individual, community, or organizational level for community resilience building.

Psychosocial Awareness Program

Our results pointed to the need to better understand all the ways in which psychosocial considerations interplay in human behavior and social interaction to improve the adequacy of disaster response. Our research, especially case studies and interviews, had shown that planners, managers, and responders had underestimated or neglected the impact of psychosocial factors, while success stories had documented the creative use of psychosocial resources. Hence we created the Psychosocial Risk Manager (PRiMer) Platform as a knowledge tool to educate people on the relevant psychosocial considerations. The PRiMer presents a framework for multi-level psychosocial considerations built as an interactive multimedia awareness program. It has a community facilitation tool designed to promote a paradigm shift in emergency planning and response. The purpose is to build community resilience and speed recovery in the face of extreme events. PRiMer was developed in an iterative fashion by the GAP-Santé research team at the University of Ottawa, based on a series of round-table consultations with a wide range of community and professional responders, from the public sector and NGOs. It provides basic psychosocial education for those tasked with planning and implementing emergency responses on how people behave before, during, and after a crisis. It aims to foster a shared understanding and common language about psychosocial risks and identifies their implications for planning and mitigation. PRiMer offers an introduction to key psychosocial considerations and concepts, drawing its content from evidence-informed research.

The research has been translated into a suite of practical tools that facilitate the embedding of psychosocial considerations into an organization's emergency planning and continuity of operations. Through a variety of instructional approaches, PRiMer promotes capacity and competency building within communities by creating a mechanism to include public engagement and participation in responding to psychosocial considerations. It demonstrates a practice method to enhance community resilience and coping capacity. It includes a web-based self study guide, a one-day interactive scenario-based workshop, a psychosocial checklist and an asset tool. The PRiMer tool has been subjected to expert validation for content. It has also been deployed in five field demonstrations in Halifax, Montreal, Ottawa, Toronto, Winnipeg, for evaluation (Lemyre, Johnson, & Corneil, 2010).

Community Asset Mapping

The collaborative process required to develop the community asset map opens communication channels among traditional and non-traditional responder groups and builds relationships and networks of support. Communities and organizations actually know little of the resources of other players in their environment. The GAP-Ville Capability Tool works on the premise that when extreme events occur, individuals and organizations have a lot to offer and will "emerge and converge" to do so. Pre-identifying these community assets will improve the ability to collaborate, coordinate, and potentially improvise when required. It is in keeping with a "strength-based approach" in building community resilience (Kretzman & McKnight, 1993). The tool aims to promote a "culture of preparedness" within communities, fostering anticipatory thinking: "What can I do? How can I help?" Its use affirms that people at the local level know their community best and can creatively respond to the challenges they face. It allows information to be collected from and used by many different groups. It utilizes a bottom-up approach to encourage greater public engagement (Colussi, 2000; Gurwitch et al., 2007).

The GAP-Ville Capability Tool starts with a face-to-face interaction to support a web-based application. It was designed to encourage greater collaboration and coordination among relevant community stakeholders at disaster planning and preparedness. It is launched at a community participation and planning workshop in which participants first learn of the process from the example of a theoretical community, GAP-Ville. They then move into characterizing their own locality and capturing its resources. The intent of the tool is not so much on identifying gaps or deficiencies in a community (although this may well become apparent in the mapping process) as on identifying the positives in the community, the common assets, and the potential for sharing.

For each community, the content is meant to be self-organizing and self-correcting in keeping with the norms of social media and wiki tools. Users mark and record community assets using an online social map, indicating contact information and locations for established or potential responder organizations. Assets can be categorized in terms of people (e.g., individuals with leadership positions and/or particular skills), equipment (vehicles, generators, communications systems), space (facilities, meeting areas, kitchens, recreation space, storage), services (mass care/shelter, transportation, child care, health care, protective services). Using Geographic Information System (GIS) from Google Maps technology, people map their community assets. An opportunity is also available to establish a Facebook link to a specific location, should one be available. Many individuals and organizations use Facebook and this offers a low-cost way to utilize its functionality in creating profiles of available assets. A wealth of information about the status of the asset is readily available, and maintaining updated profiles is therefore easier.

The capability tool is a relatively straightforward process in terms of data inputs for specific locations. Identifying interested stakeholder groups and involving them in the mapping process, however, is much more of a challenge. A well-defined process for engaging the public in this activity will enhance and extend the social networks needed to build

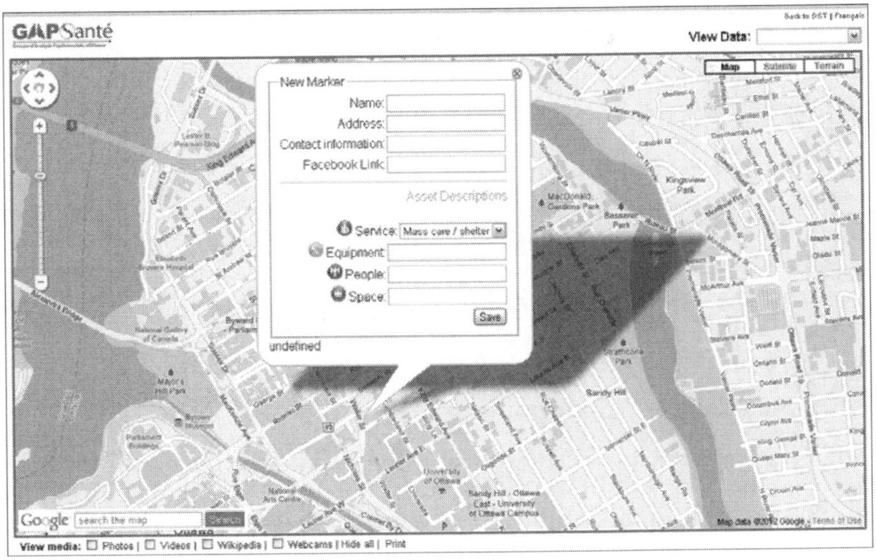

Figure 14.6 Screenshot of the GAP-Ville Capability Tool interface.

community resilience. An example of a public engagement process and how asset mapping can be utilized to build resilience can be found in the *Community Resilience Activity Book* prepared by Hutchins and Speevak-Sladowski (2009).

High-Risk Groups and Functional Needs

In a similar fashion as for assets, our results also showed that communities and organizations did not have prior knowledge of special needs. They had not planned for special populations and how to service special needs such as mobility restriction, communication deficit, and oxygen-supply-dependent people. Based on the World Health Organization Functional Needs assessment, a sister project was created, the Enhancing Resilience and Capacity for Health (EnRiCH Initiative), a community-based participatory research program which focuses on the design, implementation, and evaluation of a community intervention aimed at enhancing resilience among high-risk groups (O'Sullivan et al., 2012).

Here the term "high risk" refers to people who are at higher risk of experiencing negative impacts from a crisis or disaster due to clustering of personal and social determinants of health. These include functional needs related to communication, mobility/transportation, intellectual and emotional disabilities, and other conditions which limit an individual's ability to live independently. As part of the intervention, the team adapted the principles of asset mapping of the GAP-Ville Capability Tool into an online collaborative task using Google Docs. Following a brief face-to-face training session, the community stakeholders collaborated to populate a spreadsheet which maps the assets in the community which can assist people with disabilities and other functional needs that could limit their ability to cope with the impacts of a disaster. This tool has been pilot-tested in four communities in Canada (Truro, Nova Scotia; Gatineau, Quebec; the Region of Waterloo, Ontario; and Quebec City, Quebec).

Qualitative interviews have documented the evolution of engagement, mapping perceived resources and effective strategies to mobilize communities, while the quantitative data have indicated more networking, more exchange of critical information, and the actual construction and use of the web repertory EnRiCH High Risk Identification Tool (EHRIT) (O'Sullivan et al., 2012).

Interorganization Collaboration

Finally, our results stressed the need for collaboration between organizations beyond, and before, the mere efforts for coordination. Collaboration requires special behaviors and the sharing of higher level information on mandate and mission, plus other "cultural" norms. The In-Vivo PODS was created to provide an environment in which collaboration could be rehearsed and studied. In today's science and technology environment, most of the coordination and collaboration research is done using agent-based computer simulation and modeling. Yet, little data is being generated using real humans in group decision making and problem solving. We therefore developed an in-vivo human platform for exercises of interorganizational problem solving (Lemyre et al., 2011). The system allows for exercises involving intersectoral tasks under various scenarios, and diverse configurations of actors at different time phases. It permits testing of roles and functions that are not prescribed by standard operating procedures. Too often, exercises rehearse only the "knowns" and avoid in their script the "unknowns," or the blind spots and the emergent phenomena. Also, most simulations involve only a restrictive subset of participants and usually the associative sector and the public are left out, neglecting the largest portion of the people on scene. The PODS system offers the systematic inclusion of the third sector.

Such a platform can be used to train skills in collaboration and in stewardship. It also creates a space for virtual encounters and distance participation. Indeed a number of the stakeholders, from the government or the private sector, usually reside at a distance from the disaster. Knowledge about prior and continuous collaborative practices in major cities will enable the preparedness for higher resiliency.

CONCLUSION

There are some caveats in the resilience approach. Resilience has become a fashionable term these days to replace the mantra Prepare-Prevent-Respond-Restore of the previous decade. It may only be a repackaging of best practices that brings us back to marvel again at the basics of coping with adversity of previous generations, and civilizations across the ages. The resilience "banner" can certainly yield to noble pro-social community development; it also points to a critique of the overprofessionalization of disaster management.

States cannot afford to rely only on professional responders and emergency managers. Professionalization has also shown it can fail us in our blind spots. The revival of the orientation towards resilience also reminds us of the capabilities of lay people and the power of astute common sense, as well as the lessons to be learned of more modest countries around the globe who rebuild again and again after repeated disasters.

At another level, the current support for resilience building also grows on an insidious withdrawal of government-assisted support. The new motto might be ironic—"You're on your own": organize thyself. With this in mind, let us hope we can use our developing knowledge on resilience to increase the effectiveness and coverage of equity in well-being. The United Nations Hyogo Agreement that created the UN International Strategy for Disaster Reduction

(UN-ISDR, 2005) launched a "Building Resilience" campaign: Get your City Ready. It aims to stimulate all levels of collaborative efforts required to develop and sustain member-state resilience. In Canada, for example, the translation of these goals into a National Platform for Risk Reduction is stimulating the growth of working groups around resilient communities.

REFERENCES

Alberts, D.S., & Hayes, R.E. (1995). *Command arrangements for peace operations*. Washington, DC: NDU.

Alberts, D.S., & Hayes, R.E. (2003). *Power to the edge*. Washington, DC: CCRP.

Anaut, M. (2003). *La Résilience: Surmonter les traumatismes*. Paris: Nathan Université.

Antonovsky, A. (1987). *Unraveling the mystery of health: How people manage stress and stay well*. San Francisco: Jossey-Bass.

Bandura, A. (1979). Self-efficacy: Toward and unifying theory of behavioral change. *Psychological Review*, 84, 191–215.

Bandura, A. (1989). Human agency in social cognitive theory. *American Psychologist*, 44(9), 1175–1184.

Bandura, A. (Ed.) (1995). *Self-efficacy in changing societies*. Cambridge: Cambridge University Press.

Block, P. (1993). *Stewardship*. San Francisco: Berret-Koehler.

Bonanno, G. A., Westphal, M., & Mancini, A. (2011). Resilience to loss and trauma. *Annual Review of Clinical Psychology*, 7, 511–535.

Bowlby, J. (1978). *L'Attachement: Attachement et Perte*. Paris: PUF.

Bowlby, J. (1982). *Attachment and loss. Vol. 1: Attachment* (2nd Ed.). New York: Basic Books

Bronfenbrenner, U. (1977). Toward an experimental ecology of human development. *American Psychologist*, 32, 513–531.

Chandra, A., J. Acosta, et al. (2011). *Building community resilience to disasters: a way forward to enhance national health security*. Technical Reports. Santa Monica, CA, RAND: xxiii, 78 p.

Colussi, M.M. (2000). *The community resilience manual: A resource for rural recovery and renewal*. British Columbia: Canadian Centre for Community Renewal.

Cyrulnik, B. (2002). *Un merveilleux malheur*. Paris: Odile Jacob.

Cyrulnik, B., & Duval, P. (2006). *Psychanalyse et résilience*. Paris: Odile Jacob.

EU-ASSRBCVUL Group & Lemyre, L. (2007). Final report: Assessment of the vulnerabilities of modern societies to terrorist acts employing radiological, biological or chemical agents with the view to assist in developing preventive and suppressive crisis management strategies. Project 502 476. European Union Research Commission 6th Framework Report.

Fleury, J., & Lee, S.M. (2006). The social ecological model and physical activity in African American women. *American Journal of Community Psychology*, 37, 129–140.

Garmezy, N. (1974). The study of competence in children at risk for severe psychopathology. In A. Koupernik (Ed.), *The child and his family: Children at psychiatric risk*, vol. 3 (pp. 77–97). New York: Wiley.

Garmezy, N. (1991). Resiliency and vulnerability to adverse developmental outcomes associated with poverty. *American Behavioral Scientist*, 34, 416–430.

Garmezy, N. (1993). Vulnerability and resilience. In D.C. Funder, R.D. Parker, C. Tomlinson Keesey, & K. Widaman (Eds.), *Studying lives through time: Approaches to personality and development* (pp. 377–398). Washington, DC: American Psychological Association.

Garmezy, N., & Masten, A. (1991). The protective role of competence indicators in children at risk. In E. Cummings et al. (Eds.), *Perspective on Stress and Coping* (pp. 151–174). New Jersey: Hilldale, Erlbaum.

Gibson, S., & Lemyre, L. (2012). A multi-level analysis of preparedness behaviours. Manuscript under review.

Gurwitch, R.H., Pfefferbaum, B., Montgomery, J.M., Klomp, R.W., & Reissman, D.B. (2007). *Building community resilience for children and families.* Oklahoma City: Terrorism and Disaster Center at the University of Oklahoma Health Sciences Center.

Hutchins, R., & Speevak-Sladowski, P. (2009). Community resiliency handbook: Enhancing the role of the voluntary sector on health emergencies. Centre for Voluntary Sector Research and Development, Carleton University.

Kobasa, S.C.O. (1987). Stress responses and personality. In R.C. Barnett, L. Biener, & K. Baruch (Eds.), *Gender and stress*. New York: Free Press.

Kretzman, J.P., & McKnight, J.L. (1993). *Building communities from the inside out: A path toward finding and mobilizing a community's assets.* Chicago: ACTA Publications.

Lee, J.E.C., Gibson, S., Markon, M.P.L., & Lemyre, L. (2009). A preventive coping perspective of individual response to terrorism in Canada. *Current Psychology*, 28, 69–84.

Lemyre, L., Clément, M., Corneil, W., Craig, L., Boutette, P., Tyshenko, M.G., Karyakina, N., Clarke, R., & Krewski, D. (2005). A psychosocial risk assessment and management framework to enhance response to CBRN terrorism threats and attacks. *Biosecurity and Bioterrorism*, 3(4), 316–330.

Lemyre, L., Johnson, C., & Corneil, W. (2010). Psychosocial considerations for mass decontamination. *Radiation Protection Dosimetry*, 142, 17–23.

Lemyre, L., & Orpana, H. (2002). Integrating population health into social ecology: Role of family medicine researchers. *Canadian Family Physician*, 48, 1349–1350.

Lemyre, L., Pinsent, C., Boutette, P., Corneil, W., Riding, J., Riding, D., Johnson, C., Lemus, C., Blust, S., Kitchener, H.M., Markon, M.-P.L., & Gibson, S. (2011). *Research using in vivo simulation of meta-organizational shared decision making (SDM)*. Published report for Defence Research and Development Canada, no. CR-2011–31.

Lemyre, L., & Tessier, R. (2003). Measuring psychological stress: Concept, model, and measurement instrument in primary care research. *Canadian Family Physician*, 49, 1159–1160.

Luthar, S.S. (Ed.) (2003). *Resilience and vulnerability: Adaptation in the context of childhoodadversities*. New York: Cambridge University Press.

Masten, A.S. (2001). Ordinary magic: Resilience processus in development. *American Psychologist*, 56, 227–238.

Masten, A.S. (2006). Promoting resilience in development: A general framework for children care. In R.J. Flynn, P.M. Dudding, & J.G. Barber (Eds.), *Promoting resilience in child welfare* (pp. 3–17). Ottawa, ON: University of Ottawa Press.

Masten, A.S., Best, K., & Garmezy, N. (1990). Resilience and development: Contributions from the study of children who overcome adversity. *Developmental Psychopathology*, 2, 425–444.

Norris, F.H., Stevens, S.P., Pfeifferbaum, B., Wyche, K.F., & Pfefferbaum, R.L. (2008). Community resilience as a metaphor, theory, set of capacities and strategy for disaster readiness. *American Journal of Community Psychology*, 41, 127–150.

Norris, F.H., Tracy, M., & Galea, S. (2009). Looking for resilience: Understanding the longitudinal trajectories of responses to stress. *Social Science & Medicine*, 68, 2190–2198.

O'Sullivan, T.L., Kuziemsky, C.E., Toal-Sullivan, D., & Corneil, W. (2012). Unraveling the complexities of disaster management: A framework for critical social infrastructure to promote population health and resilience. Manuscript submitted for evaluation.

Paquet, G. (1999). *Governance through social learning.* Ottawa: University of Ottawa Press.

Paquet, G. (2005). *Gouverance: Une invitation à la subversion.* Montreal: Liber.

Paquet, G. (2011). *Gouvernance collaborative.* Montreal: Liber.

Piaget, J. (1975). *L'équilibration des structures cognitives: Problème central du développement.* Paris: Presses Universitaires de France.

Raphael, B. (1986). *When disater strikes: How individuals and communities cope with catastrophe.* New York: Basic Books.

Rutter, M. (1987). Psychosocial resilience and protective mechanisms. *American Journal of Orthopsychiatry*, 57(3), 16–331.

Rutter, M. (1993). Resilience: Some conceptual considerations. *Journal of Adolescent Health*, 14, 626–631.

Rutter, M. (2000). Resilience reconsidered: Conceptual considerations, empirical findings, and policy implications. In J.P. Shonkoff & S.J. Meisels (Eds.), *Handbook of early childhood intervention*, 2nd edition (pp. 651–682). New York: Cambridge University Press.

Selye, H. (1956). *The stress of life.* New York: McGraw-Hill.

Taylor-Powell, E., B. Rossing, et al. (1998). *Evaluating Collaboratives: Reaching the potential. Program Development and Evaluation.* Madison, Wisconsin: University of Wisconsin-Extension: 190

UN-ISDR (United Nations International Strategy for Disaster Reduction) (2005). Hyogo framework for action, 2005–2015: Building the resilience of nations and communities to disasters, international, world conference on disaster reduction, January 2005, Kobe, Hyogo, Japan. www.unisdr.org.

Werner, E.E., & Smith, R.S. (1982). *Vulnerable but invincible.* New York: McGraw Hill.

Werner, E.E., & Smith, R.S. (1992). *Overcoming the odds: High risk children from birth to adolescence.* Ithaca, NY: Cornell University Press.

15 Management Capacity and Rural Community Resilience

William L. Waugh, Jr.

INTRODUCTION

Community resilience has become the holy grail for emergency managers and policy makers concerned about risks to life and property. It is a goal driven by the needs to reduce reliance upon increasingly scarce federal resources and to encourage the exercise of local responsibility for managing risks posed by natural and unnatural hazards. It is also driven by the necessity to increase local self-reliance in the event that the nature or scale of a disaster may make it difficult or even impossible for federal and even state authorities to respond to all communities needing assistance and even for neighboring jurisdictions to share resources via formal or informal mutual assistance agreements. Demand for assistance may far outstrip federal and state capabilities, in other words. Assistance may not just be slow in coming, it may not come at all or may be very late in arriving and communities will have to fend for themselves. This is the kind of scenario that might result from a pandemic, asteroid strike, nuclear attack, and other cataclysmic event or from the cascading impacts of long-term trends such as climate change or earthquake storms. Indeed, some communities in Mississippi and Louisiana did not receive outside aid for a week or more after Hurricane Katrina made landfall in 2005.

Rural communities are particularly vulnerable because they lack the resources of urban areas and have populations that require greater support. The focus of this analysis is the impact of low management capacity on rural community resilience, in particular the lack of capacity for planning and program management necessary to manage hazards, prepare for disasters, respond to disasters, and recover quickly. The question that may not be answerable is whether rural communities will be willing to do what is necessary to reduce risks to life and property. Might culture and politics still trump social need?

Clearly, the emphasis on local self-reliance is also a response to the current economic crisis and increasing pressure to reduce budget outlays. The costs of disaster assistance have been escalating and there is pressure to reduce budget outlays in other areas to compensate for aid to communities affected by recent disasters. In fact, following Hurricane Irene and Tropical Storm Lee in 2011, conservative leaders in Congress threatened that expenditures for disaster assistance to devastated communities would require comparable cuts in other federal programs (Pear, 2011). Even without resorting to such trade-offs, there is increasing pressure to reduce spending for disaster aid and hazard mitigation. One of the ways to do that is to encourage local action to reduce vulnerabilities and to mitigate hazards. This is the change in policy direction that seems to be underway.

FEMA administrator Craig Fugate has long been a proponent of local capacity building (Fugate, 2010; Resnick, 2011), and the development of mechanisms for communities to share resources when major disasters occur has clearly changed response operations. Communities have shared resources through local mutual assistance arrangements for centuries. The development of statewide mutual assistance compacts and the national compact, the Emergency Management Assistance Compact (EMAC), have certainly increased local capabilities (Kapucu et al., 2009). Although it is difficult to generalize, most urban communities have the health care and emergency medical, transportation, communications, and emergency response systems necessary to respond to disasters although their systems may be overwhelmed by catastrophic events. Many, if not most, rural communities do not have the resources to deal with even a small disaster. In the event of a pandemic, no community may have the resources to meet the needs of its own residents and no community may have surplus resources to share.

NATURAL AND UNNATURAL HAZARDS

The increasing vulnerability of communities, urban and rural, to natural and unnatural disasters is well documented. For example, a NOAA report on demographic shifts in the U.S., including rural communities, pointed out increasing vulnerabilities along the nation's coastlines. According to the report, 50% of the American population lived within fifty miles of the coast in 2004 and the percentage is projected to rise to 70% by 2025 (Crossett, 2004). The vulnerability posed by the wind and water of coastal storms will be further exacerbated by sea level rise due to climate changes. Indeed, worldwide, population centers increasingly are located along vulnerable coastlines and two-thirds of the cities of five million or more population will be vulnerable to sea level rise and its effects (Mathez, 2009: 9). The U.S. is not alone in its task of reducing vulnerabilities of coastal communities, and the impacts of climate change will increase the levels of risk to both rural and urban communities. Charles Perrow has suggested reducing population concentrations, i.e., the sizes of towns and cities, to reduce the number of highly vulnerable population centers (Perrow, 2007), but such radical population shifts are unlikely absent catastrophic disasters.

While rural communities may have less exposure to coastal storms, simply by virtue of having fewer people and structures, they do have exposure to other manifestations of climate change. Extreme weather associated with climate change is causing deeper and more persistent droughts, as well as more intense rainfall and frequent flooding. Heat and cold waves are also becoming more frequent and devastating. Disease outbreaks are anticipated as the seasons moderate and public health officials are monitoring for subtropical diseases such as dengue fever. While many climate scientists are not ready to predict stronger and more frequent tornado outbreaks and hurricanes, they do expect that changes in weather patterns due to climate change will pose increased risk to populations around the world (Mathez, 2009).

In rural areas, the impacts upon livestock and crops are already being felt and adjustments are being made to deal with changing migratory patterns of wildlife. The Western Governors Association (WGA, 2008) has committees investigating the impact of climate change on the migratory patterns of longhorn sheep and other large animals. Migrating deer and elk herds are increasingly posing hazards along roadways in communities located in the Rocky Mountains and on the Great Plains and the migrating animals are damaging crops. Mountain lions and wolves are also following the migrations. The WGA is also seeking regional answers to declining water resources in the region due to decreased snowfall and rising temperatures.

In rural areas, the direct impact of climate change upon agriculture is perhaps a larger concern. The U.S. Global Change Research Program (2009) has concluded that warming temperatures have a negative impact upon crop growth and yields. Extreme weather, such as heavy rains and droughts, are likely to reduce crop yields. Warming, too, can encourage the growth of weeds and plant diseases, as well as insects. Crop yields decrease and the expenses associated with increased use of insect and disease control measures increase. Forage quality declines and livestock productivity declines. Higher temperatures and extreme weather events can also reduce livestock productivity. All of this is to say that agricultural production is affected by the increased carbon dioxide and warming caused by climate change and rural communities are feeling the economic impact of reduced production.

LOCAL GOVERNANCE CAPACITY AND HAZARDS

Managing risks to life and property and the environment requires resources and expertise. Cities and counties were the center of the nation's governmental and economic life well into the twentieth century. They were the closest governments to residents in largely rural communities. County offices still serve as administrative arms of state government, issuing licenses, registering deeds and deaths, collecting taxes, administering social and economic programs, and providing a forum for local politics (Waugh, 1997). The expansion of county services has been slow and has been characterized more by the "municipalization" of urban counties than the expansion of authority to all counties (Waugh & Streib, 1993). To the extent that state officials have delegated authority to county and city officials for land use planning, zoning, and the regulation of building codes, the burden of managing, including financing those activities most central to the mitigation of hazards, stretches the capacities of local offices. While all counties do not have authority to plan and regulate land use, county and municipal governments generally provide the principal forum for land use decision making. A few states, like Georgia, have mandated land use planning as a pre-condition for receiving state economic development funding. But, that state pressure to plan is the exception rather than the rule. As will be suggested later on, rural communities may well choose not to manage hazards for political and cultural reasons, as well as economic, but they still are central actors in the process of hazard mitigation.

The capabilities of rural communities to respond to emergencies and provide medical care for victims are less certain. Rural communities may be largely agricultural, reflecting the traditional notion of "rural," or some mix of rural, suburban, and urban with relatively low population density. Those on the fringes of large urban areas have access to medical care and can rely upon the emergency response capabilities of their neighbors. Circumstances differ in more isolated communities (Sears et al., 1993). Self-reliance is often assumed to be a strength of rural communities in the United States. That is the idyllic picture of the American heartland, but it is not the reality today. Undoubtedly there are still some communities characterized by strong social networks and a willingness to help neighbors in need. But, communities have changed as family farms have been displaced by agribusinesses, often owned and run by international conglomerates. Farms may be owned and run by part-time farmers who hold full-time jobs in local malls or factories. The point is simply that the image of rural America is less romantic than it was decades ago and rural communities are less resilient than they used to be. Even food and water for local consumption may be scarce commodities in regions that produce food. Clearly, it is also difficult to generalize about rural communities because all are not agricultural and all are not distant from major urban areas.

Today, many communities are not "rural" in the traditional sense of the term. They are simply small in terms of the number and/or the concentration of inhabitants. However, what they do share is reliance upon relatively small local governments, which suffer the political and administrative limitations of local governments in general and lack the financial resources of larger governments. For example, they are hampered by limited sources of revenue and limited authority to borrow (Cigler, 1993). One stereotype of local governance is the town hall meeting, a revered manifestation of direct democracy found in the Midwest, New England, and Mid-Atlantic region, which "emphasizes personal and informal organization as the vehicle for maintaining an engaged citizenry" (Sokolow, 1997: 294). Modern social and economic problems are not as easily addressed as those that communities faced fifty, a hundred, or two hundred years ago.

Small community governance has changed tremendously over the last century as the farming and ranching population has declined and towns no longer serve as centers of commercial life. Communities are disappearing because they are losing their economic purpose and cannot provide the services that residents demand (Mattson, 1997: 322–223). The size of the community affects the style of governance, with smaller communities exhibiting "caretaker governance" and "a cautious, highly personalized go-slow policy-making process." Moreover, the size of the professional staff is often is very small with only a city manager and/or planner responsible for the daily administative tasks. , (Mattson, 1997: 324–333). The capacities of small communities to address modern problems and to manage complex programs is limited.

The Disaster Mitigation Act of 2000 requires that states and localities develop hazard mitigation plans to reduce the incidence and/or impact of disasters in order to qualify for funds under the Hazard Mitigation Grant Program, but the capacities and willingness of communities to develop effective plans vary widely (Schwab, 2010). Professional emergency managers may serve several to many communities, as they do in parts of California, but, more often, the emergency management function is assigned to a fire chief or police chief or sheriff who has little or no training in the management of large-scale disasters and is even less likely to have training in the management of local hazards in order to prevent or reduce the risk of disaster. Or, it is assigned to a part-time employee or unpaid volunteer who has few resources to support programs and may or may not have training in the profession. Local emergency response may be well organized, but measures to mitigate, prepare for, and recover from disasters may be lacking because they are outside the purview of the fire chief, police chief, and sheriff, and the broader emergency management function lacks strong political and public constituencies (Waugh, 1990). Jessica Jensen (2009) points out the problem of training on NIMS when rural fire departments are largely volunteer and have few resources and limited time for training. Just as firefighters suppress fires and leave the victims to be helped by the local ARC chapter, responsibility for the larger emergency management function has been picked up by non-governmental organizations, often church groups or civic organizations, or no one at all.

The biggest proponent of rural community capacity building has been the U.S. Department of Agriculture through the Economic Research Service, Cooperative Extension Service, and the Soil Conservation Service. Capacity building has primarily been in the area of economic development, although some attention has been paid to local governance issues. Elected officials in rural communities, however, have only had marginal roles (Cigler, 1993: 100). State efforts have tended to be broadly focused, defining "rural" very loosely to capture suburban interests, and have tended to concentrate on building administrative capacity to manage state programs rather than local programs. Because county governments are administrative arms of state governments, capacity building serves to improve the administration of state programs (Waugh, 1993). The consolidation of services at the county level has helped small communities reduce costs (Cigler, 1993: 103).

The question here is how capable rural governments are to manage hazards and deal with disasters. The literature on urban/rural differences suggests that rural counties are less likely to be concerned about environmental issues, meaning that "environmentalism" is largely an urban issue. The siting of hazardous material disposal and transfer facilities, as well as prisons and electricity-generating facilities, in rural areas is cited as an example of that difference in perspective (Sears et al., 1993; Alm & Witt, 1995). In most states, local emergency management offices are creatures of the state rather than solely local offices. However, local planning and zoning commissions and other bodies responsible for land use planning, building codes and standards, and other actions to reduce the risk of disaster are decidedly local although state governments may adopt building codes and set standards for the training of building inspectors and planning officials. The extent to which elected officials receive training related to hazard management is questionable. The exceptions are floodplain management, because of federal mandates to limit development on 100-year floodplains in order to qualify for disaster assistance, and state-mandated building codes. In short, state and federal regulations do have impact because compliance is required. Citizen champions of environmental management can also sensitize a community to the risks posed by local hazards (Waugh, 2011).

RURAL GOVERNANCE AND MANAGEMENT CAPACITY

Lobao and Kraybill (2005a, 2005b, 2009) associate county government capacity with variations in poverty rates. Their focus is on rural counties. Their USDA-funded project concludes that counties with smaller and less professional staff are a statistically significant determinant of variation in adult and child poverty rates and household income. They also find that counties that are larger with more professionalized staff have lower poverty and high household income. Rural counties have aging populations with associated vulnerabilities. While the capacities of metropolitan county governments have expanded to address the needs of residents, the capacities of non-metropolitan counties have not kept up with the needs of residents. By contrast, metropolitan counties have greater capacity to support public services even though state and federal funding may be reduced. The more remote the county, the greater the likelihood that it will lack the capacity to meet resident needs. Remote populations are more likely to be older and require more public services while not generating sufficient tax revenue to support those services. Tax bases are smaller. Low population density means that the service area is greater—the population is scattered over a larger area—and therefore it is more expensive to serve.

The economic base is smaller and less diversified. The preponderance of jobs are low-skill and low-paying. Unemployment is higher than it is in more urban counties. It costs more to stimulate economic development because financial and human resources are so limited. For more remote communities, the prospect for economic development is even bleaker when companies need transportation networks and skilled workers (Waugh & Waugh, 1990). "In sum, because of attributes of the local population and economy, county governments vary in managerial and other resource capacities, scope of public services and economic development activities, and fiscal stress" (Lobao & Kraybill, 2005a: 1). Rural community demographics and economics are the root of the capacity problem, but other changes have also been taking place in rural America that have greatly reduced capacities to respond to local needs, including disaster response and recovery needs. Deregulation of bus, rail, and air service in the 1970s and 1980s means that many rural communities have no public transportation except for school buses. There are also declining numbers of food, hardware, and other retail outlets because of the development of regional centers. Mega-stores serve regional needs and small community retail stores have been disappearing. Even the closing of local post offices limits options for using the mail service to distribute pharmaceuticals or other medical supplies. Rural populations are more isolated and it is a challenge to provide basic services.

An Australian example of the impact of demographics on capacities to respond to local emergencies is the 2009 Victoria bushfires. The state had a policy of "Prepare, Stay and Defend or Leave Early," commonly referred to as "stay or go." The premise on which the policy was adopted was that residents could be prepared to defend their property and, if circumstances required, they could evacuate. When there were smaller bushfires, the policy worked and more residents were encouraged to stay home to protect their property. Unfortunately, in February 2009, the fires spread rapidly, often becoming crown fires in forested areas that were difficult to fight from ground level, and there was no time for many residents to flee. Many that did flee were caught on the road in their vehicles. Others were trapped in their homes. On "Black Saturday" and the days that followed, 173 people, including one firefighter, were killed. The investigation revealed the grim details of how residents were trapped and overcome by the flames. The defense strategy was predicated on having physically able residents with at least two adults per household. In reality, as the Royal Commission Report (2010) concluded, many residents did not have adequate plans in place regarding the defense of their property and under what circumstances they should evacuate, or the capabilities to do so. Many simply lacked such capabilities because they were elderly, disabled, living alone, and/or caring for children. The severity of events and other factors associated with the terrain and fire were also found to be problematic, but the demographic assumptions underlying the policy of "stay or go" raised serious questions about the reasonableness of the policy (Royal Commission, 2010). The demographics in rural Australia are not so different than those in rural America.

COMMUNITY RESILIENCE

Resilience is usually defined in terms of the capacity to recover quickly from catastrophic disasters and that capacity can be enhanced through the mitigation of hazards to prevent or reduce losses and preparedness efforts to improve response capabilities. In FEMA administrator Fugate's view, it is defined in terms of local capacity building, risk reduction, and preparedness (Fugate 2010). Communities need the capacity to deal with disasters with minimal outside assistance, they need to invest in mitigation programs to reduce risks and the impacts of the disasters that do occur, and they need to plan for the disasters that they cannot prevent. Having mechanisms to share critical resources can greatly expand local capabilities to deal with disasters, as well as other challenges. Admiral Thad Allen also links local preparedness to community resilience. They are basic building blocks of a civil society. They also support response and long-term recovery relationships. Trust and partnerships contribute to resiliency. He recommends that local governments provide emergency planning guidance for families when children are away from home, citing the Joplin tornado as an example. Risk-based emergency planning is needed to assure that resources are effectively allocated and the highest risks are reduced (Allen, 2011; Waugh, 2011).

Community resilience has many dimensions. Good government management, including the capacity to maintain and operate emergency management programs, is a dimension. Daniel Alesch and his colleagues (2009) have concluded that a number of factors influence a community's ability to recover from a major disaster. The severity of the disaster is one of those factors. Injuries and the loss of life, damage to the built environment, economic conditions prior to the disaster, are also factors. A large affluent community with many resources, a diverse economy, well-developed social networks, permanent institutions that are likely to be rebuilt (such as universities and military bases and prisons), and a quality of life that encourages residents to stay and rebuild is clearly more self-reliant and resilient than a community lacking those attributes. Alesch and his team found some communities in the U.S. to be very resilient and some extremely vulnerable. Despite massive damage to infrastructure and loss of life, some communities recover quickly. The differences were interventions by non-governmental organizations, insurance, and good leadership. Alesch and colleagues' recommendation is "managerial mitigation," i.e., insuring buildings and infrastructure, hiring well -trainined personnel, putting "hold harmless" provisions in place, developing mutual assistance, communicating risk, and planning for diverse events. In short, building local government management capacity and implementing preparedness programs increase the likelihood that communities will recover and recover quickly.

There are programs operating to increase managerial capacity and leadership in rural communities, such as those funded by the USDA, but enhancing rural community governance is no small task. "Governance is the process of making and carrying out decisions. In its most common use, governance refers to the management practices of governments, including cities, counties, special districts, school systems, regional governments, Indian reservations, and states. Especially in Anglo-Saxon countries, good governance denotes efficiency, effectiveness, good-value-for-the-money, and use of alternative administrative mechanisms," according to Nancy Stark, director of the Rural Policy Research Institute (Stark, n.d.). The Rural Governance Initiative includes efforts to improve decision making and implementation in rural communities by involving public, private, and non-governmental organizations, building all partnerships and other collaborative relationships, engaging citizens, and leveraging resources regionally. Local officials are under increasing pressure to address complex social and economic problems, operate with very limited resources, and have little professional staff to provide support (Stark, n.d.: 2–5). "They manage risk, do more with less, and yet 'get no respect'," according to Stark (n.d.: 5).

For Alesch and his colleagues, community characteristics and assets are critical variables in recovery from disasters. Local revenues, before and after the disaster, are important. Priority setting, resource allocation, and program management are also important. "Local government operating systems must be operational" (Alesch, Arendt, & Holly, 2009: 129). Accounting and financial management systems, adequate staffing, a "high-level problem-solving team," (135–136) helps. Some try to rebuild things as they were and some take advantage of the opportunity to implement change to reduce vulnerabilities. Change is easier if it was planned for and it is easiest if it was recommended in a strategic plan. The strategies for recovery range from re-creating what was there before to transforming the community into a very different one. Jim Schwab's report (2010) for the American Planning Association (APA) strongly recommends that communities link their future development to their hazard mitigation plans in order to reduce threats to life and property. Comprehensive plans will provide a map for hazard mitigation and encourage the community to consider risk in all of its planning. Alesch and colleagues focus on the attempt by Montezuma, Georgia to transform itself (unsuccessfully) into the kind of community it was many years before. The community lost its critical mass of businesses and could not rebuild the local economy. The attempt to revive the local economy suffered the same problems as the economy suffered before the disaster. The transformation of Greensburg, Kansas into a "green" community with a rejuvenated local business economy after being destroyed by a tornado is a largely successful example. One key is having the resources to press ahead after a devastating disaster. Another key is having the leadership to choose and implement a course of action. Recovery is driven by people, particularly people with resources, energy, and ideas.

EMERGENCY MANAGEMENT CAPACITY

If resilience is the ability to recover quickly from disaster, having adequate emergency management capacity would seem logical. However, emergency management is a relatively new profession and it is not always understood that it requires basic management skills, as well as an understanding of human behavior in crises and hazards science and other technical skills. The Certified Emergency Manager (CEM) requires training and/or education in general management as well as emergency management. Lucien Canton (2007) traces the evolution of emergency management as "management" to FEMA IS230, Principles of Emergency Management, an early training course which outlined the emergency management program functions as: laws and authority, hazard identification and risk assessment, hazard mitigation, resource management, planning, direction and control, communication and warning, operations and procedures, logistics and facilities, training, exercises evaluations and corrective actions, public education and training, and finance and administration (Canton, 2007, p. 69). These functions became the core of FEMA's Capability Assessment for Readiness (CAR) program in the 1990s. The goal was to create measures of preparedness to use in encouraging state programs to be more comprehensive. The functions became the core of the National Fire Protection Association's NFPA 1600 standard and the Emergency Management Accreditation Program (EMAP) standard. NFPA 1600 primarily provides benchmarks for private sector business continuity programs and EMAP primarily provides benchmarks for and accredits public sector emergency management programs and is pilot testing its program in college and university settings and for large federal agencies. Both sets of standards require the capacity to develop, operate, and maintain a comprehensive emergency management program. The EMAP standard requires that:

> The jurisdiction has a documented Emergency Management Program that includes an executive policy or vision statement for emergency management, a multi-year strategic plan, developed in coordination with Emergency Management Program stakeholders that defines the mission, goals, objectives, and milestones for the emergency Management Program and includes a method for implementation. (EMAP, 2012: Chapter 3.1.1)

That section of the standard goes on to state that:

> The Emergency Management Program has a documented method and schedule for evaluation, maintenance, revision and corrective actions for elements in Chapter 3 and Chapter 4 and shall conduct an evaluation of the objectives consistent with the program policies. (EMAP, 2012: Chapter 3.1.2)

Chapter 3 continues to outline the requirements for a designated office or agency with a leader with the authority to administer the program and an advisory body representing stakeholders and having "regular input" into program operation. Chapter 4 outlines program elements beginning with administrative and financial procedures to support the program and the authority to perform its responsibilities. That is, the initial concerns are that the program has legal authority, leadership, and management capacity. The remainder of Chapter 4 outlines the capabilities that should be present, including identifying hazards, assessing risk, and analyzing consequences; a hazard mitigation program; processes to prevent incidents; operational plans that encompass emergency response, continuity of operations (COOP) plans; continuity of government (COG) plans; recovery plans, with a lengthy list of emergency functions that might be performed by the office or stakeholder groups; an incident management system to provide coordination during emergencies; resource management and logistics plans; mutual aid agreements; communications and warning systems; procedures to coordinate operations with stakeholders; appropriate facilities, including an alternate emergency operations center in case the primary center is damaged or destroyed; appropriate training programs, including training needs assessment, evaluations, and record keeping and including specialized training; regular exercises, evaluations, and remedial actions; and crisis communications, public education, and public information plans. The

Table 15.1 EMAP Standards

- Program Management
- Administration and Finance
- Laws and Authorities
- Hazard Identification, Risk Assessment, and Consequence Analysis
- Hazard Mitigation
- Prevention
- Operational Planning
- Incident Management
- Resource Management and Logistics
- Mutual Aid
- Communications and Warning
- Operations and Procedures
- Facilities
- Training
- Exercises, Evaluation, and Corrective Action
- Crisis Communications, Public Education, and Information

list is long and compliance requires thorough documentation. However, not every jurisdiction faces the same hazards and stakeholder groups can take responsibility for major functions. Collaboration is not only key, but it is a requirement under the standard. The question is whether smaller jurisdictions have the capability to meet the standard and demonstrate that they have a comprehensive emergency management program. To date (as of March 2012), twenty-seven states, the District of Columbia, and six local jurisdictions are accredited. Two states and three local jurisdictions are conditionally accredited—meaning that they are in compliance in all but a very few standard areas. None of the accredited jurisdictions are small rural communities, but the EMAP Commission has decided to encourage one or more small jurisdictions to pursue accreditation and to help that jurisdiction through the process. If small communities have the wherewithal to develop and maintain a comprehensive, accredited program the EMAP Commission will have a much bigger challenge.[1]

MANAGEMENT CAPACITY AND RURAL COMMUNITY RESILIENCE

The focus of this analysis has been on the determination of management capacity, including emergency management capacity, and the state of rural community governance. Resilience is the goal. How can communities increase their capacities to minimize losses and their capacities to recover quickly from catastrophe? What are the essential variables? Some have been suggested. Dan Alesch and his colleagues (2009) focused on financial and human resources, including good community leadership. Craig Fugate (2010) mentioned local capacity building in the broad sense and in urban and less urban communities. Technology and human resources have been suggested. According to the working draft of the National Mitigation Framework, "Resilience reduces a community's vulnerability to the potential consequences from natural and man-made hazards, enabling it to absorb the impact of a disaster, response to the needs of its members, and recover in a timely and comprehensive manner" (FEMA, 2012: 5). Sustainability is a broader, more comprehensive concept. The draft goes on to point out the importance of local leadership and community-driven action to mitigate hazards, partnerships and inclusiveness, a "risk-conscious culture," and credibility and relevance to "all stakeholders" (6–7). The capabilities identified include communication, collaboration, and decentralized public involvement (15). "Regional resilience rests at the grassroots level with local government and key stakeholder organizations in partnership with state and federal government. The federal role is primarily to provide resources and assistance to localities and states consistent with policy and legal mandates" (TISP, 2011: 7).

In 2011, the Insurance Institute for Building and Home Safety (IBHS) rated coastal states in terms of their adoption of appropriate building codes, code enforcement, and contractor licensing for the types of hazards that pose risks to their communities. The states were given scores of up to fifty points for the adoption and universality of the codes, twenty-five points for the training of enforcement officials, and twenty-five points for requiring the licensing of contractors and subcontractors. Florida, Virginia, and New Jersey scored over ninety points (95, 95, and 93, respectively). Alabama, Texas, Delaware, and Mississippi scored less than twenty (18, 18, 17, and 4, respectively) (IBHS, 2012). Clearly, the management of the hazards posed by poor building codes, building standards, code enforcement, and licensing of contractors is both a function of political choice or will and management capacity. Management capacity is not the whole story when it comes to community resilience, in other words.

This analysis focuses on the capacities of rural communities to manage emergency management programs. Capacity building in rural communities begins with the question, "Capacity for what?" Different challenges require different capabilities. Do they have the capacities that Alesch and colleagues identified? While rural communities, by their very nature, have capacities that more urban communities do not, e.g., food production, they are particularly vulnerable to extremes of weather and they lack the medical, transportation, and communications resources of more urban areas (Waugh, Rotanz, & Sylves, forthcoming). Moreover, they are less likely to have the administrative capacities necessary to manage a large-scale natural or technological disaster or to manage a recovery effort without significant outside help.

As Lobao and Kraybill (2005a, 2005b) and others have concluded, the capacities of rural local governments have not kept up with the needs of their residents under normal circumstances and their aging populations have increasing needs. While measures of capacity have been lacking, there are a number of ways to assess performance and make judgments about capacity. The criteria that the Pew Center on the States' Government Performance Project is using included information collection and use, human capital development, long-term perspective on budget decisions and financial management, and infrastructure decision making and management (Barrett & Greene, 2008). Overall, in 2008, the states were given an average grade of B- for information use, a C+ for people or building human capital, a B- for budgeting, and a B- for managing infrastructure. To the extent that some local governments may fare better than their state governments in terms of their willingness to address issues of risk and their capacities to manage programs, the report's conclusions may not apply to rural jurisdictions. In fact, New York City has linked its future development to its hazard mitigation plans, including its estimation of the impact of climate change on sea level rise, rainfall, and temperature (Waugh et al., forthcoming), but interestingly, the Pew Center report noted that its score for New York State was significantly lower because of low scores for New York City's capacities.

CONCLUSION

Management capacity is viewed as essential by the USDA programs that support economic development in rural communities. Management capacity is also a requirement for emergency management programs themselves. While measures of capacity have frequently focused on the use of modern information technologies and management practices, agencies and offices may well have sufficient "low tech" capabilities to manage small programs (Waugh, 1988; Waugh & Hy, 1988; Waugh, 1990). Retired business owners and managers, military officers, teachers, and others may bring skills to local governments and community organizations that they would not otherwise have. The question is whether there are specific skills and competencies, as well as technologies, absolutely necessary to accomplish specific ends—in this case, hazard mitigation, disaster preparedness, and community redevelopment and recovery. As the Alesch report noted, local leadership and collaborative efforts can drive recovery. Collaborative leadership skills are critical to emergency management at all levels. Education and outreach are the tools to encourage public buy-in. The success of social, economic, and political recovery, dimensions of rural resilience, may also rest on local champions and collective action. Whether there is political will and support is another question. Overcoming a culture of inaction, i.e., the propensity to live with risk rather than work to reduce it, is a challenge. To the extent that emergency management capacities build upon local land use planning capacities (and vice versa), Jim Schwab's (2010) recommendation to link comprehensive planning and mitigation planning to assure that long-term development is risk-informed is a starting place. Training programs for local officials and public education can also sensitize the community to the risks posed by known hazards and, perhaps, encourage greater attention to risk reduction opportunities. To the extent that management capacities mean leadership, collaboration, accountability, and transparency, as well as financial and human resource management, more investments in training are necessary.

Most importantly, residents of communities need a sense of efficacy, a sense that they can protect themselves from environmental and human threats. They are at risk and they can do something about it. That sense of efficacy is evident in the images of townspeople filling and piling sandbags in preparation for river flooding. Shared experience builds a sense of community and successful recovery is testament to the resilience of communities.

NOTES

1. The author is a member of the EMAP Commission and is working on the selection of one or more small local governments for the Commission to assist through the accreditation process.

REFERENCES

Alesch, D., Arendt, L., & Holly, J. (2009). *Managing for long-term community recovery in the aftermath of disaster.* Fairfax, VA: Public Entity Risk Institute.

Allen, T. (2011). Community resilience: Thad Allen on creating a "community of practice". RAND audio podcast. Retrieved March 1, 2012 from http://www.rand.org/multimedia/audio/2011/12/12/creating-community-practice.html.

Alm, L.R., & Witt, S.L. (1995). Environmental policy in the intermountain west: The rural-urban linkage. *State and Local Government Review*, 27(2), 127–136.

Barret, C. & Greene, R. (2008). Measuring Performance: The State Management Report Card for 2008. *Governing*, March Issue, 24–95.

Canton, L.G. (2007). *Emergency management: Concepts and strategies for effective programs.* Hoboken, NJ: Wiley-InterScience.

Cigler, B. (1993). The special problems of rural county governments. In David R. Berman, ed., *County government in an era of change.* New York: Greenwood Press, pp. 89–106.

Crossett, K.M. (2004). *Population trends along the coastal United States: 1980–2008.* Washington, DC: National Oceanic and Atmospheric Administration, U.S. Department of Commerce.

EMAP (Emergency Management Accreditation Program). (2012). Who is accredited? Retrieved March 1, 2012 from http://www.emaponline.org/index.php?option=com_content&view=article&id=175&Itemid=139.

FEMA. (2012). National Mitigation Framework Review Package. Retrieved July 26, 2012 http://www.google.com/url?sa=t&rct=j&q=&esrc=s&source=web&cd=2&ved=0CGMQFjAB&url=http%3A%2F%2Fwww.fema.gov%2Flibrary%2Ffile%3Bjsessionid%3D9A7BEC5D10B1DE5E5A3B9337EA1EA309.WorkerLibrary%3Ftype%3DpublishedFile%26file%3Dpeo_nationalmitigationframework_20120302_v3.pdf%26fileid%3D58c40a70–6638–11e1–948f-001cc456982e&ei=no8RUKKpAYSY9QSY-YHYCA&usg=AFQjCNHEZFQarieWrFbWsqlVdQgV8fog7g&sig2=a8ItlwoC7jN4xaNcUz-6kg

Fugate, C. (2010). Opening plenary speech at the national conference of the International Association of Emergency Managers, San Antonio, TX, October 31.

IBHS (Insurance Institute for Business and Home Safety). (2011). *Rating the states: An assessment of residential building code and enforcement systems for life safety and property protection in hurricane-prone regions.* Tampa, FL: IBHS, December 31.

Jensen, J. (2009). NIMS in rural America. *Journal of Homeland Security and Emergency Management*, 27(3).

Kapucu, N., Augustine, M., Garayev, V. (2009). Interstate Partnerships in Emergency Management: Emergency Management Assistance Compact in Response to Catastrophic Disasters. *Public Administration Review*, 69(2), 297–313.

Lobao, L., & Kraybill, D.S. (2005). The emerging roles of county governments in metropolitan and nonmetropolitan areas: Findings from a national survey. *Economic Development Quarterly*, 19(3), 245–259.

Lobao, L., & Kraybill, D. (2005b). Rural county governments, economic development, and public services: The impact of decentralization over time. USDA Research, Education, and Economics Information System.

Lobao, L., & Kraybill, D.S. (2009). Poverty and local governments: Economic development and community service provision in an era of decentralization. *Growth and Change*, 40, 418–451.

Mathez, E.A. (2009). *Climate change: The science of global warning and our energy future*. New York: Columbia University Press.

Mattson, G. (1997). Small community governance: Some impediments to policy making. In J.J. Gargan, *Handbook of local government administration*. New York: Marcel Dekker Publishers.

Pear, R. (2011). In congress, storm relief is grist for dispute. *New York Times*, September 7. Retrieved from http://www.nytimes.com/2011/09/08/us/politics/08fema.html.

Perrow, C. (2007). *The next catastrophe: Reducing our vulnerabilities to natural, industrial, and terrorist disasters*. Princeton, NJ: Princeton University Press.

Resnick, B. (2011). FEMA's Craig Fugate reflects on a year of disasters. *The Atlantic*, September. Retrieved September 15, 2011 from http://www.theatlantic.com/national/print/2011/09/femas-craig-fugate-reflects-on-a-year-of-disasters.

Royal Commission Report. (2010). *The Fires and the Fire Related Deaths*. Parliament of Victoria: Government Printer of the State of Victoria.

Schwab, J.C. (2010). *Hazard mitigation: Integrating best practices into planning*. Planning Advisory Service report no. 560. Chicago: American Planning Association.

Sears, D.W., Redman, J.M., Gardner, R.L., & Adams, S.J. (1993). Designing successful state rural development strategies. In R.P. McGowan & E.J. Ottensmeyer (Eds.), *Economic development strategies for state and local governments*. Chicago: Nelson Hall Publishers.

Sokolow, A.D. (1997). Town and township government: Serving rural and suburban communities. In J.J. Gargan, *Handbook of local government administration*. New York: Marcel Dekker Publishers.

Stark, N. (n.d.). Effective rural governance: What is it? Why does it *matter?* Washington, DC: Rural Governance Initiative, Rural Policy Research Institute.

The Infrastructure Security Partnership (TISP). (2011). *Regional disaster resilience guide: 2011 edition.*

United States Global Change Research Program. (2009). *Global climate change impacts in the United States: A state of knowledge report from the U.S. global change research program*. Cambridge and New York: Cambridge University Press.

Waugh, W.L. (1988). Current Policy and Implementation Issues in Disaster Preparedness. in L.K. Comfort (ed) *Managing Disaster: Strategies and Policy Perspectives*, pp. 111–125. Durham: Duke University Press.

Waugh, W.L. (1990). Emergency management and the capacities of state and local governments. In R.T. Sylves & W.L. Waugh (Eds.), *Cities and disaster*. Springfield, IL: Charles C. Thomas.

Waugh , W.L. (1993) . Coordination or Control: Organizational Design and the Emergency Management Function . *International Journal of Disaster Prevention and Management*, 2(4), 17—31

Waugh, W.L. (1997). County government as the "administrative arm" of state government. In J.J. Gargan, *Handbook of local government administration*. New York: Marcel Dekker Publishers.

Waugh, W.L. (2011) Risk-based planning at the community level. Paper presented at the Southeastern Conference on Public Administration, New Orleans, LA, September 22–24.

Waugh, W.L., & Hy, R.J. (1988). The policymaking, administrative, and fiscal capacities of county government. *State and Local Government Review*, 20, 28–31.

Waugh, W.L., Rotanz, R., & Sylves, R.T. (forthcoming). Resilience and emergency management in New York City. In P. Barnes & A. Nakamura (Eds.), *Resilience and crisis planning in mega-cities: Issues, opportunities, and uncertainties.* Edward Elgar Publishing Ltd.

Waugh, W.L., & Streib, G. (1993). County capacity and intergovernmental relations. In D.R. Berman (Ed.), *County governments in an era of change.* Westport, CT: Greenwood Press.

Waugh, W.L., & Waugh, D.M. (1990). The political economy of seduction: Promoting business relocation & economic development in nonindustrial states. In A. Gitelson & M. Dubnick (Eds.), *Public policy and economic institutions.* Greenwich, CT: JAI Press.

Western Governors' Association (WGA). (2008). Western Wildlife Habitat Council Established. Jackson, Wyoming, June 29. Retrieved July 26, 2012 http://www.westgov.org/wga/publicat/wildlife08.pdf

16 Interorganizational Coordination
Analysis of 2011 Van Earthquake

Suleyman Celik and Sitki Corbacioglu

INTRODUCTION

Turkey is an earthquake-prone country that is located on the historically earthquake-prone Anatolian Peninsula. Local communities have lived under significant and continuous risk for thousands of years. Many destructive earthquakes occurred and even destroyed civilizations such as Hierapolis, Ephesus, and Troy. The Eastern Mediterranean sector of the Alpine-Himalayan earthquake belt exposes 98% of the nation to seismic risk. Earthquakes along with other natural and man-made disasters require an effective disaster system that can achieve timely interorganizational coordination in rapidly changing disaster environments.

Disaster environments are complex and the policy problems dealt are ill-structured. Disasters, especially earthquakes, are symmetry-breaking events that change the regular flow and order of organizational and societal life. Tens or even hundreds of organizations from different sectors with different organizational cultures assist the stricken communities in which power and telecommunications systems, local public personnel, and organizational capacity are negatively affected. Uncertainty dominates the environment through unlimited alternatives, conflicting values, unknown outcomes, and incalculable probabilities. Traditional bureaucratic systems, depending on the impacts of the earthquake, lose their capacity to follow the regular flow of work with regular means and methods that are suitable for regular times. The 1999 Marmara and 1999 Duzce earthquakes were a milestone for Turkish disaster management because they facilitated significant legal and organizational changes to improve the capacity of Turkish disaster management in the early 2000s. Earthquakes continued to occur after the 1999 earthquakes. The 2002 Sultandagi, 2003 Bingol, and 2004 Askale earthquakes were destructive ones with a considerably limited impact on local communities. Policy makers introduced a new law to fundamentally reorganize the organization structure of Turkish disaster management in 2009.

The Van earthquake occurred in 2011. It represents some similarities with the 1999 Duzce earthquake in regard to magnitude and destructive impacts.

Both earthquakes were destructive with a magnitude of 7.2. While the 2011 Van earthquake killed 604 people, the 1999 Duzce earthquake killed 761 people. This study reveals the 2011 Van earthquake response operations and compares and contrasts it with the 1999 Duzce earthquake response.

The comparison allows us to provide some insights about the results of changes in the organizational structure of the Turkish disaster system for providing intergovernmental coordination. The study analyzes the emerged informal network and its deviation from the formal disaster response system. It reveals the role of central actors and their attempts toward a coordinated disaster response.

The research questions being addressed in this study are: How did interorganizational response to the 2011 Van earthquake emerge? How did it differentiate from the 1999 Duzce earthquake? What are the leverage points for interorganizational coordination in response to disaster as revealed by the network analysis? The study attempts to address the research questions through a theoretical framework based on complex adaptive systems that involves the integration of organizational actors through sufficient and flexible organization structure, and technical and cultural capacities.

THEORETICAL FRAMEWORK

As in many areas of public management, conventional planning models in disaster management are based upon linear assumptions. According to these linear assumptions, policy actions in public management not only produce desired outcomes, but also these actions and outcomes are proportionate. In this model, an analytical approach to address the problem is: "Study the problem, develop alternatives, choose one, and move on to the next problem" (Mileti, 1999: 146). Organizations work like machines, they contain standard operating procedures and formal rules that identify responsibilities and ensure that all these procedures are reliably performed (Scott, 1992; Morgan, 1997). These organizations can be very effective where the environment is stable, problems are well-defined, and organizations can be closed to outside interference. However, in uncertain and complex environments, those machine-like organizations are likely to fail to achieve desired goals (Osborne & Gabler, 1992; Marion, 1999; Barzelay, 1992; Comfort, 1999; Axelrod & Cohen, 1999). Uncertain, complex, and socio-technical problems cannot be fully addressed by the conventional ways of problem solving (Carley, 2002).

In a traditional disaster management system, managers in a centralized system are more concerned with the intention to reduce risk and assume that execution of this intention and application of existing knowledge will result in proportional advances in hazard reduction (Turner & Pidgeon, 1997; Mileti, 1999). However, disasters create different norms, "emerging norms," from regular norms. The discrepancy between "emerging norms of disasters" and "bureaucratic norms of disaster administration" represents failures of intention (Schneider, 1995). The goal of a disaster management system is to find ways of diminishing this discrepancy as a way of dealing with the fatal consequences of disasters. A process-oriented, decentralized, flexible disaster management system with long-term mitigation plans is more likely to meet the emerging characteristics of disasters. Although local communities are more vulnerable to disasters, traditional disaster management systems put relatively less emphasis on improving the local capacity of the system. Human and material resources and the authority to mobilize these resources are often gathered at the national level. In many administrations, this appears to be a fundamental power issue between central and local organizations within the system.

The centralized system performs slowly and provides no flexibility to overcome damage from disasters during sudden disruptions. The key resolution to these problems is to foster local resiliency (Mileti, 1999). Resiliency in Mileti's words refers to the "capability of complex systems . . . to cope with changing conditions, to permanently adapt and, nevertheless, satisfy present needs" (cited inPossekel, 1999: 56). From this point of view, local resiliency represents a system where "a locality can tolerate and overcome damage, diminished productivity, and reduced quality of life from an extreme event without significant outside assistance" (Mileti 1999: 4). Local organizations should also have resiliency features that enhance both a capacity to predict the risks before they threaten the operations and the ability to survive within the circumstances in which the operations are compromised (McDonald, 2006). Therefore, to accomplish sustainable resiliency, local governments and local residents should have more power and should take more responsibility in evaluating and allocating the resources, and designing policies to reduce the impacts of disasters (Platt, 1999; Mileti, 1999).

The performance of a disaster management system depends on the system's capacity at the local level. Local conditions represent the initial conditions of the system, govern the whole disaster system, and eventually generate and characterize the choices for policies at different stages of disaster management. This requires an effective link between central and local actors. It is crucial to include local perspectives into the policy-making process to diminish disaster risk. Increasing the capacity of local components of the system improves the performance of the system. In a successful disaster management system, local governments and communities are more involved in systems dealing with future disasters.

Interdependent characteristics of the policy environment clearly show that a classic bureaucratic organization of the system is not able to provide the unity, complexity, and selectivity required for survival (Bennet & Bennet, 2000; Kapucu, 2009). A new model is needed to design an organizational system that best represents the complex characteristics of a changing organizational world.

Borrowing from evolutionary biology (Kauffman, 1993), recent theorists in organization theory have emphasized the Complex Adaptive System (CAS) as an appropriate model to meet the demand of complexities in a dynamic societal environment (Holland, 1995; Comfort, 1999; Axelrod & Cohen, 1999; Stacey, 2000, 2001). CAS theory considers systems as living organisms that merge and interact to provide the capabilities of advanced adaptability. The operation of the whole system depends on the well-functioning parts, although the whole is more than the sum of its parts. CAS is composed of self-organizing subunits that seek to maximize their own benefit, but at the same time act according to general system rules and in the context of interactions with other components and the external environment (Bennet & Bennet, 2000).

Four characteristics of CAS are recognized as fundamental for a resilient system (Kauffman, 1993; Prigogine & Stengers, 1984; Kiel, 1994; Gell-Mann, 1994; Comfort, 1994; Nicolis & Prigogine, 1989). These characteristics are: sensitive dependence on initial conditions; vulnerability to random events which may affect significantly the performance of parts of the system; potential for self-organization; and information flow among different components of the whole system.

A CAS is composed of interacting agents seeking to achieve a common goal (Cohen & Axelrod, 1999). The interaction of agents at the microlevel determines the macrolevel behavior of the system. The competing strategies used by the component agents attempt to draw the macrolevel entity, the system, to behave in a particular way (Kapucu, 2009). It is this dominant strategy that determines how the system behaves. The nature of the interaction among agents and the system's sensitivity to initial conditions generates performance adaptive to a changing environment rather than a linear relationship that produces the same behavior every time (Holland, 1995; Kauffman, 1993; Morel & Ramanujam, 1999).

At the edge of chaos, "the system has its best capacity for the processing of energy and information and is thus capable of optimizing its fitness" (Possekel, 1999: 19). This is the state where the system is able to be creative and innovative, while at the same time maintaining its identity and unity (Stacy, 2000). This state can be achieved only when information flow and connectivity among components (agents or workers) are achieved (Stacey, 2000; Comfort, 1999). Information flow and the connectivity of components are positively related. In order to increase connectivity that promotes the collective behaviors of subunits, the communication channels fostering information flow should be open. Creating these communication channels is a major task for system designers.

A disaster management system is a good illustration of CAS in practice, as many interacting organizational elements create complexity. A disaster produces unpredictable consequences that require abrupt changes in reallocation of resources in a limited time period. The unexpected consequences may also create changes in the exercise of authority in the system to meet the demands of the new environment. Disaster management systems operate at central, provincial, and local jurisdictions simultaneously. Public, private, nonprofit, and international organizations interact and collectively respond to natural disasters. The behavior and interaction among these organizations from different jurisdictions determines the success and effectiveness of the entire disaster management system. In such a complex system, acquiring and exchanging the relevant information among multiple emergency organizations from different jurisdictions is vital. If a subunit of the system fails to attain and transmit the required information, the whole system will likely fail to adapt to the requirements of the risk environments in which they operate. The success of the whole system depends on the quantity and quality of information from the risk environment to inform practice in real time (Comfort, 2007). This information is necessary for coordinated operations across emergency organizations. Quick and accurate information acquisition, processing, and dissemination decreases the uncertainty at each level of government, thereby increasing the interorganizational problem-solving capacity and the effectiveness of the overall response system (Comfort & Kapucu, 2006; Comfort & Sungu, 2001; Kapucu, 2012).

The coordination and collaboration among response organizations as well as reallocation of needed resources require a disaster management system to employ a dynamic organizational structure that can collectively respond to the rapid changes in disaster environments. The formal bureaucratic structure inhibits coordination and collaboration. A dynamic organizational structure that can self-adapt to rapidly changing conditions is necessary. A flexible organizational structure and sufficient information flow facilitate coordination between organizations and jurisdictions as well as increase organizational self-adaptation to changing conditions.

The theoretical framework of this study contains the following assumptions regarding interorganizational coordination: as interactions among organizations and jurisdictions increase, they share information and reallocate resources. As the information flow between public organizations and their managers increase, their decision-making capacity for innovative action increases. As sufficient information technology exists, the system has sufficient information infrastructure for facilitating the information flow between organizations and jurisdictions. As organizations and jurisdictions share more information and resources, their integration and coordination increases. As the information search, exchange, and distribution increases among organizations and jurisdictions, coordinated interorganizational collective action is achieved.

DISASTER MANAGEMENT IN TURKEY

The first recorded disaster policy in Turkey was formed after the Istanbul earthquake of September 14, 1519 that killed 13,000 people and destroyed

1,047 properties (Turkiye Buyuk Millet Meclisi, 1997). Policies developed until 1944 were primarily post-disaster efforts that evolved as a result of the relief and aid efforts for stricken cities and their inhabitants. Five earthquakes, which started with the 1939 Erzincan earthquake and continued until 1944, increased the public's attention to seismic risk. Along with the Erzincan earthquake, the Niksar-Erbaa, Adapazari-Hendek, Tosya, Ladik, and Bolu-Gerede earthquakes killed 43,319 people, injured 75,000 people, and demolished 200,000 residences and businesses in a span of five years. As a result, the Grand National Assembly passed in 1994 the first comprehensive mitigation law, the Law for the Measures to Be Taken before and after Earthquakes (law number 4623) (Turkiye Buyuk Millet Meclisi, 1997, 1999). This law required officials to: identify seismic risk for Turkey, develop response and relief programs in advance, and conduct geophysical examination for new development areas. Recognizing the need to do more than relief and aid and taking the first steps toward mitigation were important policy changes. Despite these steps, the implementation of mitigation principles was a great challenge. Laws and regulations without the capacity for implementation could not address the problem of seismic risk.

Disaster policies introduced between 1958 and 1999 focused on increasing the capacity of laying the legal and institutional foundations of disaster management. Organizations such as the General Directorate of Civil Defense and the General Directorate of Disaster Affairs were founded, while the Kandilli Observatory also became an earthquake research center. The reforms during the 1958–1999 period set the legal and organizational foundations of a linear policy perspective, but without creating sufficient organizational, technical, and cultural capacity to implement the laws and regulations (Corbacioglu, 2004). The 1999 Marmara and Duzce earthquakes started a new era in which the mitigation, preparedness, response, and recovery capacity of the Turkish disaster management agencies were critically questioned. The year 2009 was another turning point in Turkish disaster management because the organizational structure of the system has been significantly changed with law number 5209 and following regulations.

THE ORGANIZATIONAL STRUCTURE OF DISASTER RESPONSE BEFORE 2009

Central Organization

A formal disaster response system was organized at the central and local levels. Disaster law number 7269 and regulation number 88/12777, still in force, are concerned with the principles of emergency aid and planning. This regulation also represented the national emergency plan intended to create a planning process to coordinate all resources of public organizations ahead of time for the fastest and most effective disaster response operations (Afet Isleri Genel Mudurlugu, 1998). Along with the plan, the crisis management structure was introduced by regulation 8716 in 1996 to organize and coordinate public organizations and resources in case of crises. This central crisis management, known as the Prime Ministry Crisis Management Center (PMCM), was intended to provide the nationwide coordination of organizations and resources in case of a crisis. PMCM was authorized to take necessary measures in a disaster and assign duties to public organizations.

The Prime Ministry Crisis Management Center had three administrative bodies known as the Crisis Coordination Committee (CCC), the Crisis Evaluation and Watch Committee (CEWC), and the Secretariat. While CCC consisted of ministers, CEWC and the Secretariat involved top- and lower level bureaucrats respectively. PMCM involves other members as shown in Table 16.1.

The system involved some critical central actors, such as the General Directorate of Civil Defense of Interior Ministry and the General Directorate of Disaster Affairs of Public Works and Settlement Ministry, Turkey Emergency Management General Directorate of Prime Ministry, and the Redcrescent Society. The regulation described “crisis” as a foreign threat, widespread harmful (terrorist) attacks, natural disasters, refuge and big immigration movements, epidemics, big fires, chemical and technologic disasters such as radiation and air pollution, heavy economic crisis, and other similar situations. Regulation 8716 assigned coordination responsibility to different public agencies or ministries according to the type of crisis.

The Turkey Emergency Management General Directorate of Prime Ministry (TEMGD) was established after the 1999 earthquakes. This decision was an attempt to create a unifying agency for coordinating public organizations and resources. Although the establishment of TEMGD was an important step, it was still short in achieving the coordination of central disaster organizations and ministries, because TEMGD was given coordination responsibility

Table 16.1 Organizations in Prime Ministry Crisis Management Center

Chair: Prime Ministry or a State Ministry	
Agriculture and Forestry Ministry	National Defense Ministry
Energy and Natural Resources Ministry	National Education Ministry
Environment Ministry	National Security Council
Finance Ministry	Public Works and Settlement Ministry
Foreign Affairs Ministry	Red Crescent Society
Health Ministry	Secretary of National Security
Industry and Trade Ministry	Transportation Ministry
Interior Ministry	Labor and Social Security Ministry

without any authority over major disaster organizations and other public organizations in a bureaucratic system driven by hierarchic relations. Moreover, TEMGD, with a limited number of personnel at the central office in Ankara, did not have any local organization and had to rely on existing provincial and/or district rescue and aid committees (Corbacioglu, 2004).

Local Organization

According to the disaster law 7269 and the regulation number 12777, response and recovery operations are the responsibility of provinces and districts. Province and district rescue and aid committees must plan the response operations principally with provincial and district resources before disasters occur. The local military garrisons or military forces prepare their own plans in coordination with provincial and district plans. Provincial and district crisis management centers can ask for help from neighboring provincial and district organizations, local and regional military forces, or private organizations or individuals if local resources are inadequate for timely satisfaction of the stricken community needs. When this is the case, incoming organizations are expected to work under the coordination of local rescue and aid committees. The national disaster plan also requires central public organizations to prepare supplementary plans to aid stricken provinces and/or districts.

Provinces and districts are supposed to coordinate organizations and resources in districts and provinces through rescue and aid committees before earthquakes, in accordance with the disaster law and regulation number 88/12777. Therefore, provincial/district rescue and aid committees must prepare for disasters in advance. When crisis management is declared by the prime minister, rescue and aid committees function as the local crisis management. In case of a crisis, the rescue and relief committee functions as the local crisis management center. Rural areas including towns and small municipalities do not have a formal disaster management system. The Province/District Rescue and Aid Committee's involved organizations are presented in Table 16.2.

Table 16.2 Province/District Rescue and Aid Committee

Chair: Province Governor or Her Deputy	
Mayor	Director of Public Works and Settlement
Province Gendarmerie Commander	Director of Agriculture
Police Chief	Representative of Red Crescent Society
Director of Civil Defense	Local Military Commander
Director of National Education	

Adapted from Ergunay (1999).

The main problem with this national plan is its assumption that disasters will not occur in province centers and that province resources will principally be sufficient for response and recovery efforts. This assumption has been proved to be false by the long history of natural disasters that have hit province centers and required the help of other neighboring provinces, municipalities, as well as central organizations. The 1992 Erzincan, 1999 Marmara, and 1999 Duzce earthquakes are recent examples of such disasters. This crisis management structure was in effect to coordinate seismic response operations until 2009.

THE ORGANIZATIONAL STRUCTURE OF DISASTER RESPONSE AFTER 2009

The law number 5902 reorganized central and local Turkish disaster management in 2009. This law also attempted to address the coordination problems especially within the central government. The critical central actors of disaster management working under different ministries could not effectively coordinate with each other (Ergunay, 1999). Because agencies under different ministries cannot give orders to each other, it was a challenge to achieve coordination, especially among central organizations. Therefore, policy makers reorganized the critical actors under one roof through this law (Kapucu, 2012). The law attempted to better integrate the central and local organizations with each other. Despite this attempt, it is still questionable whether this truly represents a departure from the linear public policy tradition in Turkish disaster management.

Central Organization

The most important contribution of the policy change is the gathering of centrally operating disaster organizations under a single roof by establishing the Prime Ministry Disaster and Emergency Management Presidency known as AFAD. The new law abolished the Turkey Emergency Management General Directorate of Prime Ministry, the General Directorate of Civil Defense, and the General Directorate of Disaster Affairs. Their personnel and resources were reorganized under AFAD involving departments as follows:

- Department of Planning and Mitigation,
- Department of Response,
- Department of Recovery,
- Department of Civil Defense,
- Department of Earthquake,
- Department of Administrative Affairs.

Table 16.3 Comparison of Central Organization Structure

Before the Law Number 5902	*After the Law Number 5902*
The Prime Ministry General Directorate of Emergency Management	The Prime Ministry Disaster and Emergency Management Presidency (AFAD)
Prime Ministry Crisis Management Center Crisis Coordination Committee (CCC) (Ministers)	AFAD Disaster and Emergency High Committee (Ministers)
Prime Ministry Crisis Management Center The Crisis Evaluation and Watch Committee (CEWC) (Top-Level Bureaucrats)	AFAD Disaster and Emergency Coordination Committee (Top-Level Bureaucrats)
National Earthquake Advisory Council	AFAD Earthquake Advisory Board

As shown in Table 16.3, AFAD-involved committees are very similar to committees of the Prime Ministry Crisis Management Center. Later, the Prime Ministry Crisis Management Center was abolished by regulation 27851 in 2011. Therefore, in the framework of AFAD, the Disaster and Emergency Aid High Committee and the Disaster and Emergency High Committee, which are very similar to CEWC and CCC, were established. Finally, AFAD involved the Earthquake Advisory Board which was very similar to a previously founded but later abolished National Earthquake Council.

As clearly stated in Table 16.3, policy makers merged almost all public agencies and committees together under AFAD with the expectation of better coordination of resources and personnel. This attempt can also be viewed as coordination through hierarchy, in which central authority is preferred over coordination through information sharing and collective action in state bureaucracy.

Local Organization

The law number 5209 has also reorganized local disaster management. According to the law, Directorates of Civil Defense and Directorates of Civil Defense Search and Rescue Unit of Provincial Administrations were abolished. Their personnel and resources were transferred to two new departments of Province Special Administrations:[1] the Disaster and Emergency Aid Directorate and the Civil Defense Rescue and Search Directorate. Although these two were separate departments, the latter was made the subordinate of the former by an amendment made in 2011. As a result very similar organizations with similar names were re-established under province special administrations.

The laws and regulations have not abolished the provincial rescue and aid committees/crisis centers. However, because these committees acted as the blueprint and started functioning after disasters, in practice (Corbacioglu, 2004), policy makers introduced a new initiative, establishing the Disaster and Emergency Management Centers under Disaster and Emergency Aid Directorate in each province and if necessary in districts. Examples of these centers in the form of disaster coordination centers were established in provinces such as Istanbul, Yalova, and Sakarya after the 1999 Marmara earthquake. The new regulation attempts to establish modern coordination centers throughout Turkey with sufficient technical infrastructure.

As it is clearly stated above, these attempts were made to improve the organizational capacity of the Turkish disaster management at the central and local levels by reorganizing contemporary organizations and resources. The main mechanism to achieve this goal is a command-control-oriented coordination approach. The nature of seismic risk requires the coordination of tens or hundreds of organizations from different jurisdictions and sectors. Organizations that are typically not affiliated with disaster affairs take charge and join response and recovery operations. Therefore, putting critical organizations under one roof is only part of the problem. The more important part is how to coordinate sometimes hundreds of local, national, and international organizations through timely and valid information by establishing a self-adaptive complex disaster system.

METHOD

This is a small-n exploratory case study (Yin, 1993; Yin, 1994). The data was collected from five sets of sources: official reports, related laws and regulations, post-disaster critiques by participating organizations, analyses by researchers and national and international professional organizations, and daily news reports from the *Hurriyet* between October 23 and November 8, 2011 and the *Cumhuriyet* between November 13 and November 28, 1999. Data from the content analysis of daily newspapers was collected for twenty-one days after the Duzce earthquake and seventeen days for the Van earthquake, because a separate earthquake not an aftershock occurred in the Van province on November 9, 2009. Because the news reports focused on response to the second earthquake, and to minimize the potential for the interactions of two separate operations with each other, the content analysis for the Van earthquake ended by November 9, 2011.

The study involves two cases, the Van and Duzce earthquakes that occurred on November 12, 1999 and October 22, 2011 respectively. The reasons of comparing the response to two earthquakes are as follows: First, the two earthquakes have about the same magnitude of 7.2. Second, both earthquakes have had the highest magnitude according to the recorded ground accelerations since the 1999 Marmara earthquake, a regional disaster (METU, 2011). Third, both earthquakes heavily damaged a limited area. While they impacted the province centers, heavy damage occurred in the district centers of Duzce and Ercis along with surrounding towns. Finally, significant changes in the organization structure of Turkish disaster management occurred after the Duzce earthquake. Therefore, researching the response to the Van earthquake is useful to see the influence of these changes.

Data from the content analysis are used in the network analysis. Network analysis identifies the interactions among the actors (e.g., disaster organizations) rather than the attributes of the actors (Hanneman, 2001). We have used the UCINET software program for social network analysis (Borgatti, Everett, & Freeman, 2002). The social network analysis is based on four basic measurements of centrality data (Hanneman, 2001) used to analyze the whole disaster system network structure through the UCINET software program. These measures of centrality are as follows:

1. *Degree Centrality:* measures which organizations have the most ties to which other organizations.
2. *Closeness Centrality*: measures the distance of an organization to all other organizations in the network by focusing on the shortest distance from each organization to all others.
3. *Betweenness Centrality*: measures which organizations have a more positional advantage to the extent that they fall on the shortest pathway between other pairs of organizations.
4. *Flow Betweenness Centrality:* measures not only the shortest pathways between the two pairs of organizations but also all other alternative pathways that an organization can use to reach the others.

The content analysis of news reports for each earthquake is the primary data for network analysis. Organizations and their interactions are coded by using Microsoft Office-Excel 2003 (see Appendix 2 for sample coding). Interactions are defined as the joint acts between two or more organizations. After the initial coding, an interaction matrix was created for each earthquake. This study uses a "binary nominal level of measurement" for the matrices which include only the names of interacting organizations, the direction of interactions, and the frequency of interactions. An interaction between two organizations is coded as 1. If an interaction does not exist between two organizations, it is coded as 0. If the same organizations interact a second time, the previously assigned value of 1 is increased to 2. Later, the matrices were imported from Microsoft Office-Excel 2003 into the UCINET software program. By using one of the transformation methods in the UCINET program, the data were normalized. Normalization procedures help to standardize the data, especially when it is collected from different respondents who may use different scales of measurement (Borgatti et al., 2002). This network analysis explored the patterns of interactions and information exchange among disaster organizations during the disasters.

Using network analysis to analyze the interactions is important in identifying influential actors, their roles, and the informal organizational structure in response operations. Although only domestic public and nonprofit organizations with interactions are included in the network analysis, all public, nonprofit, and private organizations mentioned in the news reports are identified regardless of any interaction.

1999 DUZCE EARTHQUAKE

The 1999 Duzce earthquake caused heavy damage in the Duzce District, especially the city of Duzce and Kaynasli Municipality.[1] Duzce was located in Northwest Turkey and bordered Bolu, Zonguldak, and Sakarya provinces. The city was on the Ankara-Istanbul Highway and geographically close to other provinces and districts that were stricken by the 1999 Marmara earthquake, about three months prior.

The Duzce earthquake occurred at 7:02 pm (local time) on November 12, 1999 with a magnitude of 7.2. The earthquake killed 761 people and injured 4,948 people (Basbakanlik Kriz Yonetim Merkezi, 2000). The Duzce fault which is a segment of the North Anatolian fault line caused the earthquake (Erdik, 2000). This fault was 70 km long and extended between Akyazi and Kaynasli (Duzce Valiligi, 2000). The other two faults in the area are Hendek (50 km long) and Cilimli (13 km long) and did not break. The Marmara earthquake had already ruptured 30 km of the Duzce fault. The 1999 Duzce earthquake created 40–45 km of surface fault rupture between the Efteni Lake and Pirahmentler (Duzce Valiligi, 2000; Erdik, 2000). The epicenter of the earthquake was 6 km south of the city of Duzce (Erdik, 2000).

While most of the casualties were in the city of Duzce and Kaynasli, the fire and smoke caused by traditional heaters significantly increased the number of deaths and injuries, especially in Kaynasli. As a result, at least fifty people died by fires alone in Kaynasli. Deaths due to smoke and fire were also reported in Duzce (Corbacioglu, 2004).

Two earthquakes less than three months apart created heavy damage in the region. The Duzce earthquake caused heavy damage in the city of Bolu, the city of Duzce, and the Kaynasli municipality, as well as impacting surrounding provinces (Basbakanlik Kriz Yonetim Merkezi, 2000). The Duzce earthquake caused the collapse of many buildings already weakened by the Marmara earthquake.

Both the Marmara and Duzce earthquakes significantly damaged Duzce's economy by heavily damaging 3,228 businesses and slowing down economic development (Duzce Valiligi 2000; Devlet Planlama Teskilati, 2001). The estimated cost of the earthquake was $1 billion U.S. dollars (Gulkan, 2000).

As was the case in the Marmara earthquake, the Duzce earthquake immediately knocked out power and telecommunications systems. While the Turkish Amateur Radio Club was the primary means of communication on the first day, limited communication was available by the third day (Comfort & Sungu, 2001). Both telecommunications and power were largely restored in Duzce by the fifth day (Comfort & Sungu, 2001). Damage to the water and sewage systems caused delays for water distribution for the first few days in Kaynasli and Duzce (Kara Kuvvetleri Komutanligi Egitim ve Doktrin Komutanligi, 2000). The most significant damage on highways occurred on a section of the Trans-European Motorway (TEM) between Bolu and Duzce. Bolu viaducts #1 and #2, Bolu Bridge, and Bolu Tunnel located in the last segment of the TEM were strongly affected by the earthquake (Erdik, 2000). Additionally, collapses blocked inner-city roads in Kaynasli and Duzce (Corbacioglu, 2004).

2011 VAN EARTHQUAKE

The Van earthquake occurred in the Van province bordering Iran and neighboring the provinces of Agri, Mus, Bitlis, Hakkari, Siirt, and Sırnak in Eastern Turkey. The earthquake essentially hit an area by Lake Van including Van (pop. 370,000) and Ercis (pop. 75,000) cities as well as some related towns.

The earthquake, with a magnitude of 7.2 on the Richter scale, epicentered between the cities of Ercis and Van on October 23, 2011 at 10:41 am (GMT). Because there are different estimations, the exact epicenter of the earthquake is unclear. However, it is north of the city ofVan and east of Lake Van. Both field observations and aftershocks indicate that it was at most a distance of 30 km from Van's city center (METU, 2011; CEDİM, 2011). The number of aftershocks following the Van earthquake was more than any recorded earthquake in Turkey. There were 114 aftershocks with magnitudes 4.0–4.9 and seven aftershocks with magnitudes greater than 5.0 throughout the week after the earthquake, and a daily mean of 180 aftershocks occurred as of December 9, 2011 (AFAD, 2011a). The earthquake took place east of the North and East Anatolian faults, on the Bitlis suture zone (European-Mediterranean Seismological Center, 2011). Regarding moment magnitude, the Van earthquake is among the first ten earthquakes of the last 110 years and the third earthquake according to the recorded ground accelerations after the 1999 Marmara and Duzce earthquakes in Turkey (METU, 2011).

The earthquake impacted the Van province with heavy damage to its Ercis District, and was felt by surrounding districts and the provincial centers (MTA, 2011). Both Van and Ercis are adjacent to Van Lake and suggest soft soil conditions. Along with Ercis and Van cities, towns adjacent to Van Lake were seriously damaged. A total of 604 people died and more than 2,000 people were injured in Ercis, Van, and nearby towns (AFAD, 2011b; METU, 2011). Twenty-eight thousand properties collapsed or were heavily damaged (MEU, 2011). The most heavily impacted dwellings were found in Ercis in which about 30% of buildings were inhabitable (CEDIM, 2011; *Hurriyet*, November 1, 2011). The estimations indicate total economic losses ranging from $555 million USD to $2.2 billion USD (CEDIM, 2011). The natural gas system, water supply systems, and power and telecommunications were affected but became functional in twenty-four hours (CEDIM, 2011). The road between Van and Ercis cities was also affected due to collapse and cracklings (METU, 2011).

Another earthquake, not an aftershock, with a magnitude of 5.7 occurred on November, 9 2011, southeast of Van Lake, in the city of Edremit in the Van province. The earthquake killed forty people and caused heavy damage to 7,000 properties. Because it is a separate event, this study has not covered it.

EMERGENCE OF AN INTERORGANIZATIONAL NETWORK

Table 16.4 provides information about organizations that participated in the Van response system. According to the table, national organizations (74, or 70%) have played a more important role in response operations than international organizations (30, or 30%).

The findings show that public organizations are the heart of disaster management for the Van response system. Approximately 73% of all organizations are national and international public organizations while 27% of them are nonprofit and private organizations. The larger section of public organizations is international public organizations. Thirty, or 29% of all organizations that participated in the Van response system, are public international organizations, while 21% and 14% of all organizations are public-central and public-provincial organizations, respectively. National nonprofit (10%) and private organizations (16%) played significant roles in the response system as well. Only 2, or 2%, of 105 organizations are public-district organizations, whereas municipal organizations (8%) were involved more in response activities.

In summary, national and international public organizations constitute a very high proportion of the Van response operations. Public provincial, nonprofit, and private organizations are more involved in the Van response operations than municipal and district organizations. According to the data (Corbacioglu, 2004; Celik, 2006), national organizations in the Duzce earthquake (141, or 77.1%) played a more important role in response operations than international organizations (42, or 22.9%) compared to the Van earthquake. The findings of the Duzce response system show that public organizations played a relatively more important role than other private and nonprofit organizations. Approximately 83% of all organizations are national and international public organizations while only 16.4% of them are nonprofit and private organizations. This accounts for 10% more public organizations compared to the Van response system. Similar to the Van

Table 16.4 Types of Organizations in the 2011 Van-Ercis Earthquake

National Organization	N	%
Public Central	22	21
Public Provincial	15	14
Public District	2	2
Public Municipal	8	8
Nonprofit	10	10
Private	17	16
Total National	74	70
International Organizations		
Public International	30	29
Private International	1	1
Total International	31	30
Total	105	100

N = Number of organizations; % = Percentage of organizations.
Sources: *Hurriyet* news reports, October 23–November 8, 2011.

response system, among the national public organizations, the larger section is public-central organizations. Fifty-one, or 27.9% of all organizations that participated in the Duzce response system, are public-central organizations, while 25.7% and 7.1% of all organizations are public-province and public-municipal organizations, respectively. Only nine, or 4.9%, of 96 organizations are public-district organizations. Domestic private organizations are less involved (3.8%) in the Duzce response operations than the Van response operations (16%). National nonprofit organizations (10%) are relatively more involved in the Van response operations than Duzce response operations (7.7%). There are no international nonprofit organizations in the Van response system while there are nine international nonprofit organizations that constitute roughly 5% of the Duzce response system.

National public organizations constitute a very high proportion of both the Duzce and Van response operations. Among public organizations, public-central organizations, public-provincial, and public-district organizations are involved relatively more in the Duzce response operations than in the Van response operations. On the other hand, involvement of public international, domestic nonprofit, and private organizations in the Van response operations is considerably higher than in the Duzce response operations.

Figure 16.1 shows the organizations in the 2011 Van response network that emerged following the earthquake. Organizations have different positions depending on how connected they are to each other. The informal network indicates that AFAD, Van governorate, and Ercis District are at the center of activity. The centrality analyses provide more information about the roles and relations of actors.

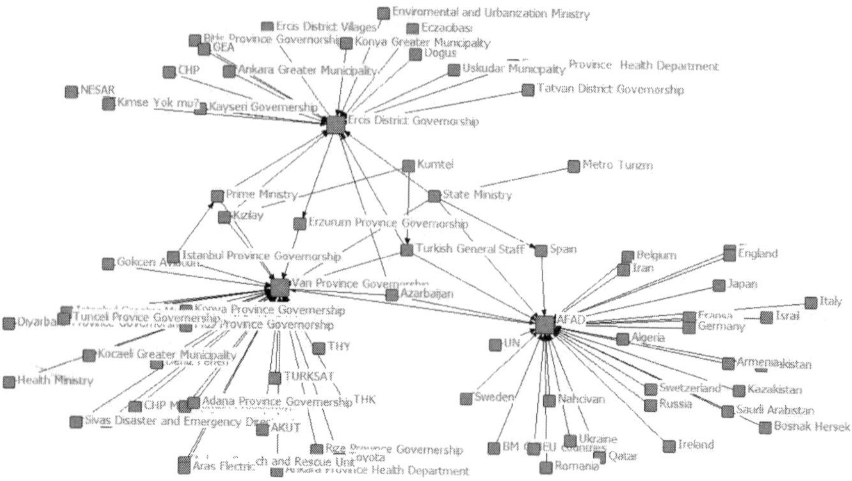

Figure 16.1 Interorganizational network in response to 2011 Van earthquake.

Degree Centrality

According to Freeman's degree centrality, the total number of degrees indicates how many connections one organization has with others in the network. In-degree data indicate how many ties an organization receives from others. If the organization receives an interaction, its in-degree value gets bigger. On the other hand, out-degree data indicate how many ties an organization sends to others. Thus, if the organization initiates interactions, and ties with others, its out-degree value gets higher.

Table 16.5 shows that AFAD, Van Province Governorship, and Ercis District Governorship are the central organizations during the Van response. The Erzurum Province Governorship (actor #8), Kizilay (actor #5), AFAD (actor #1), State Ministry (actor #6), and the Turkish General Staff (actor #7) are the most influential organizations in terms of out-degree centrality. The Van Province Governorship (actor #2), Ercis District Governorship (actor #3), and AFAD (actor #1) are the most influential organizations in terms of in-degree and total degree centrality measurements.

Kizilay and Kimse Yok Mu were two national nonprofit organizations that were active and influential. Freeman's degree centrality shows that provincial governorships and one public-central organization (AFAD) have larger in-degree and out-degree centrality than other organizations. Van Governorship, AFAD, and Ercis District Governorship operated as centers for collecting incoming assistance and distributing it to needy communities.

The degree centrality analysis of the Duzce response network (Celik, 2006) indicates that in the Duzce operations provincial and local organizations have higher in-degree centrality while central organizations have higher out-degree centrality. On the other hand, in the Van response system, there are both provincial and central organizations that have a higher degree of in- and out-degree centrality.

Both analyses present descriptive statistics of the two response networks (Celik, 2006). According to these statistics, on average, organizations have a degree centrality of 1.495, and 1.662, respectively, for the Duzce and Van response network. The ranges of variability for in-degree (24 versus 13) and out-degree (62 versus 59) are larger in the Van response network than in the Duzce response network. This shows that some organizations played a substantive role during the Van earthquake in terms of initiating (e.g., Erzurum Province Governorship) and receiving interactions (e.g., Van Governorship). In addition to these statistics, network centralization measures explore the

Table 16.5 Freeman's Degree Centrality for the Van Response System

Actor #	*OutDegree*	*InDegree*	*NrmOutDeg*	*NrmInDeg*
8 Erzurum Province Governorship	24.000	1.000	1.316	0.055
5 Kizilay	10.000	1.000	0.548	0.055
1 AFAD	7.000	30.000	0.384	1.645
6 State Ministry	6.000	1.000	0.329	0.055
7 Turkish General Staff	4.000	1.000	0.219	0.055
3 Ercis District Governorship	3.000	28.000	0.164	1.535
9 Prime Ministry	3.000	1.000	0.164	0.055
10 Azerbaijan	2.000	1.000	0.110	0.055
41 Kimse Yok mu?	2.000	0.000	0.110	0.000
35 Istanbul Province Governorship	2.000	0.000	0.110	0.000
45 Kumtel	2.000	0.000	0.110	0.000
2 Van Province Governorship	0.000	62.000	0.000	3.399
DESCRIPTIVE STATISTICS				
1 Mean	1.662	1.662	0.091	0.091
2 Std Dev	2.930	8.315	0.161	0.456
3 Sum	128.000	128.000	7.018	7.018
4 Variance	8.587	69.133	0.026	0.208
5 SSQ	874.000	5536.000	2.627	16.640
6 MCSSQ	661.221	5323.221	1.987	16.000
7 Euc Norm	29.563	74.404	1.621	4.079
8 Minimum	0.000	0.000	0.000	0.000
9 Maximum	24.000	62.000	1.316	3.399
10 N of Obs	77.000	77.000	77.000	77.000

Network Centralization (Out-degree) = 1.241%
Network Centralization (In-degree) = 3.352%
Sources: *Hurriyet* news reports, October 23–November 8, 2011.

degree of centralization in the networks. Freeman's graph centralization measure indicates that the Van response network has 1.24% in-degree, and 3.35% out-degree centrality, while the Duzce response network has 1.79% in-degree, and 8.96% out-degree centrality. Although network centralization statistics do not show highly centralized response systems, these data support evidence of the hypothesis that the Van response network is less concentrated in terms of both in-degree and out-degree centrality.

Closeness Centrality

Freeman's degree centrality indicates an actor's immediate ties. If one actor has more direct ties with others, the actor could be the most central actor in the network. By focusing on the shortest distance from each organization to all others, however, closeness centrality measures the distance of an

Table 16.6 Closeness Centrality for the Van Response System

Actor #	*Farness*	*nCloseness*
1 AFAD	211.000	36.019
2 Van Province Governorship	212.000	35.849
6 State Ministry	223.000	34.081
7 Turkish General Staff	224.000	33.929
10 Azerbaijan	226.000	33.628
5 Kizilay	252.000	30.159
9 Prime Ministry	253.000	30.040
8 Erzurum Province Governorship	254.000	29.921
3 Ercis District Governorship	254.000	29.921
57 Spain	267.000	28.464
35 Istanbul Province Governorship	269.000	28.253
DESCRIPTIVE STATISTICS		
1 Mean	287.632	26.672
2 Std Dev	26.646	2.724
3 Sum	21860.000	2027.038
4 Variance	710.022	7.422
5 SSQ	6341588.000	54628.371
6 MCSSQ	53961.684	564.109
7 Euc Norm	2518.251	233.727
8 Minimum	211.000	23.171
9 Maximum	328.000	36.019
10 N of Obs	76.000	76.000

Sources: *Hurriyet* news reports, October 23–November 8, 2011.

organization to all other organizations in the network. According to Table 16.6, during the Van response actions, AFAD (actor #1) is the most central organization in terms of closeness centrality. Van Province Governorship (actor #2), State Ministry (actor #6), and Turkish General Staff (actor #7) are the other most central organizations in the response network. Erzurum Province Governorship (actor #8) and Kizilay (actor #5) lost their positional advantage compared to degree centrality measurements. The same as the Duzce response network (Celik, 2006), closeness centrality and degree centrality analyses produce similar findings.

Betweenness Centrality

Betweenness centrality measures an organization's positional advantage based on its location along the shortest pathway between other pairs of actors in the response network. According to Table 16.7, the Ercis District Governorship (actor #3) was the most central organization which falls on the

Table 16.7 Betweenness Centrality for the Van Response System

Actor #	*Betweenness*	*nBetweenness*
3 Ercis District Governorship	94.000	1.649
1 AFAD	27.000	0.474
10 Azerbaijan	25.500	0.447
8 Erzurum Province Governorship	7.500	0.132
6 State Ministry	7.000	0.123
9 Prime Ministry	5.000	0.088
7 Turkish General Staff	3.500	0.061
5 Kizilay	2.500	0.044
DESCRIPTIVE STATISTICS		
1 Mean	2.234	0.039
2 Std Dev	11.385	0.200
3 Sum	172.000	3.018
4 Variance	129.608	0.040
5 SSQ	10364.000	3.190
6 MCSSQ	9979.792	3.072
7 Euc Norm	101.804	1.786
8 Minimum	0.000	0.000
9 Maximum	94.000	1.649
10 N of Obs	77.000	77.000

Network Centralization Index = 1.63%
Sources: *Hurriyet* news reports, October 23–November 8, 2011.

shortest pathway between other pairs of actors, while it was the sixth most central organization in terms of degree centrality and ninth most central organization in terms of closeness centrality. AFAD (actor #1), Azerbaijan (actor #10), and Erzurum Province Governorship (actor #8) are the other central organizations with the highest degree of betweenness centrality.

The betweenness centrality statistics (Celik, 2006) indicate that there is more variation in the Duzce response network (from 0 to 255.17) than the Van response network (from 0 to 94). The overall network centralization is a little higher (1.90%) in the Duzce response network (1.90%) than in the Van response network (1.63%). Although this data shows that the Duzce response system had more intermediary actors, both response systems needed more agents to provide links among disaster organizations.

Table 16.8 FlowBetweenness Centrality for the Van Response System

Actor #	*FlowBet*	*nFlowBet*
1 AFAD	50.000	0.877
2 Van Province Governorship	0.000	0.000
3 Ercis District Governorship	36.667	0.643
4 Ercis District Villages	0.000	0.000
5 Kizilay	1.000	0.018
6 State Ministry	2.000	0.035
7 Turkish General Staff	1.000	0.018
8 Erzurum Province Governorship	0.000	0.000
9 Prime Ministry	2.000	0.035
10 Azerbaijan	38.667	0.678
DESCRIPTIVE STATISTICS		
1 Mean	1.714	0.030
2 Std Dev	8.157	0.143
3 Sum	132.000	2.316
4 Variance	66.542	0.020
5 SSQ	5350.000	1.647
6 MCSSQ	5123.714	1.577
7 Euc Norm	73.144	1.283
8 Minimum	0.000	0.000
9 Maximum	50.000	0.877
10 of Obs	77.000	77.000

Network Centralization Index = 0.858%
Sources: *Hurriyet* news reports, October 23–November 8, 2011.

Flow Betweenness Centrality

Flow betweenness centrality measures both the shortest pathways between two pairs of organizations and all other alternative pathways which an actor can use to reach the others. According to Table 16.8, AFAD (actor #1) is the most central actor. After that Azerbaijan (actor #10) and Ercis District Governorship (actor #3) have the most advantaged positions in terms of flow betweenness centrality. Interestingly, the analysis shows that Erzurum and Van governorships do not play any notable role in terms of the flow betweenness centrality measurement.

The descriptive statistics of flow betweenness centrality of both response networks, Van and Duzce, reveal similar results to the betweenness centrality findings. The variation in the Van response network is lower (50) than that of the Duzce response network (258). Likewise, network centralization is smaller (0.858%) in the Van response network than in the Duzce network (2.19%). The network centralization data of the flow betweenness centrality is smaller than the betweenness centrality based on the shortest pathway.

IMPROVING INTERORGANIZATIONAL COORDINATION IN SEISMIC RESPONSE

The centrality analysis and comparison of both response systems indicate that the main central organization in the formal structure is also the most influential actor in the informal network. As revealed by the network analysis, the Prime Ministry Crisis Management Center is the most central organization in the Duzce response operations whereas AFAD is the most central organization in the Van operations. This means that despite the recent changes in organizational structure in 2009, dominance of the central organizations continues in the Turkish disaster management system. Erzurum, Van, and Ercis district governorships were also central organizations, as was the Bolu Crises Management Center and the Duzce Crises Management Center in Duzce operations. However, the analysis shows that these organizations receive or distribute the assistance. Therefore, the systems in both operations lack sufficient intermediary actors that connect organizations to each other and create more inclusive networks. This also means that the Turkish disaster management system is not sufficiently strong at the local level, and risk-prone communities are still essentially depending on outside assistance.

The network analysis indicates that international assistance is an increasingly important part of the emergency system. A total of thirty-one organizations, 30% of all involved, are international organizations. Therefore, the formal structure needs to incorporate non-domestic actors into the system and rethink the level of international participitation within the emergency system. Moreover, many nonprofit actors as well as private and public organizations of different jurisdictions are involved in response and recovery operations, although they are not part of the official response system at the central and local levels. Therefore, the emerged informal network significantly differentiates from the formal organizational structure. These findings indicate that a linearly designed formal response system cannot be sufficient for responding to disasters. Command-control-oriented bureaucracies fail to control the system and its actors. However, complex conditions require a dynamic organizational structure that requires the integration of critical public, private, and nonprofit actors with each other.

Such a dynamic organizational structure can happen through approaching complex policy problems from a complexity perspective. Therefore, a complex adaptive disaster system that can adapt to the changing conditions of stricken communities is still needed after the 2011 Van earthquake to more effectively coordinate interorganizational collective action. Critical aspects in such organizational structure are going beyond the introduction of laws and regulations to develop the organizational, technical, and cultural capacity of the system before an earthquake strikes. Sufficient information flow through a flexible organizational structure can help an emerging informal network to respond to the needs of a stricken community faster and in a more coordinated way.

CONCLUSION

Turkey has been vulnerable to seismic risk for thousands of years and still struggles to find a better way to respond to stricken communities especially since the 1940s. The disaster management system has significantly developed from the recovery perspective to effective mitigation, preparedness, and response stages. The Marmara and Duzce earthquakes of 1999 represented a milestone in developing the organizational and technical capacities of the disaster system. However, the critical problem within the system has remained basically intact: a linear public policy approach and command-control-oriented bureaucratic model. Despite improvements, the central actors are still influential, while local capacity is insufficient. Moreover, as revealed by the informal response networks of two disaster response systems in this chapter, the emerging conditions of disasters create a dynamic organizational structure. Many public, private, and nonprofit organizations create an emerged informal network that is critically different from the formal disaster organization. Such a result indicates that the linearly designed formal structure cannot function effectively in response to symmetry-breaking earthquakes. Despite changes in the central and local organizational structures of the system since 2009, similar problems exist in 2011 Van response operations.

The disaster system needs intermediary actors that can coordinate organizations from different jurisdictions operating within the stricken communities. The integration of nonprofit and international actors to the response system is also a critical point of improvement. Moreover, the content analysis along with the findings of network analysis indicate that Turkish disaster response and recovery systems need to develop its organizational, technical, and cultural capacity by changing its governing values from a linear-bureaucratic mind-set towards complex self-adaptive response through the complex adaptive system perspective.

NOTES

1. Province Special Administration is a type of local government that operates in each province besides municipalities. Province Special Administration is a mix of elected representatives and appointed bureaucrats. Its head, the top manager, is the governor of the province administration that represents the central government in the provincial level. However, the decision-making body, the assembly, consists of only elected members.
2. Civil Defense Rescue and Search Directorate exists in only eleven provinces.
3. Kaynasli became a district of Duzce province on December 9, 1999. The local organization of Turkish central government essentially involves province and district governments. Provinces involve districts that are composed of small towns and municipalities. Kaynasli by the time of the earthquake was a small inhabitancy with a municipal administration. Following the earthquake, the Duzce District, then part of the Bolu province, became a province as Kaynasli became a district of Duzce province.

REFERENCES

AFAD (Prime Ministry Disaster and Emergency Management Presidency). (2011a). Van earthquake on October 23, 2011. Press release, December 13, 2011. Retrieved from http://www.afetacil.gov.tr/.

AFAD (Prime Ministry Disaster and Emergency Management Presidency). (2011b). October 23, 2011 and November 9, 2011 Van earthquakes. Press release, November 28, 2011. Retrieved from http://www.afetacil.gov.tr/.

Afet Isleri Genel Mudurlugu. (1998). *Kanunlar ve yonetmelikler* [Laws and regulations]. Ankara: Bayindirlik ve Iskan Bakanligi.

Axelrod, R., & Cohen, D.M. (1999). *Harnessing complexity.* New York: Free Press.

Barzelay, M. (1992). *Breaking through bureaucracy: A new way for managing in government.* Berkeley: University of California Press.

Basbakanlik Kriz Yonetim Merkezi. (2000). *Depremler 1999* [1999 earthquakes]. Ankara: Basbakanlik.

Bennet, A., & Bennet, D. (2000). Characterizing the next generation knowledge organization. *Knowledge and Innovation: Journal of the KMCI*, 1(1), 8–42.

Borgatti, S.P., Everett, M.G., & Freeman, L.C. (2002). *Ucinet for Windows: Software for social network analysis.* Harvard, MA: Analytic Technologies.

Carley, K.M. (2002). Computational organizational science and organizational engineering. *Simulation Modeling Practice and Theory*, 10(5–7), 253–269.

CEDIM (Center for Disaster Management and Risk Reduction Technology). (2011). Comparing the current impact of the Van Earthquake to past earthquakes in

Eastern Turkey. Report no. 4. Retrieved from http://www.cedim.de/download/CEDIMForensicEQAGTurkeyVanEQ_Report4.pdf.

Celik, S. (2006) Socio-technical approaches to complex phenomena: An analysis of the Turkish disaster management system under stress. Unpublished dissertation. Graduate School of Public and International Affairs, University of Pittsburgh, Pittsburgh, PA.

Comfort, L.K. (1994). Risk and resilience: Inter-organizational learning following the Northridge earthquake of 17 January 1994. *Journal of Contingencies and Crisis Management*, 2(3), 157–170.

Comfort, L.K. (1999). *Shared risk: Complex systems in seismic response*. Amsterdam: Pergamon.

Comfort, L.K. (2007). Crises management in hindsight: Cognition, communication, coordination, and control. In Administrative failure in the wake of Katrina, special issue, *Public Administration Review*, S1, 88–96.

Comfort, L.K., & Kapucu, N. (2006). Interorganizational coordination in extreme events: The World Trade Center attacks, September 11, 2001. *Natural Hazards*, 39(2), 309–327.

Comfort, L.K., & Sungu, Y. (2001). Organizational learning from seismic risk. In U. Rosenthal, L.K. Comfort, & A. Boin (Eds.), *From crisis to contingencies: A global perspective* (pre-published copy). Chicago: Charles C. Springer, Inc.

Corbacioglu, S. (2004). Intergovernmental collective action in complex environments: Toward a self-adaptive Turkish disaster management system. Unpublished dissertation. Graduate School of Public and International Affairs, University of Pittsburgh. Pittsburgh, PA.

Cumhuriyet daily newspaper. (1999). Istanbul, Turkey, November 13–28, 1999 issues.

Devlet Planlama Teskilati. (2001). *Duzce ili raporu* [Duzce province report]. Ankara: Devlet Planlama Teskilati.

Duzce Valiligi. (2000). *Yeni kent, yeni yasam: Duzce* [New city, new life: Duzce]. Ankara: Yargici Ofset.

Erdik, M. (2000). Report on 1999 Kocaeli and Duzce (Turkey) earthquakes. Retrieved from http://www.koeri.bound.edu.tr/depremmuh/Kocaelireport.pdf.

Ergunay, O. (1999) A Perspective of Disaster Management in Turkey: Issues and Prospects. Urban Settlements and Natural Disasters, Proceedings of UIA Region II Workshop, June 1999, pp.1–9 .

European-Mediterranean Seismological Center. (2011). Earthquake details. Retrieved from http://www.emsc-csem.org/Earthquake/202/Earthquake-M7–2-Eastern-Turkey.

Gell-Mann, M. (1994). *The quark and the jaguar: Adventures in the simple and the complex*. New York: WH Freeman and Co.

Gulkan, P. (2000). Recent natural disasters in Turkey: An overview of the national technological capacity and its utilization. Unpublished essay. Middle East Technical University, Ankara.

Hanneman, R.A. (2001). *Introduction to social network methods*. Riverside: University of California. Retrieved from http://faculty.ucr.edu/~hanneman/nettext/.

Holland, J.H. (1995). *Hidden order*. Reading, MA: Addison-Wesley.

Hurriyet daily newspaper. (2011). Istanbul, Turkey, October 23–November 8, 2011 issues.

Kapucu, N. (2009). Interorganizational coordination in complex environments of disasters: The evolution of intergovernmental disaster response systems. *Journal of Homeland Security and Emergency Management*, 6(1), 1–26.

Kapucu, N. (2012). *The network governance in response to acts of terrorism: Comparative analyses*. New York: Routledge.

Kara Kuvvetleri Komutanligi Egitim ve Doktrin Komutanligi. (2000). *Depremler ve alinmasi gereken dersler* [Earthquakes and lessons need to be taken]. Ankara.

Kauffman, S.A. (1993). *The origins of order: Self-organization and selection in evolution*. New York: Oxford University Press.

Kiel, L.D. (1994). *Managing chaos and complexity in government*. San Francisco: Jossey-Bass Publishers.

Marion, R. (1999). *The edge of organization*. Thousand Oaks, CA: Sage Publications.

McDonald, N. (2006). Organizational resilience and industrial risk. In E. Hollnagel, D.D. Woods, & N. Leverson (Eds.), *Resilience engineering, concepts and precepts*. Brookfield, VT: Ashgate Publishing Company.

METU (Middle East Technical University Earthquake Engineering Research Center). (2011). Field observations about Mw 7.2-October 23, 2011 Van earthquake seismic and structural damage. Retrieved from http://www.imo.org.tr.

MEU (The Ministry of Environment and Urbanism). (2011). http://www.cevresehircilik.gov.tr/turkce/sayfa.php?Sayfa=bakanlikbasin&ID=716.

Mileti, D.S. 1999. *Disasters by design: A reassessment of natural hazards in the united centrals*. Washington, DC: Joseph Henry Press.

Morel, B., & Ramanujam, R. (1999). Through the looking glass of complexity: The dynamics of organizations as adaptive and evolving systems. *Organization Science*, 10(3), 278–293.

Morgan, G. (1997). *Images of organization*. Thousand Oaks, CA: Sage Publications.

MTA (General Directorate of Mineral Research and Exploration). (2011). 2011 Van earthquake information note. Retrieved from http://www.mta.gov.tr/v2.0/deprem/BilgiNotu_2011–10_Tabanli-Van.pdf.

Nicolis, G., & Prigogine, I. (1989). *Exploring complexity: An introduction*. New York: W.H. Freeman.

Osborne, D., & Gabler, T. (1992). *Reinventing government*. Reading, MA: Addison Wesley.

Platt, R.H. (1999). *Disaster and democracy: The politics of extreme natural events*. Washington, DC: Island Press.

Prigogine, I., & Stengers, I. (1984). *Order out of chaos*. New York: Bantam Press.

Possekel, A.K. (1999). Living with the unexpected: Linking disaster recovery to sustainable development. New York: Springer; Berlin: Heidelberg.

Schneider, S.K. (1995). *Flirting with disaster: Public management in crisis situations*. Armonk, NY: M.E. Sharpe.

Scott, W.R. (1992). *Organizations: Rational, natural, and open systems*. Englewood Cliffs, NJ: Prentice Hall.

Stacey, R.D. (2000). *Complexity and management*. London: Routledge.

Stacey, R.D. (2001). *Complex responsive processes in organizations*. London: Routledge.

TBMM (Turkiye Buyuk Millet Meclisi). (1997). *Dogal afetlerde can ve mal kaybinin en aza indirilmesi icin alinmasi gereken tedbirlerin belirlenmesine iliskin meclis arastirmasi* [Turkey Grand National Assembly special committee on natural disasters]. Ankara.

TBMM (Turkiye Buyuk Millet Meclisi). (1999). *Deprem arastirma raporu* [Earthquake investigation report]. Report nos. 10/66, 67, 68, 69, 70. Ankara.

Turner, B.A., & Pidgeon, N.F. (1997). *Man-made disasters*. 2nd edition. Butterworth-Heinemann.

Yin, R.K. (1993). *Applications of case study research*. Thousand Oaks, CA: Sage Publications.

Yin, R.K. (1994). *Case study research*. Thousand Oaks, CA: Sage Publications.

17 Community Capacity and Interorganizational Networks for Disaster Resilience

Comparison of Rural and Urban Counties

Fatih Demiroz, Naim Kapucu, and Rebecca Dodson

INTRODUCTION

Communities in the U.S. face a variety of natural and man-made disasters and Florida is one of the most hazard-prone states in the country. Building disaster-resilient communities is a complex and multi-dimensional policy issue. The effectiveness of disaster response and recovery, and the adaptive capacities that a community has are the fundamental aspects of disaster resiliency. The overall effectiveness of the emergency management phases (preparedness, mitigation, response, and recovery) relies on the effectiveness of cross-sector and interorganizational networks that are designed to work towards these goals.

Several chapters in this book cited the importance of collaborative relationships and stakeholder involvement in disaster management (see Chapters 2, 5, and 11). This chapter approaches emergency management from a similar perspective. Stakeholder involvement is discussed using a resource dependence approach within the context of collaborative public management. Social network analysis (SNA) tools are used for analyzing stakeholder involvement. We compare stakeholder networks of a rural and an urban county because the availability of resources and the level of interdependence among organizations vary significantly in both counties. In this chapter, we examine the structural differences between stakeholder networks of Orange (urban) and Sumter (rural) counties and impact of resource dependence. This chapter is a modest attempt to integrate a resource dependence approach to emergency management. The results we present are descriptive and we are aware that this work is far from being comprehensive enough. Nonetheless, we hope that it will be a good starting point for future research focusing on resource dependence and social network analysis in the emergency management setting.

This chapter consists of five sections. The first section presents a brief review of the literature on network effectiveness, resource dependence, and emergency management networks. The second section provides general information about the potential hazards in Central Florida and background information about the emergency management systems in Orange and Sumter counties. The third section explains the methodology that was followed for data collection and analysis. The fourth section discusses the results of the analysis and compares the networks of the two counties, and the last section concludes the findings of the chapter.

NETWORKS FOR DISASTER RESILIENCE

The effectiveness of emergency management networks has a direct impact on community disaster resilience. There is extensive literature on the effectiveness of interorganizational networks; however, the use of social network analysis on this topic is relatively limited (Cross, Borgatti, & Parker, 2002; Harris & Clements, 2007). Social network analysis (SNA) tools are helpful in examining the structure of organizational networks, identifying "points of misalignment and accelerating collaboration in the right places, and determining whether certain organizations and functions are achieving the connectivity required for desired results" (Kapucu & Demiroz, 2011: 557).

Network structure (particularly for policy networks) is shaped by the internal motives of organizations and external forces in the environment (Henry, 2011; Podolny & Page, 1998). As Henry (2011) notes, some policy networks are formed as a result of formal mandates. For example, intergovernmental relations (i.e., among local, state, and federal government agencies) are good examples for mandated networks in emergency management. However, various other organizations take important roles in emergency management even though they are not mandated to do so. These organizations join the network in different stages of the emergency management process and make networks appear different than they were planned. This situation, Henry (2011) highlights, is an outcome of the internal motives of policy networks. Willingness to reach additional resources can be an important example of the internal motives. On the flip side of the coin, external conditions can force the way organizations join a network informally. An example of external constraints for network structure is the institutional rules and formal boundaries that set the stage for a policy problem. That means the rules and regulations may determine who can join the network and how they can do it.

The interdependence of organizations is a key attribute and an internal motive of service delivery networks (Thomson, Perry, & Miller, 2009). Level of interdependence is a predictor of relationship building in network settings (Gulati & Gargiulo, 2001). Scholars such as Gulati and Gargiulo (2001) and Podolny and Page (1998) mention economic factors such as profit making, transaction costs, and resource dependence are important components of interdependence.

In emergency management, interdependent relations are spelled out in official documents such as NRF and local comprehensive emergency management plans (CEMPs). For example, a local CEMP assigns a certain number of organizations for a specific emergency support function (e.g., transportation, mass care, sheltering) either in a primary or a supporting role. These organizations are expected to work together (and also with the local emergency management organization) for accomplishing that specific task. In this context, the level of collaboration among partnering organizations serves as an indicator of effectiveness of that network. Furthermore, compliance with the planned network is an indicator of level of collaboration (Kapucu & Demiroz, 2011).

Having organizations work collaboratively according to a plan is a challenging task. Organizations usually have informal connections among each other at some level and these relationships can be transformed to formal ones (Ansell & Gash, 2007). Nonetheless, not all informal ties become formal policy relationships. There are several reasons this. For example, in local governments where availability of resources is a challenge (e.g., rural communities) transforming informal relationships into institutionalized ties might be problematic. Despite interorganizational social capital helps partnering organizations lower transaction costs, that may not necessarily be enough for some public organizations and nonprofits for which emergency management (or being involved in a collaborative network for this purpose) is not an organizational goal. In a situation where organizations lack the resources to be involved in emergency management, the local emergency management agency serves as a network administrative organization (Provan & Kenis, 2008) to coordinate the network via the resources it owns and the legal authority it bears.

In this chapter we argue that the structure of the network is impacted by the capacity of organizations to transform their informal relationships into formal policy networks. This is significantly impacted by the availability of resources in the network. Furthermore, the effectiveness of the emergency management network is impacted by its structure. SNA was used to visualize the differences between an urban county, where there are more resources available for emergency management, and a rural county where resources are relatively scarce.

CONTEXT OF THE STUDY

Why Central Florida?

Much of Florida remains rural. Of the 406 cities, 89% (363) have populations less than 50,000 and 80% have less than 25,000. Among the 67 counties, 46% (31) have a population of less than 50,000. Central Florida houses a considerable amount of rural and small communities. Furthermore, urban areas such as Orlando are part of the region as well. This makes the region a good representation of the entire state. Second, the state of Florida is one of the nation's most at-risk states. The Central Florida region is subject to a majority of the natural hazards that threaten Florida (Kapucu, 2012). Some of the major hazards prevalent in the region are mentioned below.

Hurricanes in Central Florida

While the response to Hurricane Katrina uncovered many problems, it also made us question our capacity to collect and process information for managing hurricanes (Comfort, 2006; Fitzpatrick, 1999; Kapucu & Özerdem, 2012; Schneider, 1995). Florida is a prominent example for observing how local and state governments deal with hurricanes. For example, in 2004, four major hurricanes struck Florida in rapid succession, three of which hit Central Florida. Hurricane Charley made landfall on August 13, Hurricane Frances on September 5, Hurricane Ivan on September 16, and Hurricane Jeanne on September 25. Collectively, they killed 117 people and caused more than $45 billion in estimated damages (Newman, 2004). One can say that the state of Florida managed these hurricanes well. Unsurprisingly, scholars such as Waugh (2006) mention Florida's emergency management system as one of the best in the nation and suggest that others might consider the state as an example for best practices. Alongside hurricanes, the Central Florida region is also prone to other natural hazards such as wildfires and tornados. The unpredictability of wildfires and tornados are important to analyze because they provide a different emergency management context than those of hurricanes.

Wildfires in Central Florida

Florida is known for its hot climate and remarkable flora, but this also makes wildfires another natural threat for people living in Central Florida. An increase in population and the proximity between wildlife and communities elevates the importance of wildfire resiliency. A document prepared by the Florida Department of Agriculture and Consumer Services, Division of Forestry (FDOF) indicates that approximately 5,500 wildfires occur in Florida every year (FDOF, 2004a). According to the Division of Forestry data, 1,897 fires have occurred within the Orlando District, which includes Seminole, Orange, Osceola, and Brevard counties, since 2000 and burned more than 120,000 acres in these incidences. Also, Orange County Fire Rescue Department, alone, responded to a total of 590 wildfires in 2007 (UCF OEM&RS, 2011).

There have been more than sixty significant wild land fires in Florida since 1981 (FDOF, 2004b). The Everglades fires in the 1970s and the Palm Coast and Black Friday fires in 1985 affected tens of thousands of acres of natural land. The 1985 Palm Coast and Black Friday fires resulted in 250 and 400 home losses respectively (FDOF, 2004a; FBRM, n.d.), and the 1985 Palm Coast fire was the first wildlife/urban interface fire in the history of Florida (FBRM, n.d.). Florida experienced other dangerous wildfires in 1998, which forced more than 100,000 residents to move from their homes. As the Florida Division of Forestry states, wildfires cannot be managed with one single program (FDOF, 2004c), and as the Florida Department of Community Affairs, Bureau of Recovery and Mitigation (FBRM) notes, fire protection systems are very fragmented.[1]

Tornadoes in Central Florida

Florida's climate is suitable for tornados in the summer (June–September) and spring (February–April) seasons (FDEM, 2002, 2009). Tornadoes in the summer season are more frequent and are usually a result of tropical cyclones. In the spring, tornadoes become more powerful by the presence of the jet stream. The National Climatic Data Center (n.d.) reports that fifty-eight fatal tornados have occurred in Florida since 1950. These fifty-eight tornados caused 185 deaths and 1,641 injuries in total, along with $2.1 billion in property damage and $2.5 million in crop damage. The state of Florida experienced the most deadly tornado outbreak in 1998 in Central Florida. Three tornadoes took place in the region between February 22 and 23, which killed forty-two people and injured 256 (*St. Petersburg Times*, 2007; Tornado Project, 2000). During the 1998 tornadoes, 500 buildings were destroyed in Lake County and 625 houses were destroyed in Seminole County. The 1998 tornadoes were the ninth biggest single fatalities-related weather event in the state (NOAA, n.d.). The most recent deadly tornado happened in February 2007 in Central Florida, which killed twenty-one people, damaged 1,500 houses and resulted in $218 million in damages (Forbes, 2007; Simmons & Sutter, 2008).

The state of Florida has lost 204 lives due to tornadoes and 1,918 people have sustained tornado-related injuries since the first year of recording in 1884 (NOAA, 2007). Finally, the 2007 Central Florida tornado event made the local and national news immediately and it was well known to policy makers and the public simultaneously (Collins & Kapucu, 2008).

METHODS

In general terms, the chapter compares the effectiveness of Orange and Sumter counties' informal and formal (i.e., emergency preparedness and response) networks. The data for this study comes from a survey that was a part of a research project funded by the U.S. Department of Agriculture and was administered in eleven counties in Central Florida.[2] The participants were organizations that were identified as a partner in the emergency management systems of study counties. They were identified in each county's comprehensive emergency management plan.

Orange and Sumter counties are compared using two different types of data. The first type of data is relational (i.e., network data) and identifies the structure of organizational relations in the dataset. Three types of relations are examined which are friendship, disaster preparedness, and disaster response. Participating organizations were given a roster of organizations that are a part of the emergency management system in their county, and asked whom they know, whom they work with for disaster response, and whom they work with for disaster preparedness. The second type of data is responses to open-ended questions in the survey.

UCINET software was used for analyzing network data. UCINET (Borgatti, Everett, & Freeman, 2002) is a software commonly used for analyzing relational data. The results of the survey are analyzed using Freeman's degree centrality measure. Furthermore, information about each county's emergency management background was collected. This information comes from personal communication with emergency management officials from each county, county websites, and other county documents.

There are several reasons for choosing Orange and Sumter counties for comparison. First, Orange County is the largest urban county in the region while Sumter county is a remarkably rural county. Further, both counties have sufficient survey response rates which make them better choices than other counties in the region. The response rates for Sumter and Orange counties are 31.58% and 37.88% respectively and these are the highest response rates for rural and urban counties in the study. The following section provides background information about the emergency management systems at Sumter and Orange counties.

Emergency Management at Sumter County

Sumter County was created in 1853 and prior to its foundation was part of Marion County. The county seat is the city of Bushnell. The county covers an area of 545.73 square miles. Sumter County's population was 97,700 in 2010 with 171.2 persons per square mile as of 2010 (Census Bureau, 2011). The majority of the population (90%) lives in unincorporated areas. The county population was 53,545 in 2000; this constitutes an 83.17% population increase between 2000 and 2010 (Sumter County, n.d.). According to the Census Bureau (2010), the 2009 median household income was estimated at $46,586 in Sumter County. Per capita income was $30,253 in 2009 (Office of Economic and Demographic Research, 2011). Major industries that generate employment are construction (24.8%); trade, transportation, and utilities (17.1%); public administration (10.4%); and education and health services (9.6%) 2009 (Sumter County Chamber of Commerce, 2011).

Emergency management was first introduced to Sumter County though the use of a civil defense director in the 1950s. The first full-time emergency management director was brought onto the staff around 1997 (J.

Wright, personal communication, July 2011). Sumter County Emergency Management (SCEM) operates as a division under the Sumter County Sheriff's Office. The Emergency Management Division states its mission as "ensuring Sumter County is prepared to respond to emergencies, recover from them, and mitigate their impact" (Sumter County Sheriff's Office, 2005). SCEM employs four full-time employees and one contracted labor employee. The annual budget may vary but is estimated at a yearly average of $375,000 (J. Wright, personal communication, July 2011).

Emergency operations in Sumter County call for a coordinated effort and gradual response by personnel and equipment from municipal, county, and other disaster support agencies in preparation for, and in response to, local disasters. If the county's resources are overwhelmed, assistance will be requested from the State Emergency Operations Center (SEOC) (SCEM, 2011). In the event of a disaster, the Emergency Operations Center "will be activated for all incidents requiring a significant dedication of resources and/or extraordinary inter-agency coordination outside the realm of normal, day to day emergency situations responded to by law enforcement, fire, and EMS agencies" (SCEM, 2011).

Emergency Management at Orange County

Orange County was originally named Mosquito County when created in 1824 and was later changed to Orange County in 1845. Originally the name was related to the large production of oranges in the area; today there are no orange groves and only a few packing and distribution centers (County Florida, n.d.). Currently Orange County's population is expanding. In 2010 the population was 1,145,956, with an additional seasonal population of 47.78 million domestic visitors and 3.675 million international visitors (Census Bureau, 2012; Visit Orlando, 2012). Due to population expansion steps have been taken which consider future land use and the preservation of rural settlements that are still thriving.

Orange County recognizes that rural areas are being lost quickly to urban sprawl and an increase in population to the area. From 2000 to 2010 there was a 27.8% increase in population (Census Bureau, 2012). To address these issues, "the designation of Rural Settlements on the Future Land Use Map of the Orange County 1990–2010 Comprehensive Policy Plan, the development of related future land use policies, protection of some Rural Settlements included in Joint Planning Area Agreements with municipalities, and creation of Preservation Districts" have been created (OC Planning, 2003).

Emergency management was first introduced to Orange County though the implementation of a volunteer civil defense director in October of 1950 (R. Plummer, personal communication, July 2011). The Orange County Office of Emergency Management (OCOEM) is a branch of Orange County Fire Rescue which is under the guidance of Orange County's Public Safety

Division (Orange County, 2011). OCOEM is comprised of seven employees and two student interns. The annual budget may vary but is estimated at a yearly average of $1.39 million (R. Plummer, personal communication, July 2011). Considering the potential hazards, Orange County is susceptible to a wide range of geological and man-made threats. Within the Local Mitigation Strategy five hazards were identified as having a high level of vulnerability and probability and affecting the entire county. These are: flooding, lightening, hurricanes and tropical storms, and wild land fires (OCOEM, 2009). During a disaster the county mayor assumes the responsibility of director of emergency management. These duties include declaring a Local State of Emergency, ordering evacuations and curfews, entering into contracts, and receiving and distributing resources, to name a few (OCOEM, 2008).

OCOEM's role in a disaster is to "coordinate the activities of the County's Emergency Response Team (OCERT) prior to and after a disaster" (OCOEM, 2008). These duties include: maintaining the Planning Program which includes the creation of all guiding documents, training, and exercises; the Homeland Security Program which coordinates OCERT; the Mitigation Program; the Citizens Corps Council which educates and trains local citizens; the Hazardous Materials Program; the Sheltering Program; the Grants Program; and maintaining the Communications and Warning system for the county (OCOEM, 2008). The following section presents the analysis and findings of the SNA survey for Sumter and Orange counties.

ANALYSIS AND FINDINGS

Network analysis findings present an opportunity to compare the networks of Orange and Sumter counties. Because of a relatively low response rate in both counties, the networks are analyzed at the individual level (i.e., ego networks) rather than from a socio-centric view. Organizations' ego networks are compared based on their connectedness and structure. Also, similar organizations (e.g., Emergency Management Office, Health Department) are compared according to structural equivalence. This section presents first, an overview of the friendship, preparedness, and response networks in Orange and Sumter counties. Second, individual organizations' ego networks are compared alongside organizational attributes.

In the friendship network, Orange County Office of Emergency Management (OCOEM) has the highest number of outgoing ties. OCOEM identified all organizations in the roster as a partner that they know. Orange County 311 (OC 311), the Animal Services Division (OCAnmlSD), the Sherriff's Office (OCSO), and Orange County Board of County Commissioners (OCBCC) hold the second through fifth places in the friendship network respectively.

Characteristics of the preparedness network are significantly different than the friendship network. Naturally, OCOEM is the most connected actor in the network. It has the highest number of outgoing ties and the organization identified almost every other organization in the roster as their partner in disaster preparedness. OCBCC and OCSO have the second highest number of outgoing ties in the network which are followed by OCUD and OCHRD.

Table 17.1 Out-degree Measures of Organizations that Responded to Survey in Orange County

Friendship Network			*Preparedness Network*			*Response Network*		
	OutDeg	*nOutDeg*		*OutDeg*	*nOutDeg*		*OutDeg*	*nOutDeg*
OCOEM	64	81.01	OCOEM	64	81.01	OCOEM	51	64.55
OC 311	58	73.41	OCBCCC	42	53.16	OCBCCC	43	54.43
OCAn-mlSD	57	72.15	OCSO	42	53.16	OCSO	42	53.16
OCSO	54	68.35	OCUD	37	46.83	OCHRD	35	44.30
OCBCCC	54	68.35	OCHRD	37	46.83	OCUD	34	43.03
OFD	51	64.55	OC 311	30	37.97	OCCFU	30	37.97
OCNPRD	49	62.02	OCISS	30	37.97	PD	25	31.64
OCHFSD	48	60.75	OCNPRD	24	30.38	OCISS-ES	23	29.11
VC	43	54.4	OCPAO	23	29.11	OCCD	22	27.84
OCSMD	42	53.16	OCCD	22	27.84	OCNPRD	20	25.31
OCCFU	36	45.57	OCHFSD	21	26.58	OCGMD	18	22.78
OCCD	31	39.24	OCGMD	17	21.51	OCHFSD	14	17.72
OCPS	31	39.24	OFD	16	20.25	OFD	14	17.72
OUC	30	37.97	CoAFD	15	18.98	OC 311	13	16.45
OCUD	29	36.70	OCPS	12	15.19	OCPS	12	15.19
FAoF	27	34.17	OUC	11	13.92	OUC	11	13.92
OCPAO	26	32.91	OCAn-mlSD	10	12.65	OCPSTS	11	13.92
SA	26	32.91	PD	9	11.39	OCAn-mlSD	10	12.65
OCPSTS	25	31.64	VC	8	10.12	VC	9	11.39
OOCEA	23	29.11	OCPSTS	6	7.59	SA	9	11.39
PD	20	25.31	SA	3	3.79	OOCEA	7	8.86
OCISS-ES	19	24.05	OOCEA	2	2.53	OCSMD	2	2.53
OCISS	18	22.78	OCSMD	2	2.53	OCPAO	1	1.26
OCGMD	17	21.51	OCISS-ES	1	1.26			
CoAFD	11	13.92						
OCHRD	11	13.92						

In the response network, the composition of the network is different than previous results. Orange County Office of Emergency Management (OCOEM) has the highest number of outgoing ties and Orange County Board of County Commissioners (OCBCC) follows it. It is important to note that OCBCC participated in the survey even though it was not identified as an official partner for disaster response in the CEMP. The Orange County Sherriff's Office (OCSO) has the third highest number of outgoing ties which has only one link less than OCBCC. Orange County Human Resources Division (OCHRD) and Orange County Consumer Fraud Unit (OCCFU) hold the fourth and fifth places in terms of number of outgoing connections.

The network analyses of Sumter County's data present somewhat different results than of Orange County. In the friendship network, Sumter County Health Department (SCHD) has the highest number of outgoing ties and Sumter County Emergency Management (SCEM) follows it. Sumter County School District (SCSD), Sumter County Public Works Departments (SCPW), and Amateur Radio Emergency Services (ARES) are the other actors in the network that have a relatively high number of connections to others. In addition to these actors, Sumter County Housing (SCH) and the University of Florida Extension (UFE) identified a significant number of actors in the network as their friends. Sumter County Public Works Department (SCPW) has the highest number of outgoing ties with SCEM holding second place. UFE, SCHD, and SCBCC hold third, fourth, and fifth places respectively.

Table 17.2 Out-degree Measures of Organizations that Responded to Survey in Sumter County

Friendship Network			*Preparedness Network*			*Response Network*		
	OutDeg	*nOutDeg*		*OutDeg*	*nOutDeg*		*OutDeg*	*nOutDeg*
SCHD	54	90	SCPW	49	81.66	SCEM	44	73.33
OsCEM	52	86.66	SCEM	38	63.33	SCHD	42	70
SCEM	45	75	UFE	35	58.33	SCBCC	18	30
SCSD	37	61.66	SCHD	23	38.33	SCSD	12	20
UFE	36	60	SCBCC	18	30	ARES	11	18.33
SCH	29	48.33	ARES	16	26.66	SCCST	1	1.66
SCBCC	18	30	SCSD	16	26.66			
SCPW	8	13.33	OsCEM	10	16.66			
ARES	8	13.33	SCCSAC	7	11.66			
SCCSAC	6	10	SCCST	5	8.33			
SCCST	3	5						

For emergency response, SCEM has the highest number of outgoing ties (44) which is followed by SCHD (42). There is a significant drop in the number of connections for actors in third place. SCBCC has eighteen, SCSD has twelve, and ARES has eleven connections in the network. There is another sharp decline from ARES to Sumter County Community Services-Transit (SCCST). SCCST has only one outgoing tie. ARC has the highest number of incoming ties (six), followed by the Sherriff's Office (five). The remaining actors have similar values of incoming ties as can be understood from the standard deviation for in-degree statistics. SCHD and SCEM are the most central actors in the network.

Descriptive statistics for the responding organizations in Orange and Sumter counties are presented in Table 17.3. The average number of ties per responder in Orange County is 34.61 for the friendship network but it declines to 20.16 in the preparedness network, and falls down to 19.82 in the response network. The standard deviation for friendship and preparedness networks are quite similar while the response network has a lower standard deviation. Interestingly, the number of responses declines in preparedness and response networks. In response to the friendship network, twenty-six organizations responded to the survey, however, three out of twenty-six organizations did not identify others as a partner for disaster preparedness. FAoF responded only to friendship network questions, CoAFD and OCISS responded to friendship and preparedness network questions, and OCCFU responded to friendship and response network questions.

Table 17.3 Descriptive Statistics for Responding Organizations

		Friendship		*Preparedness*		*Response*	
		OutDeg	*Nrm OutDeg*	*OutDeg*	*Nrm OutDeg*	*OutDeg*	*Nrm OutDeg*
Orange	Mean	34.61	43.82	20.16	25.53	19.82	25.10
	s.d.	15.64	19.80	15.83	20.04	13.61	17.24
	Max	64.00	81.01	64.00	81.01	51.00	64.56
	Min	11.00	13.92	1.00	1.27	1.00	1.27
	Range	53.00	67.09	63.00	79.75	50.00	63.29
	N of Obs	26.00	26.00	24.00	24.00	23.00	23.00
Sumter	Mean	26.90	44.85	21.70	36.17	21.33	35.56
	s.d.	19.19	31.98	14.53	24.22	23.30	29.43
	Max	54.00	90.00	49.00	81.67	44.00	73.33
	Min	3.00	5.00	5.00	8.33	1.00	1.67
	Range	51.00	85.00	44.00	73.33	43.00	71.67
	N of Obs	11.00	11.00	10.00	10.00	6.00	6.00

In Sumter County, the average number of ties per respondent is 26.9 for the friendship network with a standard deviation of 19.19. The average number of ties is slightly more than twenty-one ties per organization for preparedness and response networks. However, standard deviations for preparedness and response networks are significantly different from each other. The standard deviation for the preparedness network is 14.53 while it is 23.3 for the response network. SCPW and SCH have responded only to friendship network questions, OsCEM, SCCSAC, and UFE have responded to friendship and preparedness network questions, and SCPW responded only to preparedness network questions.

Orange and Sumter counties have their unique network structures and participants. Based on the responses to the survey, there are four organizations that are common in both of the networks within the counties. These organizations are the County Emergency Management Office, county public schools, Health Department, and Board of County Commissioners. The primary reason for having four comparable organizations is the relatively low response rate. In the case of a complete response, a greater number of organizations could be compared. When the four organizations' ego networks are compared from a structural equivalence standpoint, notable differences between similar organizations are identified. The organizations are-compared according to their normalized out-degree values. Normalized values show the percentage of the number of ties that an actor has with respect to the entire network. The comparison of four organizations for Orange and Sumter counties is presented in Table 17.4.

Emergency management offices in both counties have similar types of ego networks. Both organizations have connections with a majority of the organizations in the roster. Orange County Office of Emergency Management (OCOEM) has ties with 98.46% of the organizations in the roster for friendship and preparedness networks, but this falls to 78.46% for the response network. Sumter County Emergency Management (SCEM) has identified 80.35%, 67.85%, and 78.57% of the organizations as a partner for friendship, preparedness, and response networks respectively.

Orange County Health and Family Services Department (OCHFSD) has a comparatively small role in disaster preparedness and response. OCHFSD has identified 73.84% of the organizations in the roster as a friend but only 32.30% of these organizations were identified as a partner for preparedness and 21.53% were identified for response. However, Sumter County Health Department (SCHD) has a distinctive role in both preparedness and response. SCHD has identified 96.43% of the organizations in the roster as a friend. The percentage of partners in the roster falls to 41.07% for preparedness but it is again rises to 75% in disaster response. This means SCHD has responsibilities in disaster response which are almost equivalent to SCEM.

Orange County Public Schools (OCPS) and Sumter County School District (SCSD) have similar network structures in terms of the number of

Table 17.4 Comparison of Similar Organizations in Both Counties

		Friendship		*Preparedness*		*Response*	
		OutDeg	*Nrm OutDeg*	*OutDeg*	*Nrm OutDeg*	*OutDeg*	*Nrm OutDeg*
Emergency Management	Orange	64	98.46	64	98.46	51	78.46
	Sumter	45	80.35	38	67.85	44	78.57
Health Department	Orange	48	73.84	21	32.30	14	21.53
	Sumter	54	96.42	23	41.07	42	75.00
Public Schools	Orange	31	54.38	12	18.18	12	18.18
	Sumter	37	66.07	16	28.57	12	20.00
Board of County Commissioners	Orange	54	94.73	42	63.63	43	54.43
	Sumter	18	31.57	18	31.57	18	30.00

connections and the distribution of ties in friendship, preparedness, and response networks. SCSD has more responsibilities than OCPS in all networks. SCSD has identified 66.07% of the organizations in the roster, while OCPS has identified 54.38%, in the friendship network. In the preparedness network, SCSD and OCPS have identified 28.57% and 18.18% of the organizations as a partner respectively. In the response network, they have identified almost an equal percentage of organizations as a partner. SCSD has identified 20% while OCPS has identified 18.18% of the organizations as a partner for disaster response.

The Board of County Commissioners in Orange and Sumter counties have responsibilities at different weights. County commissioners in Orange County have a greater share in disaster preparedness and response than the Sumter County commissioners. Orange County Board of County Commissioners (OCBCC) has identified 94.73% of the organizations in the roster as a friend whereas this ratio is 31.57% for Sumter County Board of County Commissioners (SCBCC). OCBCC has also identified 63.63% and 54.43% of the organizations as a partner for disaster preparedness and response respectively. SCBCC, on the other hand, has identified 31.57% and 30% of the organizations in the roster as a partner for disaster preparedness and response respectively.

The analysis of comparable organizations implies that both county emergency management organizations have similar positions in disaster preparedness and response networks. Nevertheless, Sumter County Health Department has a significant role particularly in disaster response. Orange County BCC is significantly involved in preparedness and response while Sumter County BCC has a comparatively small role in preparedness and response. Also the attributes of the top five organizations show that Orange and Sumter networks have different characteristics. Orange County

Sherriff's Office (OCSO), Utilities Department (OCUD), and Human Resource Department (OCHRD) have a notable role in preparedness and response in Orange County. In Sumter County, the Public Works Department (SCPW), Health Department, and School District have important roles in the network. Remaining organizations in the network have a very small number of ties in disaster preparedness and response when compared to these three organizations.

From a resource dependence perspective it is safe to state that organizations in counties with greater resources have more flexibility to be involved in relationships with other organizations for disaster preparedness and management purposes. Although organizations identify others as a friend, this relationship does not always transform into partnerships. Qualitative data collected in this study support these discussions. In one of the open-ended questions, participants were asked about their opinion with regard to the obstacles of building disaster-resilient communities. The majority of the participants who replied to this question from Orange County (eleven out of sixteen responses) and one participant from Sumter County (one out of six responses) highlighted a lack of funding. One of the respondents stated that "ignorance and hypocrisy (expecting resilience while refusing to pay for it)" is an obstacle for resilient communities and continued that emergency managers are "practitioners with starved budgets who, despite becoming more and more efficient year after year, simply cannot keep up with rising professional standards and the additional risk caused by crumbling infrastructure, climate change, urban sprawl, and growing poverty." Another respondent emphasized the lack of concern for disasters in other sectors and mentioned that "eliminating community apathy toward disasters" is necessary and added that the "business community needs to be better prepared." Similar to these comments, a third participant underlined that "getting the cooperation of other organizations in the community" is an obstacle for building disaster-resilient communities. Additionally, four participants underlined apathy towards disasters, two underlined a loss of institutional memory in organizations, and another two underlined lack of trust in the community as other obstacles to disaster resiliency.

Involvement in a collaborative network requires substantial resources. Lack of funding for emergency management business appears to be the primary reason for highly centralized networks in preparedness and response actions. Organizations which are inherently responsible for emergency management incur most of the burden while others are likely to participate in emergency management efforts if they have resources to do so. Additionally, apathy and complacency prevent individuals and community organizations from taking responsibility for emergency management processes. Loss of institutional memory and lack of trust between individuals and organizations are significant inhibitors of collaboration in emergency management networks. Connections between individuals are oftentimes the starting point for interorganizational collaborative relationships, and emergency managers can be looked at as the coordinators of this process. These individuals also constitute the institutional memory of their organizations. Loss of individuals who maintain collaborative relationships between organizations result in the disappearance of ties between organizations; budget cuts and departmental reorganization can contribute to these issues. Building new relationships with other organizations requires significant trust. Lack of trust thereof leads to an emergency management network in which most of the organizations are tied to a limited number of organizations who have relatively high resources, are trustworthy, and keep up with relations (no loss of institutional memory). This also explains the significant difference between the friendship networks and work-related networks (i.e., preparedness and response).

Another finding of the survey is the discrepancy between the list of survey participants and the roster of emergency management organizations. This happened due to three reasons: 1) the responders did not print the name of the organization they represent completely (e.g., UFE), 2) the responders were listed in other emergency management plans such as the local mitigation strategy but not the CEMP, 3) the responders were part of the emergency management system without being mentioned in an official document. The Board of County Commissioners for both counties are prominent examples of the last example case. Because emergency managers are accountable to elected officials, it is understandable to have them in their network. That may lead to a lack of institutionalized relationships between partners. Consequently, the life span of these relationships is questionable.

When the two counties are compared, Orange County has less centralized work networks than Sumter County. This can be explained with the availability of resources in urban counties. Sumter County has more centralized networks for friendship, preparedness, and response than Orange.

Planned and Perceived Networks

As mentioned by Kapucu and Demiroz (2011), the convergence of actual emergency management networks with planned networks is an indicator of effectiveness of that network. Following this idea, we compare planned networks for Orange and Sumter counties based on the CEMPs, with the networks based on survey results. This is a modest attempt to compare two networks based on their structure and there is a need for a theoretical basis to assess the effectiveness of survey-based networks. In these comparisons we are aiming to make a methodological presentation for future researchers that would focus on the effectiveness of emergency management networks.

Similar to our previous analysis, we focus on ego networks rather than socio-centric networks due to response rate issues. In our analysis, we connected the organizations that are located in the same ESF process. That means a link is drawn between an organization and the other organizations

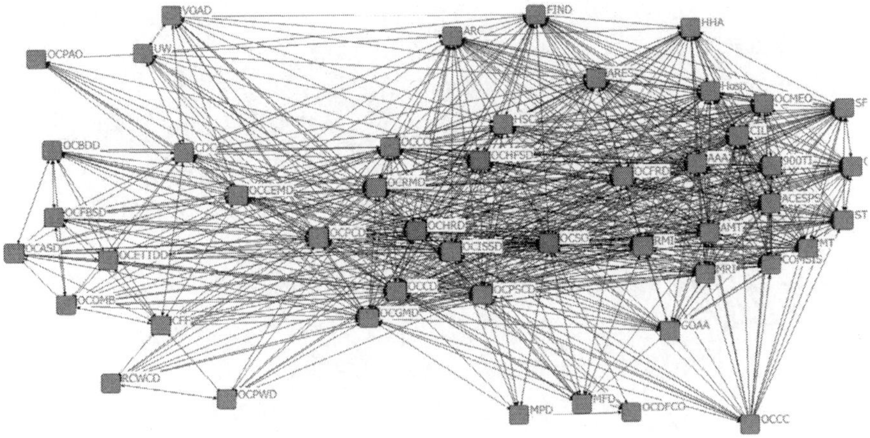

Figure 17.1 Ego network of OCHRD and OCSO based on Orange County CEMP.

that are mentioned in the same support function. Based on this analysis, we discovered the networks of organizations in both counties. For the sake of simplicity and response rate issues, we choose OCSO and OCHFSD from Orange County, and SCHD and SCEM from Sumter County for analyzing ego networks. Figures 17.1 and 17.2 compare planned and survey-based ego networks.

There is an observable difference between the planned and survey-based networks. Both organizations are strongly connected to their partners in

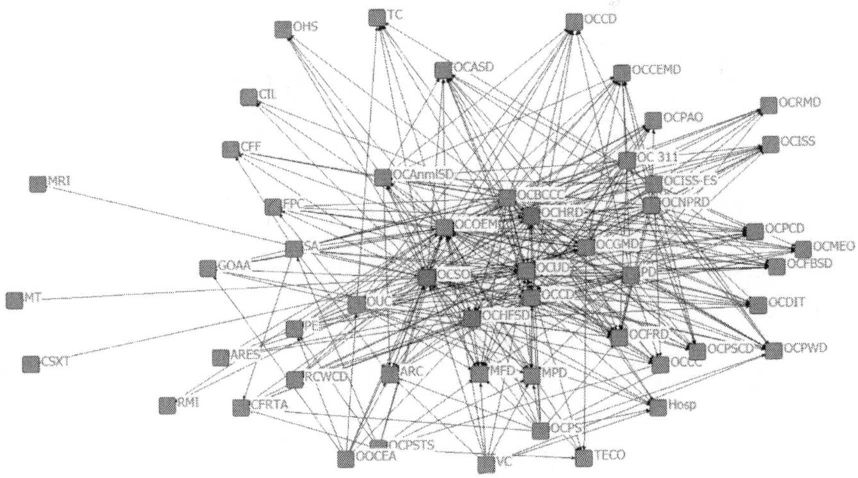

Figure 17.2 Ego network of OCHRD and OCSO based on the Orange County survey response.

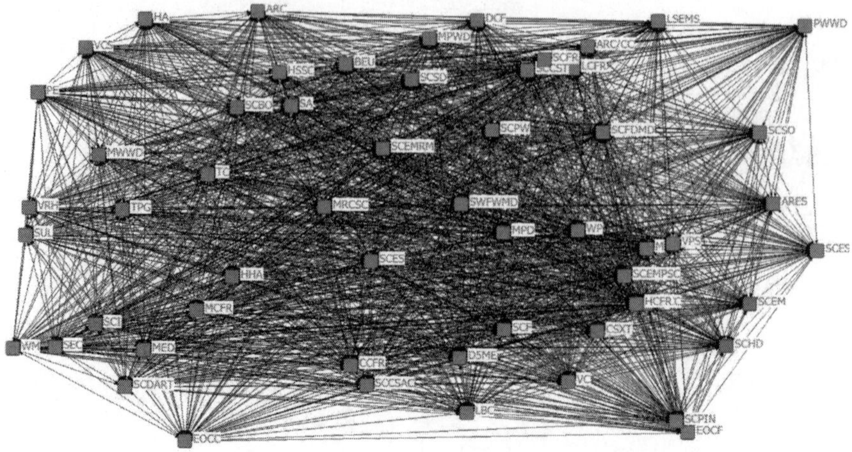

Figure 17.3 Ego network for SCEM and SCHD based on the Sumter County CEMP.

the plan and there is a decline in the level of connectedness in the survey. Nonetheless, the difference is not significant when compared to organizations in Sumter County. The planned and survey-based ego networks for SCHD and SCEM are presented in Figures 17.3 and 17.4.

Sumter County CEMP consists of seventeen ESFs. In three of these support functions, all of the organizations were given either primary or supporting roles. That requires all of the organizations to be connected to each other. Consequently, Figure 17.3 shows a network where all possible connections

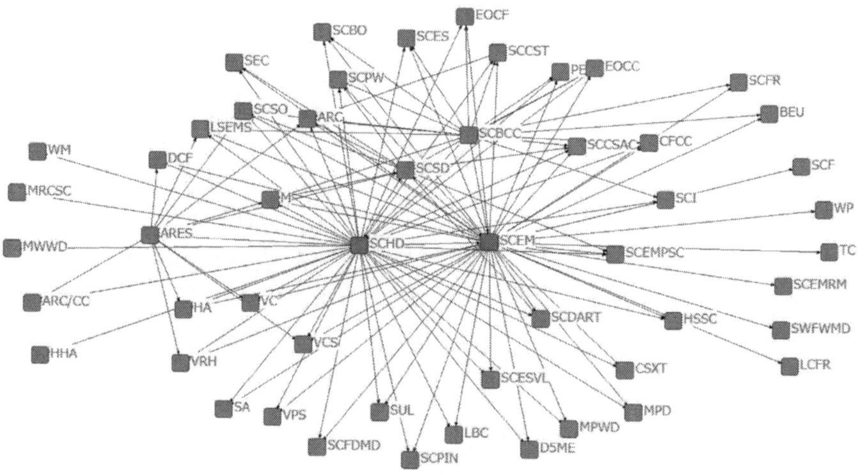

Figure 17.4 Ego network for SCEM and SCHD based on the Sumter County survey response.

are actualized (even though it is an ego network). On the other hand, survey-based ego networks shown in Figure 17.4 depict a significantly different network structure. SCEM and SCHD are in the center of the network and they are serving as a bridge for connecting almost all other organizations in the network. The difference between the planned and survey-based networks could partially be explained with the response rate problem, yet the way the ESFs were structured appears to be the primary reason for this.

CONCLUSION

This paper presented a comparison of the emergency management networks of Orange and Sumter counties from Central Florida. Three different networks were analyzed using survey data: friendship, preparedness, and response. Both counties have centralized networks, meaning a majority of the connections in the network concentrate on a limited number of actors. Nevertheless, Orange County has less centralized networks when compared to Sumter County. Responses to open-ended questions highlighted a lack of funding, apathy towards disasters, loss of institutional memory, and a lack of trust as inhibitors of community disaster resiliency from an organizational perspective. These inhibitors also shape the emergency management networks of each county. Due to a lack of resources, the Sumter County network is centralized to a limited number of organizations while Orange County has a relatively abundant amount of organizations to share the burden within the network. Another finding of the study shows that there is discrepancy between the list of organizations that were identified in the CEMPs and the organizations that participated in the survey. The participants who were not listed in the CEMPs are also part of the emergency management system; however, their roles were not institutionalized (i.e., formalized). The sustainability of the relationships with these organizations is subject to weaken over time unless they are institutionalized.

ACKNOWLEDGEMENTS

We would like acknowledge the U.S. Department of Agriculture as the funder of this project (Award No. 2010–67023–21698). Also, we greatly appreciate input of Orange County Emergency Manager Dave Freeman and Sumter County Emergency Manager Dave Casto.

NOTES

1. The three types of fire protection systems include city fire departments which are not responsible for fires out of their districts; forestry departments which are not responsible for structural fires; and rural fire protection

districts which provide full-time protection (FBRM, n.d.). Unfortunately, due to low population density, rural areas often cannot afford fire protection districts.

2. U.S. Department of Agriculture Award No. 2010–67023–21698.

REFERENCES

Ansell, C., & Gash, A. (2007). Collaborative governance in theory and practice. *Journal of Public Administration Research and Theory*, 18(4), 543–571.

Borgatti, S.P., Everett, M.G., & Freeman, L.C. (2002). *Ucinet for Windows: Software for social network analysis*. Harvard, MA: Analytic Technologies.

Census Bureau. (2000). *United States Census 2000*. Retrieved December 6, 2009 from http://www.census.gov/main/www/cen2000.html.

Census Bureau. (2010). *Estimates for Florida counties, 2009, in small area income and poverty estimates*. Retrieved July 22, 2011 from http://www.census.gov/cgi-bin/saipe/saipe.cgi?year=2009&type=county&table=county&submit=States%20%26%20Counties&display_data=Display%20Data&state=12&areas=12119.

Census Bureau. (2011). *Sumter County, Florida*. Retrieved July 22, 2011 from http://quickfacts.census.gov/qfd/states/12/12119.html.

Census Bureau. (2012). *Orange County, Florida*. Retrieved May 14, 2012 from http://quickfacts.census.gov/qfd/states/12/12095.html.

Collins, M.L., & Kapucu, N. (2008). Early warning systems and disaster preparedness and response in local government. *Disaster Prevention and Management: An International Journal*, 17(5), 587–600.

County Florida. (n.d.). *Orange County Florida*. Retrieved May 2011 from http://orange.county-florida.com/.

Cross, R., Borgatti, S.P., & Parker, A. (2002). Making invisible work visible: Using social network analysis to support strategic collaboration. *California Management Review*, 44(2), 25–46.

FBRM (State Of Florida Department of Community Affairs Bureau of Recovery & Mitigation). (n.d.). *Wildfires*. Retrieved October 30, 2009 from http://www.sjcemergencymanagement.org/wildfires.html#Wild.

FDEM (Florida Division of Emergency Management). (2002). *Florida hazard watch: Tornadoes*. Retrieved November 12, 2009 http://www.floridadisaster.org/bpr/EMTOOLS/severe/tornadoes.html.

FDEM (Florida Division of Emergency Management). (2009). Retrieved November 15, 2009 from http://www.floridadisaster.org/bpr/EMTOOLS/Severe/tornadoes.htm.

FDOF (Florida Division of Forestry). (2004a). *Wildfire mitigation in Florida*. Retrieved October 30, 2009 from http://www.fl-dof.com/wildfire/wf_pdfs/Wildfire_Mitigation_in_FL.pdf.

FDOF (Florida Division of Forestry). (2004b). *Wildland fire*. Retrieved October 30, 2009 from http://www.fl-dof.com/wildfire/stats_significant_fires.html.

FDOF (Florida Division of Forestry). (2004c). *Wildland fire and fire protection*. Retrieved October 30, 2009 from http://www.fl-dof.com/wildfire/index.html.

Fitzpatrick, P. (1999). *Natural disasters: Hurricanes*. Santa Barbara, CA: ABC-CLIO.

Forbes, G. (2007). *Deadly Florida tornado outbreak*. Retrieved November 16, 2009 from http://www.weather.com/blog/weather/8_11687.html.

Gulati, R., & Gargiulo, M. (2001). Where do interorganizational networks come from? *American Journal of Sociology*, 104(5), 1439–1493.

Harris, J., & Clements, B. (2007). Using social network analysis to understand Missouri's system of public health emergency planners. *Public Health Reports*, 122, 488–498.
Henry, A.D. (2011). Ideology, power, and the structure of policy networks. *Policy Studies Journal*, 39(3), 361–383.
Kapucu, Naim. (2012). Disaster Resilience and Adaptive Capacity in Central Florida, US, and in Eastern Marmara Region, Turkey. *Journal of Comparative Policy Analysis: Research & Practice*. 14(3): 202–216.
Kapucu, N., Arslan, T., & Demiroz, F. (2010). Collaborative emergency management and national emergency management network. *Disaster Prevention and Management*, 19(4), 452–468.
Kapucu, N., & Demiroz, F. (2011). Measuring performance for collaborative public management using network analysis methods and tools. *Public Performance Management Review*, 34(4), 551–581.
Kapucu, N., & Özerdem, A. (2012). *Managing emergencies and crises*. Burlington, MA: Jones & Bartlett Learning.
National Climatic Data Center. (n.d.) *Storm events for Florida*. Retrieved November 12, 2009 from http://www4.ncdc.noaa.gov/cgi-win/wwcgi.dll?wwevent~storms.
Newman, J. (2004). Numbers tell story of season's destruction. *Orlando Sentinel*, November 29. A10.
NOAA (National Oceanic and Atmospheric Administration). (2007). *Known Florida tornado deaths 1882–2000*. Retrieved from http://www.srh.noaa.gov/mlb/PDFs/florida_tornado_death_stats.pdf.
NOAA (National Oceanic and Atmospheric Administration). (n.d.). *Event summary*. Retrieved November 16, 2009 from http://www.nws.noaa.gov/om/data/pdfs/ctrlfl.pdf.
OCOEM (Orange County Office of Emergency Management). (2008). *Orange County comprehensive emergency management plan*.
OCOEM (Orange County Office of Emergency Management). (2009). *Orange County local mitigation strategy*.
Office of Economic and Demographic Research. (2011). Sumter County. Retrieved July 22, 2011 from http://edr.state.fl.us/Content/area-profiles/county/sumter.pdf.
Orange County Planning Division. (2003). *Rural resettlement study*. Retrieved 3 May, 2009 from http://www.orangecountyfl.net/Portals/0/Resources/Internet/DEPARTMENTS/Growth_Management/Planning/compplan/ruralsettlement.pdf.
Orlando Sentinel. (2007). February 2–9 issues.
Podolny, J., & Page, K. (1998). Network forms of organization. *Annual Review of Sociology*, 24, 57–76.
Provan, K.G., & Kenis, P. (2008). Modes of network governance: Structure, management, and effectiveness. *Journal of Public Administration Research and Theory*, 18(2), 229–252.
SCEM (Sumter County Emergency Management). (2011) *Sumter County comprehensive emergency management plan*.
Schneider, S.K. (1995). *Flirting with disaster: Public management in crisis situations*. Armonk, NY: M.E. Sharpe.
Simmons, K.M., & Sutter, D. (2008). Manufactured home building regulations and the February 2, 2007 Florida tornadoes. *Natural Hazards*, 46(3), 415–425.
St. Petersburg Times. (2007). Storm Q&A. February 2. Retrieved November 16, 2009 from http://www.sptimes.com/2007/02/02/State/Storm_QA.shtml.
Sumter County. (n.d.). *Sumter County, Florida population table*. Retrieved July 22, 2011 from http://sumtercountyfl.gov/DocumentView.aspx?DID=2271.

Sumter County Chamber of Commerce. (2011). *Sumter for business.* Retrieved July 22, 2011 from http://www.sumterchamber.org/sumter-business.asp.

Sumter County Sheriff's Office. (2005). *Sumter County emergency management.* Retrieved July 22, 2011 from http://www.sumtercountysheriff.org/emergency-management.asp.

Thomson, A.M., Perry, J.L., & Miller, T.K. (2009). Conceptualizing and measuring collaboration. *Journal of Public Administration Research and Theory*, 19(1), 23–56.

Tornado Project. (2000). *Most important US tornadoes by states.* Retrieved November 16, 2009 from http://www.tornadoproject.com/alltorns/worstts.htm#FL.

UCF OEM&RS (University of Central Florida Office of Emergency Management and Natural Resources). (2011). *University of Central Florida wildfire response plan.* Retrieved May 14, 2012 from http://emergency.ucf.edu/Plans/UCF%20Wildfire%20Response%20Plan%202011.pdf.

Visit Orlando. (2012). *Visitor volumes.* Retrieved May 14, 2012 from http://corporate.visitorlando.com/research-and-statistics/orlando-visitor-statistics/visitor-volumes/.

Wasserman, S., & Faust, K. (1994). *Social network analysis: Methods and applications.* Cambridge: Cambridge University Press.

Waugh, W.L. (2006). The political costs of failure in the Katrina and Rita disasters. *Annals of the American Academy of Political and Social Science*, 604, 10–25.

18 Emerging Research in Disaster Resiliency and Sustainability

Implications for Policy and Practice

Naim Kapucu, Christopher V. Hawkins, and Fernando I. Rivera

Disaster resiliency has gained significant attention among scholarly and practitioner communities in the U.S. and abroad. The chapters in this book provide different definitions of the concept of resiliency and offer individual, organizational, and community-level analyses. In addition, the book empirically examines disaster resiliency from an interdisciplinary perspective and applies different methods of analysis (e.g., quantitative, qualitative, and network analyses). The chapters identify a range of factors, including policy tools , and governance mechanisms that are important for improving disaster resiliency.

In this chapter we briefly note some key findings and their implications for policy, practice, and research. We do not intend to repeat the details of each chapter, but rather, highlight important results related to improving disaster resiliency. We recommend readers to visit the chapters for a comprehensive understanding of the issues discussed in the book.

The design and planning of resilient communities are highly complex and require interdisciplinary perspectives. In the first section of the book we learned the need to create resilient communities from a "whole community" perspective. This view emphasizes the importance of stakeholder involvement in the design and implementation of disaster-resilient communities. The section also highlights the importance of understanding the vulnerabilities and perceptions unique to each community. The key role that nonprofit and other community-based organizations can have in building disaster-resilient communities is also discussed.

The second section of the book details the importance of planning processes and regulations for disaster resiliency. Empirical findings suggest that for communities to become resilient, environmental and ecological factors must be considered in long-term planning. This, of course, is not easily accomplished. There is a great deal of uncertainty that surrounds long-term comprehensive planning decisions. Nonetheless, communities can improve their resiliency by employing inclusive planning processes, maintaining key environmental features, and recognizing the importance of regulations that mitigate disaster impacts. This section recommends being adaptive and context-sensitive in long-term planning.

Finally, the last section of the book provides an overview of the importance of networks in building disaster-resilient communities. We learned that stakeholders from multi-jurisdictions and multi-sectors need to be involved in the planning and implementation for disaster resiliency. Cross-sector (public, private, and nonprofit, including faith-based) and inter-organizational coordination play a significant role, especially in disaster response and recovery.

The individual chapters in this book contribute to disaster resiliency from both discipline-specific and interdisciplinary perspectives. The ultimate goal of this research is to put forward a clear and concise framework to study disaster resilience. The development of this framework is important for bridging the gap between practice and research. The framework, called the *adaptive resiliency framework*, is presented below.

The framework was created during facilitated discussions among academics and practitioners at the Disaster Resiliency Workshop at the University of Central Florida. After the workshop, Robert Deyle, with contributions from William Butler, further developed and detailed the framework. The framework presented in Figure 18.1 is designed to provide guidance for future empirical work on resiliency. A key element of the framework is integrating learning and adaption into the traditional phases of disaster

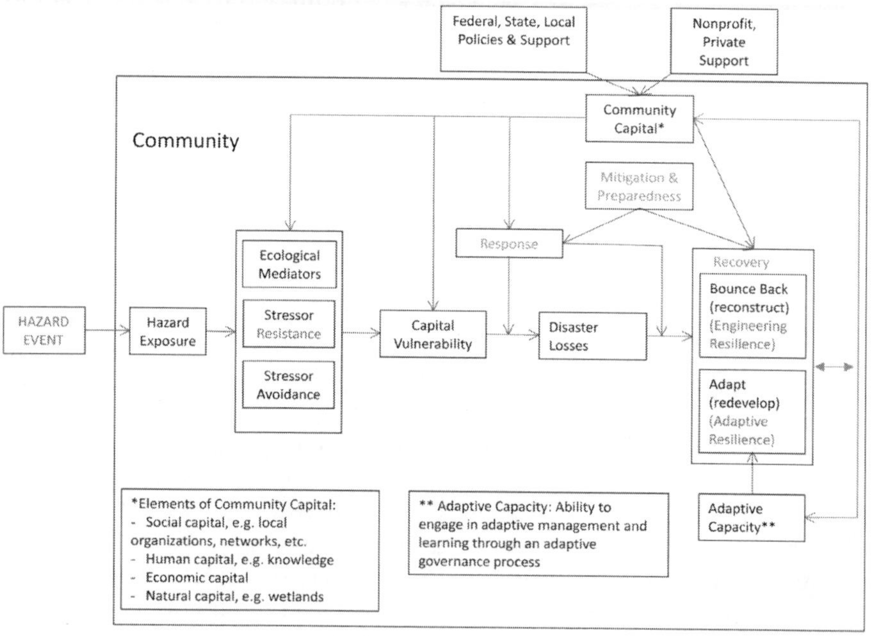

Figure 18.1 Adaptive resiliency framework.

management (mitigation, preparedness, response, and recovery) in response to a stressor (hazard event). The framework can be applied to either urban or rural contexts.

During the development of the framework, the distinction between disaster resilience and hazard resilience was raised by several of the workshop participants. As suggested by some at the workshop, a way to address the question of whether or not preparedness, response, and mitigation are elements of community resilience might be to differentiate disaster resilience from hazard resilience.

If we define the stressor as the losses that occur as a result of a disaster, then the conventional definition of resilience is the ability to "bounce back" (engineering resilience). This confines the term "resiliency" to the community's behavior after the disaster. From this perspective individual- and community-level activities focus on returning social, economic, and environmental conditions to their previous state ("back to normal"). However, if we apply the adaptive resiliency concept (as defined in Chapter 10) to disaster losses as the stressor, then disaster resilience is the ability to adapt through the redevelopment of the community in ways that reflect the community's values and goals, and its evolving understanding of external forces with which it must contend. This also incorporates the evolving understanding of how to move in a desirable direction to sustain positive community structures and functions and remove the negative ones. This refocuses attention to the "community capital" that can be brought to bear on preparing/planning for and implementing activities during the recovery process. This capital is partly influenced by local socio-economic conditions and by external forces such as state and federal-level policies and resources. Support from nonprofit and community-based organizations are also critical elements of community capital.

From this perspective, where the community chooses an adaptive rather than engineering resilience approach, disaster resilience is considered a function of the community's adaptive capacity. This capacity helps the community engage in adaptive management and continuous learning through an adaptive governance process. In this framework, there is a reciprocal relationship between adaptive capacity and community capital. Adaptive learning enhances community capital and helps develop local capacity. This in turn influences disaster resilience through mitigation and preparedness that enables a more effective response to and recovery from disasters. We also expect that lessons learned from response and recovery experiences will help with the design of more resilient communities in the future.

We can also take an outward view, defining the stressor as an external force, i.e., the hazard event, such as a hurricane. From this conceptual vantage point, all community attributes and behaviors that affect disaster losses, as well as the recovery process, are considered elements of community resilience. The determinants of a community's hazard resilience are its community capital as well as its hazard exposure, its ecological mediators, and mitigation initiatives. To minimize disaster losses, it is important for communities to understand the vulnerability of their capital and its ability to prepare and implement both response and recovery activities. Based on the adaptive resiliency framework, adaptive capacity determines community resilience. In this manner, the reciprocal relationship between adaptive capacity and community capital feeds back to hazard resilience through mitigation, preparedness, response, and recovery.

This framework can provide guidance to policy, practice, and research. Resiliency has been highlighted by many policy documents at international, national, and local levels, as discussed by Tierney in her foreword to this book. The policy documents, besides encouraging resiliency, should provide community-specific, community-measurable outcomes that can be used to demonstrate progress and accountability for practitioners at different levels of government. We expect that the research presented in this book, as well as future empirical research based on our adaptive capacity framework, will provide guidance for policy makers in developing effective policies, programs, and functioning governance structures. For these initiatives to be successful, they should include all community stakeholders in the design and implementation of effective disaster resiliency policies.

All disasters are local. The rising expectations of disaster resilience place burdens on local government responsible for managing disasters. Emergency managers must learn to partner with other organizations, community members, and jurisdictions. In short, risks are shared and a variety of community resources should be leveraged under stressful environments produced by disasters. This is essential, especially for rural communities, when faced with budget constraints and deciding on the distribution of scarce resources.

Disasters are complex in nature. Complex problems require interdisciplinary solutions. We emphasize the importance of interdisciplinary research on disaster resiliency in this book. We are certain that the framework provides guidance for resiliency research, particularly interdisciplinary community-based research from local and international perspectives. The chapters presented in this book provide some examples for both. In the future, more interdisciplinary, theory-driven, and empirically robust research is needed to fully understand the concept of resiliency.

Appendices

APPENDIX 1 ORGANIZATIONAL DEFINITIONS FOR COMMUNITY EMERGENCY MANAGEMENT ACTIVITY

COAD: Community Organization Active in Disaster (also known as Collaborating Organization Active in Disaster or Community Inter-Agency Disaster Organization) is defined in this chapter as a partnership of community-based and governmental organizations that collaborates, communicates and coordinates response and recovery efforts in a predetermined geographical area. Its defining features are (a) local activity, (b) broad representation from public, private and not-for-profit agencies, and (c) the ability to participate in all four phases of emergency management, including planning and mitigation as well as response and recovery.

VOAD: A Voluntary Organization Active in Disaster is an independent charitable or religious organization with a mission related to disaster response and recovery. Examples include the American Red Cross, Salvation Army, Mennonite Disaster Service, and the American Radio Relay League. VOADs that operate at national or state levels may be integrated into federal or state emergency planning documents and mass care may be coordinated through nonprofit coalitions such as the National Voluntary Organizations Active in Disaster (NVOAD.org) or state-level VOAD coordinating organizations.

LTRC: A Long Term Recovery Committee is a coordinating body of VOADs, social service agencies, public officials and other community organizations organized to respond to a prior disaster. Their mission may be temporary and informal, focusing on case management and service coordination. However, some LTRCs become permanent and some are absorbed into a COAD subcommittee.

LEPC: Local Emergency Planning Committee, a partnership of community-based, business and government organizations which work jointly to prevent and plan responses to industrial and chemical accidents (U.S. EPA 2008).

CERT: Community Emergency Response Teams are a program of Citizen Corps through FEMA's Individual and Community Preparedness Division. CERTs educate citizens about disaster preparedness and train them in basic disaster response skills. These individuals then serve as volunteers to supplement or support professional first responders. www.citizencorps.gov/cert

Citizen Corps Councils: Created post 9/11 by Presidential Order, Citizens Corps Councils support public education and emergency training, and play a role in the whole community planning by providing input to local emergency plans.

APPENDIX 2 DATA SOURCES AND MEASUREMENT OF VARIABLES

Category and Item	*Source Mean (s.d.)*	*Measure*
	BULDING CODE ENFORCEMENT DATASET	
Code Provision	Analysis of state legislation and regulations	Except for Alaska, which has unique provisions, each state's building code provisions were coded as one of three types: (1) provisions applying only to state facilities or none at all; (2) provisions that enable local application of building codes; and (3) provisions that require local building codes and enforcement.
Enforcement Measures		
Enforcement Effort	Survey 3.52 (.66) Alpha = .73	Mean of respondent rating of effort on a 1 (low) to 5 (high) scale put into each of five tasks: public relations, surveillance, plan checking, inspection, legal prosecution, technical assistance, public awareness.
Code Budget Priority	Survey 2.70 (.82)	Respondent rating of budget priority of the building code function over the past twelve months on a scale of 1 (low) to 5 (high).
Perceived Capacity	Survey 3.63 (.79) Alpha = .70	Mean of respondent rating of capacity on a 1 (low) to 5 (high) scale of adequacy of non-personnel budget, staffing, technical expertise, and authority for enforcing codes.
State Technical Assistance	Survey 1.23 (.84) Alpha = .80	Sum of uses of state technical assistance each coded as 1 = yes, 0 = no: information or advice about code enforcement; and participated in training opportunities offered by the state.
Enforcement Environment		
Interest Demand	Survey 33.36 (26.86)	Mean percentage of seven groups actively advocating strong code enforcement based on respondent rating: architect or engineer groups, chamber of commerce, environmental groups, general contractors association, historic preservation groups, homebuilder association, neighborhood groups. Scaled to fit a range of 0 to 100.
Interest Opposition	Survey 20.34 (21.82)	Mean percentage of the same groups actively opposing strong code enforcement. Scaled to fit a range of 0 to 100.
Ratio Demand to Opposition	Computed 16.50 (27.47)	Balance of support to opposition.

(*continued*)

CRITICAL INFRASTRUCTURE PROTECTION DATASET

Stakeholder Participation Type	Coding of sector participation lists	Coding of participation lists for each CIP sector according to four types of participants: (1) individual businesses or firms; (2) industry associations; (3) other professional associations; or (4) other entities (e.g., research experts, government agencies). Initial intercoder agreement for this coding was 96% with 100% after discussion.
Engagement Index	Coding of industry and association websites 2.20 (.29)	Coding of website entries for each association as having: (1) no mention of CIP issues; (2) a lesser concern with passing mention in the relevant listing; (3) an intermediate concern as one of a half dozen to dozen issues or many agenda items; and (4) a central concern that shows up as a priority issue. The agreement rate for a sample of fifty of these after coder discussion was 83% with eventual resolution of all but two disagreements, resulting in an overall 96% agreement.
Percent Engaged	Calculated from engagement index. 50 (.05)	Percentage of relevant associations that were coded as having engagement scores of 3 or 4.

APPENDIX 3 EARTHQUAKE MAPS OF TURKEY

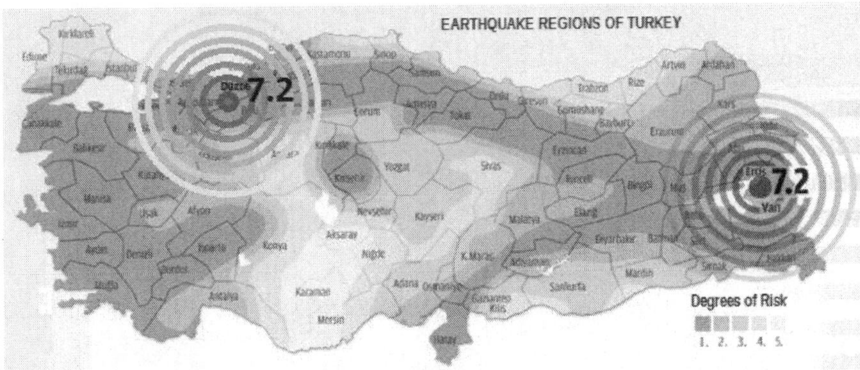

APPENDIX 4 EXAMPLE OF CODING

Date	*Organization*	*Source of funding*	*Transaction*	*Interacted organization*
October 23, 2011	Turkish General Staffing	Public-state	Three transportation aircrafts sent for rescue operations	AFAD (The Prime Ministry Disaster and Emergency Management Presidency)

Bibliography

Abell, R.A. (2000). *Freshwater ecoregions of North America: A conservation assessment.* Washington, DC: Island Press.

Adger, W.N. (2000). Social and ecological resilience—are they related? *Progress in Human Geography*, 24(3), 347–364.

Adger, W.N., Agrawala, S., Mirza, M.M.Q., Conde, C., O'Brien, K., Pulhin, J., Pulwarty, R., Smit, B., & Takahashi, K. (2007). Assessment of adaptation practices, options, constraints and capacity. In M.L. Parry, O.F. Canziani, J.P. Palutikof, P.J. van der Linden, & C.E. Hanson (Eds.), *Climate change 2007: Impacts, adaptation and vulnerability; Contribution of Working Group II to the fourth assessment report of the Intergovernmental Panel on Climate Change.* New York: Cambridge University Press.

Adger, W.N., Brooks, N., Bentham, G., Agrnew, M., & Eriksen, S. (2004). *New indicators of vulnerability and adaptive capacity.* Technical report no. 7. Tyndall Centre for Climate Change Research. Norwich, UK: East Anglia University.

Adger, W.N., Hughes, T.P., Folke, C., Carpenter, S.R., & Rockstrom, J. (2005) Social-ecological resilience to coastal disasters. *Science*, 309, 1036–1039.

Adger, W.N., & Vincent, K. (2005). Uncertainty in adaptive capacity. *Comptes Rendues Geoscience*, 337, 399–410.

Adler, P.S., & Kwon, S. (2002). Social capital: Prospects for a new concept. *Academy of Management Review*, 27(1), 17–40.

AFAD (Prime Ministry Disaster and Emergency Management Presidency). (2011a). Van earthquake on October 23, 2011. Press release, December 13, 2011. Retrieved from http://www.afetacil.gov.tr/.

AFAD (Prime Ministry Disaster and Emergency Management Presidency). (2011b). October 23, 2011 and November 9, 2011 Van earthquakes. Press release, November 28, 2011. Retrieved from http://www.afetacil.gov.tr/.

Afet Isleri Genel Mudurlugu. (1998). *Kanunlar ve yonetmelikler* [Laws and regulations]. Ankara: Bayindirlik ve Iskan Bakanligi.

Agranoff, R., & McGuire, M. (2003). *Collaborative public management: New strategies for local governments.* Washington, DC: Georgetown University Press.

Agrawal, A. (2008). The role of local institutions in adaptation to climate change. Paper prepared for the Social Dimensions of Climate Change, Social Development Department, World Bank, Washington, DC, March 5–6.

Alberts, D.S., & Hayes, R.E. (1995). *Command arrangements for peace operations.* Washington, DC: NDU.

Alberts, D.S., & Hayes, R.E. (2003). *Power to the edge.* Washington, DC: CCRP.

Aldrich, D.P. (2010). Fixing recovery: Social capital in post-crisis resilience. *Journal of Homeland Security*, 6, 1–10.

Alesch, D., Arendt, L., & Holly, J. (2009). *Managing for long-term community recovery in the aftermath of disaster.* Fairfax, VA: Public Entity Risk Institute.

Alexander, D. (1993). *Natural disasters.* New York: Chapman & Hall.

Ali, A., Angelov, P., & Hutchison, D. (2011). Towards an adaptive resilience strategy for future computer networks. In *Proceedings of the 11th UK workshop on computational intelligence* (pp. 201–206). Manchester, UK: University of Manchester School of Computer Science. Retrieved February 10, 2012 from http://ukci.cs.manchester.ac.uk/files/Proceedings.pdf.

Allen, H., & Katz, R. (2009). Demography and public health emergency preparedness: Making the connection. *Population Research and Policy Review*, 29, 527–539.

Allen, K.M. (2006). Community-based disaster preparedness and climate change adaptation: Local capacity building in the Philippines. *Disasters*, 30(19), 81–101.

Allen, T. (2011). Community resilience: Thad Allen on creating a "community of practice". RAND audio podcast. Retrieved March 1, 2012 from http://www.rand.org/multimedia/audio/2011/12/12/creating-community-practice.html.

Alm, L.R., & Witt, S.L. (1995). Environmental policy in the intermountain west: The rural-urban linkage. *State and Local Government Review*, 27(2), 127–136.

Alter, C., & Hage, J. (1993). *Organizations working together.* Newbury Park, CA: Sage.

Anaut, M. (2003). *La Résilience: Surmonter les traumatismes.* Paris: Nathan Université.

Anderson, J.L., Hilborn, R.W., Lackey, R.T., & Ludwig, D. (2003). Watershed restoration: Adaptive decision making in the face of uncertainty. In R.C. Wissmar & P.A. Bisson (Eds.), *Strategies for restoring river ecosystems: Sources of variability and uncertainty in natural and managed systems* (pp. 203–232). Bethesda, MD: American Fisheries Society.

Angus, L. (2004). Our direction for recovery in New Zealand. In S. Norman (Ed.), *New Zealand Recovery Symposium 2004.* Wellington, New Zealand: Ministry of Civil Defense and Emergency Management. Retrieved March 1, 2012 from http://www.civildefence.govt.nz/memwebsite.NSF/Files/Recovery/$file/Recovery_symposium2004.pdf#page=91.

Ansell, C., & Gash, A. (2007). Collaborative governance in theory and practice. *Journal of Public Administration Research and Theory*, 18(4), 543–571.

Antonovsky, A. (1987). *Unraveling the mystery of health: How people manage stress and stay well.* San Francisco: Jossey-Bass.

AP (Associated Press). (2012). Tornadoes kill at least 14 across southern Indiana. March 2. Retrieved April 16, 2012 from http://www.wthitv.com/dpps/weather/us_wx_news/authorities-report-extreme-tornado-damage-in-indiana-nt12-tvw_4091754.

Arendt, R.G. (1996). *Conservation design for subdivisions: A practical guide to creating open space networks.* Washington, DC: Island Press.

Arendt, R.G. (1999). Growing greener: Conservation subdivisions design. *Planning Commissioners Journal*, 33, 7–14.

Arlikatti, S., Lindell, M.K., & Prater, C.S. (2006). Risk area accuracy and hurricane evacuation expectations of coastal residents. *Environment and Behavior*, 38, 226–247.

Arnold, C.L., & Gibbons, J.C. (1996). Impervious surface coverage: The emergence of a key environmental indicator. *Journal of the American Planning Association*, 62(2), 243–258.

ARPC (Apalachee Regional Planning Council). (2010). *Statewide regional evacuation study program*, vol. 1–2, *Apalachee region technical data report.* Blountstown, FL: ARPC.

Auf der Heide, E. (1989). *Disaster response: Principles and preparation and coordination.* St. Louis, MO: The C.V. Mosby Company.

Axelrod, R., & Cohen, D.M. (1999). *Harnessing complexity.* New York: Free Press.

Baker, E.J. (1991). Hurricane evacuation behavior. *International Journal of Mass Emergencies and Disasters*, 9, 287–310.

Bandura, A. (1979). Self-efficacy: Toward and unifying theory of behavioral change. *Psychological Review*, 84, 191–215.

Bandura, A. (1989). Human agency in social cognitive theory. *American Psychologist*, 44(9), 1175–1184.

Bandura, A. (Ed.) (1995). *Self-efficacy in changing societies.* Cambridge: Cambridge University Press.

Barret, C. & Greene, R. (2008). Measuring Performance: The State Management Report Card for 2008. *Governing*, March Issue, 24–95.

Barzelay, M. (1992). *Breaking through bureaucracy: A new way for managing in government.* Berkeley: University of California Press.

Basbakanlik Kriz Yonetim Merkezi. (2000). *Depremler 1999* [1999 earthquakes]. Ankara: Basbakanlik.

Beatley, T. (1998). The vision of sustainable communities. In J.R. Burby (Ed.), *Cooperating with nature: Confronting natural hazards with land-use planning for sustainable communities.* Washington, DC: Joseph Henry Press.

Beatley, T. (2009). *Planning for coastal resilience: Best practices for calamitous times.* Washington, DC: Island Press.

Beever, J.W., Gray, W., Trescott, D., Cobb, D., Utley, J., Hutchinson, D., Gibbons, J., . . . & Ott, J. (2009). *City of Punta Gorda adaptation plan.* Technical report 09–4. Ft. Myers, FL: Southwest Florida Regional Planning Council and Charlotte Harbor National Estuary Program.

Bendick, R., Dahlin, K.M., Smoliak, B.V., Kumler, L., Jones, S.J., Aktipis, A., . . . & Scott, D. (2010). Choosing carbon mitigation strategies using ethical deliberation. *Weather, Climate and Society*, 2, 140–147.

BenDor, T., Brozovic, N. & Pallathucheril, V.G. (2007). Assessing the socioeconomic impacts of wetland mitigation in the Chicago region. *Journal of the American Planning Association*, 73(3), 263–282.

Bengston, D.N., Fletcher, O.J., & Nelson, K.C. (2003). Public policies for managing urban growth and protecting open space: Policy instruments and lessons learned in the United States. *Landscape and Urban Planning*, 69, 271–286.

Bennet, A., & Bennet, D. (2000). Characterizing the next generation knowledge organization. *Knowledge and Innovation: Journal of the KMCI*, 1(1), 8–42.

BENS. (2007). *Getting down to business: An action plan for public-private disaster response coordination; The report of the business response task force.* Washington, DC: Business Executives for National Security.

Benton, J.E., Byers, J., Cigler, B.A., Klasse, K.A., Menzel, D.C., Salant, T.J., . . . & Waugh, W.L. (2008). Service challenges and governance issues confronting American counties in the 21st century: An overview. *State and Local Government Review*, 40(1), 54–68.

Bergal, J., Shipley Hiles, S., & Koughan, F. (2007). *City adrift: New Orleans before and after Katrina.* Baton Rouge: Louisiana State University Press.

Berke, P., & Beatley, T. (1992). *Planning for earthquakes: Risk, politics and policy.* Baltimore: Johns Hopkins University Press.

Berke, P.R., & Campanella, T.J. (2006). Planning for postdisaster resiliency. *Annals of the American Academy of Political and Social Science*, 604, 192–207.

Berke, P.R., & Conroy, M.M. (2004). Are we planning for sustainable development? An evaluation of 30 comprehensive plans. *Journal of the American Planning Association*, 66, 21–33.

Berke, P., & Godschalk, D. (2009). Searching for the good plan: A meta-analysis of plan quality studies. *Journal of Planning Literature*, 23(3), 227–240.

Berke, P., Roenigk, D., Kaiser, E., & Burby, R. (1996). Enhancing plan quality: Evaluating the role of state planning mandates for natural hazard mitigation. *Journal of Environmental Planning and Management*, 39, 79–96.

Berkes, F., Colding, J., & Folke, C. (2003). *Navigating social ecological systems: Building resilience for complexity and change.* Cambridge: Cambridge University Press.

Birkland, T.A., Burby, R.J., Conrad, D., Cortner, H., & Michener, W.K. (2003). River ecology and flood hazard mitigation. *Natural Hazards Review*, 4(1), 46–54.

Birkland, T., & Waterman, S. (2008). Is federalism the reason for policy failure in Hurricane Katina? *Publius: The Journal of Federalism*, 38, 692–714.

Birkmann, J. (2006). Measuring vulnerability to promote disaster-resilient societies: Conceptual frameworks and definitions. In United Nations, *Measuring vulnerability to natural hazards: Towards disaster resilient societies.* Tokyo: United Nations University.

Block, P. (1993). *Stewardship.* San Francisco: Berret-Koehler.

Block, P. (2008). *Community: The structure of belonging.* San Francisco: Berrett-Koehler.

Bloodworth, D., Kevorkian, C., Rumbaut, E., & Chiou-Tan, F. (2007). Impairment and disability in the astrodome after Hurricane Katrina: Lessons learned about the needs of the disabled after large population movements. *American Journal of Physical Medicine and Rehabilitation*, 86(9), 770–775.

Bolin, B. (2006). Race, class, ethnicity, and disaster vulnerability. In H. Rodriguez, E.L. Quarantelli, & R. Dynes (Eds.), *Handbook of disaster research.* United States of America: Springer.

Bolin, R. (1986). Disaster impact and recovery: A comparison of black and white victims. *International Journal of Mass Emergencies and Disasters*, 4(1), 35–50.

Bolin, R. (1994). *Household and community recovery after earthquakes.* Boulder, CO: University of Colorado.

Bolin, R.C., & Stanford, L. (1991). Shelter, housing and recovery: A comparison of U.S. disasters. *Disasters*, 15, 24–34.

Bolton, P.A., Liebow, E.B., & Olson, J.L. (1993). Community conflict and uncertainty following a damaging earthquake: Low-income Latinos in Los Angeles, California. *Environmental Professional*, 15, 240–247.

Bonanno, G. A., Westphal, M., & Mancini, A. (2011). Resilience to loss and trauma. *Annual Review of Clinical Psychology*, 7, 511–535.

Borgatti, S.P., Everett, M.G., & Freeman, L.C. (2002). *Ucinet for Windows: Software for social network analysis.* Harvard, MA: Analytic Technologies.

Borgatti, S.P., Mehra, A., Brass, D.J., & Labianca, G. (2009). Network analysis in the social sciences. *Science*, 323(892), 892–895.

Boris, E.T., & Steuerle, C.E. (Eds.) (2006). *After Katrina: Public expectation and charities' response.* Washington, DC: Urban Institute.

Bouwen, R., & Taillieu, T. (2004). Multi-party collaboration as social learning for interdependence: Developing relational knowing for sustainable natural resource management. *Journal of Community & Applied Social Psychology*, 14(3), 137.

Bowlby, J. (1978). *L'Attachement: Attachement et Perte.* Paris: PUF.

Bowlby, J. (1982). *Attachment and loss. Vol. 1: Attachment* (2nd Ed.). New York: Basic Books.

Bowlby, J. (1992). L'avènement de la psychiatrie développementale a sonné. *Devenir*, 4, 10–21.

Boyd, E., & Folke, C. (2011). Conclusions: Adapting Institutions and Resilience. In: Boyd, E. and C. Folke (eds.). *Adapting Institutions: Governance, Complexity and Social-Ecological Resilience* pp 264–280. Cambridge University Press, Cambridge, UK.

Boyd, E., & Folke, C. (2012). *Adapting institutions: Governance, complexity and social-ecological resilience*. Cambridge: Cambridge University Press.

Braden, J.B., & Johnston, D.M. (2004). Downstream economic benefits from storm-water management. *Journal of Water Resources Planning and Management*, 130(6), 498–502.

Brass, D., Galaskiewicz, J., Greve, H., & Tsai, W. (2004). Taking stock of networks and organizations: A multilevel perspective. *Academy of Management Journal*, 47(6), 795–817.

Brennan, M.A., & Flint, C.G. (2007). Uncovering the hidden dimensions of rural disaster mitigation: Capacity building through community emergency response teams. *Southern Rural Sociology*, 22, 111–126.

Brezonik, P.L., & Stadelmann, T.H. (2002). Analysis and predictive models of stormwater runoff volumes, loads and pollutant concentrations from watersheds in the twin cities metropolitan area, Minnesota, USA. *Water Resources*, 36, 1743–1757.

Brody, S.D. (2003a). Measuring the effects of stakeholder participation on the quality of local plans based on the principles of collaborative ecosystem management. *Journal of Planning Education and Research*, 22, 407–419.

Brody, S.D. (2003b). Are we learning to make better plans? A longitudinal analysis of plan quality associated with natural hazards. *Journal of Planning Education and Research*, 23, 191–201.

Brody, S.D., Carrasco, V., & Highfield, W.E. (2006). Measuring the adoption of local sprawl reduction planning policies in Florida. *Journal of Planning Education and Research*, 25(3), 294–310.

Brody, S.D., Groverb, H., Lindquist, E., & Vedlitz, A. (2010). Examining climate change mitigation and adaptation behaviours among public sector organisations in the USA. *Local Environment*, 15, 591–603.

Brody, S.D., & Highfield, W.E. (2005). Does planning work? Testing the implementation of local environmental planning in Florida. *Journal of the American Planning Association*, 71, 159–175.

Brody, S.D., Highfield, W.E., & Kang, J.E. (2011). *Rising waters: Causes and consequences of flooding in the United States*. Cambridge: Cambridge University Press.

Brody, S. D., Highfield, W.E., Ryu, H., & Spanel-Weber, L. (2007). Examining the relationship between wetland alteration and watershed flooding in Texas and Florida. *Natural Hazards*, 40(2), 413–428.

Brody, S.D., Kang, J.E., & Bernhardt, S. (2010). Identifying factors influencing flood mitigation at the local level in Texas and Florida: The role of organizational capacity. *Natural Hazards*, 52, 167–184.

Brody, S.D., Zahran, S., Grover, H., & Vedlitz, A. (2008). A spatial analysis of local climate change policy in the US: Risk, stress, and opportunity. *Landscape and Urban Planning*, 87, 33–41.

Brody, S.D., Zahran, S., Highfield, W., Bernhardt, S., & Vedlitz, A. (2009). Policy learning for flood mitigation: A longitudinal assessment of CRS activities in Florida. *Risk Analysis*, 29(6), 912–929.

Brody, S.D., Zahran, S., Highfield, Grover, H., & Vedlitz, A. (2008). Identifying the impact of the built environment on flood damage in Texas. *Disasters*, 32(1), 1–18.

Brody, S.D., Zahran, S., Praveen, M., Grover, H., & Highfield, W.E. (2007). The rising costs of floods: Examining the impact of planning and development decisions on property damage in Florida. *Journal of the American Planning Association*, 73(3), 331–345.

Bronfenbrenner, U. (1977). Toward an experimental ecology of human development. *American Psychologist*, 32, 513–531.

Brouwer, R., & Nhassengo, J. (2006). About bridges and bonds: Community responses to the 2000 floods in Mabalane District, Mozambique. *Disasters*, 30(2), 234–255.

Brower, R., Choi, S., Jeong, H., & Dilling, J. (2009). Forms of inter-organizational learning in emergency management networks. *Journal of Homeland Security and Emergency Management*, 6(1).

Brudney, J.L., & Gazley, B. (2009). Planning to be prepared: An empirical examination of the role of voluntary organizations in county government emergency planning. *Public Performance and Management Review*, 32(3), 372–399.

Bruneau, M., Chang, S., Eguchi, R., Lee, G., O'Rourke, T., & Reinhorn, A. (2003). A framework to quantitatively assess and enhance the seismic resilience of communities. *Earthquake Spectra*, 19(4), 733–752.

Bruneau, M., & Reinhorn, A. (2006). Overview of the concept of resilience. Paper no. 2040. In *Proceedings of the 8th U.S. National Conference on Earthquake Engineering*, San Francisco, April 18–22.

Bryson, J.M., Crosby, B.C., & Stone, M.M. (2006). The design and implementation of cross-sector collaborations: Propositions from the literature. Special issue, *Public Administration Review*, 66, 44–55.

Bulkeley, H., & Betsill, M.M. (2003). *Cities and climate change: Urban sustainability and global environmental governance*. New York: Routledge.

Bullock, A., & Acreman, M. (2003). The role of wetlands in the hydrological cycle. *Hydrology and Earth System Sciences*, 7(3), 358–389.

Burby, R.J. (1998). Natural hazards and land use. In R.J. Burby (Ed.), *Cooperating with nature: Confronting natural hazards with land-use planning for sustainable communities*. Washington, DC: Joseph Henry Press.

Burby, R.J. (2001). Flood insurance and floodplain management: The U.S. experience. *Journal of Environmental Hazards*, 3(3), 111–122.

Burby, R.J. (2003). Making plans that matter: Citizen involvement and government action. *Journal of the American Planning Association*, 69(1), 33–49.

Burby, R.J. (2005). Have state comprehensive planning mandates reduced insured losses from natural disasters? *Natural Hazards Review*, 6(2), 67–81.

Burby, R.J. (2006). Hurricane Katrina and the paradoxes of government disaster policy: Bringing about wise government decisions for hazardous areas. *Annals American Academy of Political and Social Sciences*, 604(1), 171–191.

Burby, R.J., Beatley, T., Berke, P.R., Deyle, R.E., French, S.P., Godschalk, D.R., Kaiser, E.J., . . . & Platt, R.H. (1999). Unleashing the power of planning to create disaster resistant communities. *Journal of the American Planning Association*, 65(3), 247–258.

Burby, R.J., Deyle, R.E., Godschalk, D.R., & Olshansky, R.B. 2000. Creating hazard resilient communities through land-use planning. *Natural Hazards Review*, 1(2), 99–106.

Burby, R.J., French, S.P., & Cigler, B.A. (1985). *Floodplain land use management: A national assessment*. Boulder, CO: Westview Press.

Burby, R.J., French, S.P., & Nelson, A.C. (1998). Plans, code enforcement, and damage reduction: Evidence from Northridge Earthquake. *Earthquake Spectra*, 14, 59–72.

Burby, R.J., & May, P.J. (1998). Intergovernmental environmental planning: Addressing the commitment conundrum. *Journal of Environmental Planning and Management*, 41(1), 95–110.

Burby, R.J., May, P.J., & Patterson, R.C. (1998). Improving compliance with regulations: Choices and outcomes for local governments. *Journal of the American Planning Association*, 64(3), 324–334.

Burt, R. (1997). The contingent value of social capital. *Administrative Science Quarterly*, 42, 339–365.

Cabinet Office of Japan. (2011). White paper on disaster management. In Japanese. Retrieved from http://www.bousai.go.jp/hakusho/H23_zenbun.pdf.

Cabinet Office of Japan, Fire and Disaster Management Agency, & Japan Meteorological Agency. (2011). *Analysis results of interview survey on evacuation behaviors for the 2011 Great East Japan Earthquake (residents)*. Retrieved from http://www.bousai.go.jp/jishin/chubou/higashinihon/7/1.pdf.

Cabinet Office of the United Kingdom. (2011). Strategic national framework on community resilience. Retrieved March 5, 2012 from http://www.cabinetoffice.gov.uk/sites/default/files/resources/Strategic-National-Framework-on-Community-Resilience_0.pdf.

California Resiliency Alliance. (2011). About us. Retrieved March 1, 2012 from http://www.caresiliency.org/page/about-us.

Calkin, D., Gebert, K., Jones, J., & Neilson, R. (2005). Forest service large fire area burned and suppression expenditure trends, 1970–2002. *Journal of Forestry*, 103(4), 179–183.

Callaway, J.M. (2004). Adaptation benefits and costs: Are they important in the global policy picture and how can we estimate them? *Global Environmental Change*, 14, 273–282.

Camerer, C.F., & Kunreuther, H. (1989). Decision processes for low probability risks: Policy implications. *Journal of Policy Analysis and Management*, 8(4), 565–592.

Campanella, T.J. (2006). Urban resilience and the recovery of New Orleans. *Journal of the American Planning Association*, 72(2), 141–146.

Canton, L.G. (2007). *Emergency management: Concepts and strategies for effective programs*. Hoboken, NJ: Wiley-InterScience.

CAO. (2007). Bousaihakusho [White paper on disaster preparedness]. In Japanese. Retrieved June 26, 2008 from http://www.bousai.go.jp/hakusho/h19/BOUSAI_2007/html/index.htm.

CAO. (2008). Cabinet Office: Disaster management in Japan, Director General for Disaster Management. In Japanese.

Carley, K.M. (2002). Computational organizational science and organizational engineering. *Simulation Modeling Practice and Theory*, 10(5–7), 253–269.

Carley, K. & Harrald, J. (1997). Organizational Learning Under Fire: Theory and Practice. *American Behavioral Scientist*, 40(3), 310–332.

Carroll, M.S., Higgins, L.L., Cohn, P.J., & Burchfield, J. (2006). Community wildfire events as a source of social conflict. *Rural Sociology*, 71(2), 261–280.

Caruson, K., & MacManus, S.A. (2007). Structuring homeland security grants: Florida's local finance officials evaluate the funding process. *Journal of Public Finance and Management*, 7, 191–226.

Caruson, K., & MacManus, S.A. (2008). Disaster vulnerabilities: How strong a push toward regionalism and intergovernmental cooperation? *The American Review of Public Administration*, 38, 286–306.

CEDIM (Center for Disaster Management and Risk Reduction Technology). (2011). Comparing the current impact of the Van Earthquake to past earthquakes in Eastern Turkey. Report no. 4. Retrieved fromhttp://www.cedim.de/download/CEDIMForensicEQAGTurkeyVanEQ_Report4.pdf.

Celik, S. (2006) Socio-technical approaches to complex phenomena: An analysis of the Turkish disaster management system under stress. Unpublished dissertation.

Graduate School of Public and International Affairs, University of Pittsburgh, Pittsburgh, PA.

Celik, S., & Corbacioglu, S. (2011). Turk afet yönetim sisteminin reorganizasyonu baglamında basbakanlık afet ve acil durum yönetimi başkanlığını yapılandırma çalısmalarının eleştirel analizi. *E-akademi*, 111. Retrieved from http://www.e-akademi.org/arsiv.asp?sayi=110.

Census Bureau. (2000). *United States Census 2000*. Retrieved December 6, 2009 from http://www.census.gov/main/www/cen2000.html.

Census Bureau. (2010). *Estimates for Florida counties, 2009, in small area income and poverty estimates*. Retrieved July 22, 2011 from http://www.census.gov/cgi-bin/saipe/saipe.cgi?year=2009&type=county&table=county&submit=States%20%26%20Counties&display_data=Display%20Data&state=12&areas=12119.

Census Bureau. (2011). *Sumter County, Florida*. Retrieved July 22, 2011 from http://quickfacts.census.gov/qfd/states/12/12119.html.

Census Bureau. (2012). *Orange County, Florida*. Retrieved May 14, 2012 from http://quickfacts.census.gov/qfd/states/12/12095.html.

Chamlee-Wright, E. (2010). *The cultural and political economy of recovery*. London: Routledge.

Chamlee-Wright, E., & Storr, V.H. (2009). Club goods and post-disaster community return. *Rationality and Society*, 21(4), 429–458.

Chandra, A., J. Acosta, et al. (2011). *Building community resilience to disasters: a way forward to enhance national health security*. Technical Reports. Santa Monica, CA, RAND: xxiii, 78 p.

Chang, H., & Franczyk, J. (2008). Climate change, land-use change, and floods: Toward an integrated assessment. *Geography Compass*, 2(5), 1549–1579.

Chang, S.E., & Shinozuka, M. (2004). Measuring improvements in the disaster resilience of communities. *Earthquake Spectra*, 20, 739–755.

Chattopadhyay, P., Glick, W.H., & Huber, G.P. (2001). Organizational actors in response to threats and opportunities. *The Academy of Management Journal*, 44(5), 937–955.

Chavis, D., & Wandersman, A. (1990). Sense of community in the urban environment: A catalyst for participation and community development. *American Journal of Community Psychology*, 18(1), 55–81.

Chisholm, D. (1989). *Coordination without hierarchy*. Berkeley: University of California Press.

Christmann, G., & Ibert, O. (forthcoming). Vulnerability and resilience in a socio-spatial perspective: A social-scientific approach. In Vulnerability and resilience in a socio-spatial perspective, special issue, *Raumordnung und Raumforschung*.

Christmann, G., Ibert, O., Kilper, H., & Moss, T. (2011). Vulnerabilitaet und Resilienz in Sozio-Raeumlicher Perspektive: Begriffliche Klaerungen und Theoretischer Rahmen. IRS working paper no. 44, Institute for Regional Development and Structural Planning (IRS), Erkner, Germany. Retrieved April 26, 2012 from http://www.irs-net.de/publikationen/veroeffentlichungen/working-papers/index.php.

Christoplos, I., Mitchell, J., & Liljelund A. (2001). Re-framing risk: The changing context of disaster mitigation and preparedness. *Disasters*, 25(3), 185–198.

Chu, K. (2007). Small businesses struggle to recover from Katrina. *USA Today*, August 28. Retrieved August 29, 2007 from http://www.usatoday.com/money/smallbusiness/2007–08–28-katrina-finances_N.htm.

Cigler, B.A. (1993). Challenges Facing Fiscal Federalism in the 1990s. *Political Science and Politics*, 26, 181–186.

Cigler, B. A. (1993). The special problems of rural county governments. In David R. Berman, ed., *County government in an era of change*. New York: Greenwood Press, pp. 89–106.

City Mayors. (2010). *The largest US cities: Cities ranked 1 to 100*. Retrieved April 27, 2012 from http://www.citymayors.com/gratis/uscities_100.html.

City of Seattle. (2011). Seattle City Council emergency preparedness initiative to launch Monday. News advisory, June 9. Retrieved April 30, 2012 from http://www.seattlechannel.org/news/detail.asp?ID=11798&Dept=41.

COAD-Orange County. (2012). Mission and purpose statement. Retrieved from http://egov.ocgov.com/ocgov/Info%20OC/Departments%20&%20Agencies/Emergency%20Operations%20Center%20-%20Orange%20County%20Disaster%20Preparedness/Volunteer/COAD-OC.

Coleman, J.S. (1988). Social capital in the creation of human capital. *The American Journal of Sociology*, 94, S95–S120.

Coles, E., & Buckle, P. (2004). Developing community resilience as a foundation of effective disaster recovery. In S. Norman (Ed.), *New Zealand Recovery Symposium 2004*. Wellington, New Zealand: Ministry of Civil Defense and Emergency Management.

Collins, M.L., & Kapucu, N. (2008). Early warning systems and disaster preparedness and response in local government. *Disaster Prevention and Management: An International Journal*, 17(5), 587–600.

Collins, P.H. (2009). *Black feminist thought*. 2nd edition. New York: Routledge.

Colussi, M.M. (2000). *The community resilience manual: A resource for rural recovery and renewal*. British Columbia: Canadian Centre for Community Renewal.

Comerio, M.C. (1998). *Disaster hits home: New policy for urban housing recovery*. Berkeley: University of California Press.

Comfort, L.K. (1994). Risk and resilience: Inter-organizational learning following the Northridge earthquake of 17 January 1994. *Journal of Contingencies and Crisis Management*, 2(3), 157–170.

Comfort, L.K. (1999). *Shared risk: Complex systems in seismic response*. Amsterdam: Pergamon.

Comfort, L.K. (2006). Cities at risk: Hurricane Katrina and the drowning of New Orleans. *Urban Affairs Review*, 41(4), 501–516.

Comfort, L.K. (2007). Crises management in hindsight: Cognition, communication, coordination, and control. In Administrative failure in the wake of Katrina, special issue, *Public Administration Review*, S1, 88–96.

Comfort, L.K., Birkland, T.A., Cigler, B.A., & Nance, E. (2010). Retrospectives and prospectives on Hurricane Katrina: Five years and counting. *Public Administration Review*, 70(5), 669–678.

Comfort, L.K., Dunn, M., Johnson, D., Skertich, R., & Zagorecki, A. (2004). Coordination in complex systems: Increasing efficiency in disaster mitigation and response. *International Journal of Emergency Management*, 2(1–2), 62–80.

Comfort, L., & Haase, T. (2006). Communication, coherence, and collective action: The impact of Hurricane Katrina on communications infrastructure. *Public Works Management Policy*, 10(4), 328–343.

Comfort, L., & Kapucu, N. (2006). Inter-organizational coordination in extreme events: The World Trade Center attacks, September 11, 2001. *Natural Hazards*, 39(2), 309–327.

Comfort, L.K., & Sungu, Y. (2001). Organizational learning from seismic risk. In U. Rosenthal, L.K. Comfort, & A. Boin (Eds.), *From crisis to contingencies: A global perspective* (pre-published copy). Chicago: Charles C. Springer, Inc.

Comparing the current impact of the Van Earthquake to past earthquakes in Eastern Turkey. Report No 4. Retrieved From.http://www.cedim.de/download/CEDIMForensicEQAGTurkeyVanEQ_Report4.pdf

Conroy, M.M., & Berke, P. 2004. What makes a good sustainable development plan? An analysis of factors that influence principles of sustainable development. *Environment and Planning*, 36, 1381–1396.

Cooper, C., & Block, R. (2006). *Disaster! Hurricane Katrina and the failure of homeland security*. New York: Henry Holt and Company.

Cooper, M. (2011). Battered Vermont looks first to its roads. *New York Times*, September 2. A13.

Corbacioglu, S. (2004). Intergovernmental collective action in complex environments: Toward a self-adaptive Turkish disaster management system. Unpublished dissertation. Graduate School of Public and International Affairs, University of Pittsburgh. Pittsburgh, PA.

Costanza, R., Pe´rez-Maqueo, O., Luisa, M., Sutton, P., Anderson, S.J., & Mulder, K. (2008). The value of coastal wetlands for hurricane protection. *Ambio*, 37(4), 241–247.

County Florida. (n.d.). *Orange County Florida*. Retrieved May 2011 from http://orange.county-florida.com/.

Cox, R.S., & Perry, K.M.E. (2011). Like a fish out of water: Reconsidering disaster recovery and the role of place and social capital in community disaster resilience. *American Journal of Community Psychology*, 48, 395–411.

Coyne, C.J., & Lemke, J.S. (2012). Lessons from *The cultural and political economy of recovery*. *American Journal of Economics and Sociology*, 71(1), 215–228. Retrieved March 1, 2012 from http://www.ccoyne.com/The_Cultural_and_PE_of_recovery.pdf.

Cross, R., Borgatti, S.P., & Parker, A. (2002). Making invisible work visible: Using social network analysis to support strategic collaboration. *California Management Review*, 44(2), 25–46.

Crossett, K.M. (2004). *Population trends along the coastal United States: 1980–2008*. Washington, DC: National Oceanic and Atmospheric Administration, U.S. Department of Commerce.

Crossett, K.M., Culliton, T.J., Wiley, P.C., & Goodspeed, T.R. (2004). Population trend along the coastal United States: 1980–2008 [electronic version]. Retrieved October 20, 2006 from www.oceanservice.noaa.gov.

Cumhuriyet daily newspaper. (1999). Istanbul, Turkey, November 13–28, 1999 issues.

Cuotter, D.A. (2004). Communities of work: Rural restructuring in local and global contexts. *Rural Sociology*, 69, 581–584.

Cutter, S.L. (1996). Vulnerability to environmental hazards. *Progress in Human Geography*, 20(4), 529–539.

Cutter, S.L., Barnes, L., Berry, M., Burton, C., Evans, E., Tate, E., & Webb, J. (2008). A place-based model for understanding community resilience to natural disasters. *Global Environmental Change*, 18(4), 598–606.

Cutter, S.L., Boruff, B.J., & Shirley, W.L. (2003). Social vulnerability to environmental hazards. *Social Science Research*, 84(2), 242–261.

Cutter, S.L., Burton, C.G., & Emrich, C.T. (2010). Disaster resilience indicators for benchmarking baseline conditions. *Journal of Homeland Security and Emergency Management*, 7(1), 1–22.

Cutter, S.L., & Emrich, C.T. (2006). Moral hazard, social catastrophe: The changing face of vulnerability along the hurricane coasts. *Annals of the American Academy of Political and Social Science*, 604, 102–112.

Cutter, S.L., & Finch, C. (2008). Temporal and spatial changes in social vulnerability to natural hazards. *Proceedings of the National Academy of Sciences USA*, 105, 2301–2306.

Cutter, S.L., Johnson, L.A., Finch, C., & Berry, M. (2007). The U.S. Hurricane coasts: Increasingly vulnerable. *Environment*, 49(7), 8–20.

Cutter, S., Osman-Elasha, B., Campbell, J., Cheong, S.-M., McCormick, S., Pulwarty, R., Supratid, S., & Ziervogel, G. (2012). Managing the risks from climate extremes at the local level. In C.B. Field, V. Barros, T.F. Stocker, D. Qin, D.J.

Dokken, K.L. Ebi, M.D. Mastrandrea, . . . & P.M. Midgley (Eds.), *Managing the risks of extreme events and disasters to advance climate change adaptation: A special report of Working Groups I and II of the Intergovernmental Panel on Climate Change (IPCC)* (pp. 291–338). Cambridge: Cambridge University Press.

Cyrulnik, B. (2002). *Un merveilleux malheur.* Paris: Odile Jacob.

Cyrulnik, B., & Duval, P. (2006). *Psychanalyse et résilience.* Paris: Odile Jacob.

Dalton, L.C., & Burby, R.J. (1994). Mandates, plans and planners: Building local commitment to development management. *Journal of the American Planning Association*, 60, 444–461.

Dash, N., Peacock, W.G., & Morrow. B.H. (1997). And the poor get poorer: A neglected Black community. In W.G. Peacock, B. H. Morrow, & H. Gladwin (Eds.), *Hurricane Andrew: Ethnicity, gender and the sociology of disasters* (pp. 206–225). London and New York: Routledge Press.

Davis, T. (2006). *A failure of initiative: The final report of the select bipartisan committee to investigate the preparation for and response to Hurricane Katrina.* Washington, DC: Select Bipartisan Committee to Investigate the Preparation for and Response to Hurricane Katrina. Retrieved March 3, 2012 from http://www.katrina.house.gov.

DeVita, C.J. (2007). Preparing for the next disaster. In C.J. DeVita (Ed.), *After Katrina: Shared challenges for rebuilding communities* (pp. 27–31). Washington, DC: Urban Institute.

Devlet Planlama Teskilati. (2001). *Duzce ili raporu* [Duzce province report]. Ankara: Devlet Planlama Teskilati.

Deyle, R.E., Bailey, K.C., & Matheny, A. (2007). *Adaptive response planning to sea level rise in Florida and implications for comprehensive and public-facilities planning.* Tallahassee: Florida Planning and Development Lab, Florida State University.

Deyle, R.E., Chapin, T.S., & Baker, E.J. (2008). The proof of the planning is in the platting: An evaluation of Florida's hurricane exposure mitigation planning mandate. *Journal of the American Planning Association*, 74(3), 349–370.

Deyle, R.E., French, S.P., Olshansky, R.B., & Paterson, R.G. (1998). Hazard assessment: A factual basis for planning and mitigation. In R.J. Burby (Ed.), *Cooperating with nature: Confronting natural hazards with land-use planning for sustainable communities* (pp. 119–166). Washington, DC: Joseph Henry Press.

DHS (Department of Homeland Security). (2006). *National infrastructure protection plan.* Washington, DC: DHS.

DHS (Department of Homeland Security). (2009). *National infrastructure protection plan.* Washington, DC: DHS.

DHS (Department of Homeland Security). (2010). *Critical infrastructure partnership advisory council* Annual. Washington, DC: DHS.

DHS (Department of Homeland Security). (2011). Presidential policy directive 8 (PPD-8): National preparedness. Retrieved from http://www.dhs.gov/xabout/laws/gc_1215444247124.shtm.

Diduck, A. (2010). The learning dimension of adaptive capacity: Untangling the multi-level connections. In D. Armitage & R. Plummer (Eds.), *Adaptive capacity and environmental governance.* New York: Springer-Verlag.

Dirks, K.T., & Ferrin, D.L. (2001). The role of trust in organizational settings. *Organization Science*, 12(4), 450–467.

Dombeck, M.P., Williams, J., & Wood, C. (2004). Wildfire policy and public lands: Integrating scientific understanding with social concerns across landscapes. *Conservation Biology*, 18(4), 883–889.

Donner, W., & Rodriguez, H. (2008). Population composition, migration and inequality: The influence of demographic changes on disaster risk and vulnerability. *Social Forces*, 87(2), 1089–1114.

Donoghue, E. (2007). Social science constructs in ecosystem assessments: Revisiting community capacity and community resiliency. *Society & Natural Resources*, 20, 899–912.

Drabek, T.E. (1987). Emergent structures. R.R. Dynes, B. DeMarchi, & C. Pelanda (Eds.), *Sociology of disasters: Contributions of sociology to disaster research*. Milan, Italy: Franco Angeli.

Drabek, T.E. (2003). *Strategies for coordinating disaster responses*. Boulder, CO: Institute of Behavioral Science, University of Colorado.

Drabek, T.E., & McEntire, D.A. (2002). Emergence phenomena and multiorganizational coordination in disasters: Lessons from the research literature. *International Journal of Mass Emergencies and Disasters*, 20(2), 197–224.

Dronkers, J., Misdorp, R., Schröder, P.C., Carey, J.J., Spradley, J.R., Vallianos, L., . . . & James, E. (1990). *Strategies for adaptation to sea level rise: Report of the Coastal Zone Management Subgroup to the Intergovernmental Panel on Climate Change response strategies working group*. The Netherlands: Ministry of Transport and Public Works. Retrieved from http://papers.risingsea.net/federal_reports/IPCC-1990-adaption-to-sea-level-rise.pdf

Duzce Valiligi. (2000). *Yeni kent, yeni yasam: Duzce* [New city, new life: Duzce]. Ankara: Yargici Ofset.

Dyer, C.L., & McGoodwin, J.R. (1999). "Tell them we're hurting": Hurricane Andrew, the culture of response, and the fishing peoples of South Florida and Louisiana. In A. Oliver-Smith & S.M. Hoffman (Eds.), *The angry earth: Disaster in anthropological perspective* (pp. 213–231). New York: Routledge.

Dynes, R. (2002). *The importance of social capital in disaster response*. Newark, DE: Disaster Research Center.

Dynes, R. (2003). Finding order in disorder: Continuities in the 9–11 response. *International Journal of Mass Emergencies and Disasters*, 21, 9–23.

Eaton, L. (2007). In New Orleans, plan to raze low-income housing draws protest. *New York Times*, December 14. Retrieved January 17, 2011 from http://www.nytimes.com/2007/12/14/us/nationalspecial/14orleans.html.

Edwards, F.L. (2006). Mitigation versus prevention: A view from the local government level. Paper presented at the ASPA National Conference, Washington, DC.

Edwards, F.L., & Goodrich, D.C. (2011). *Continuity of operations/continuity of government for state-level transportation organizations*. San Jose, CA: Mineta Transportation Institute.

Edwards, M.M., & Haines, A. (2007). Evaluating smart growth implications for small communities. *Journal of Planning Education & Research*, 27, 49–64.

Eikenberry, A., Arroyave, V., & Cooper, T. (2007). Administrative failure and the international NGO response to Hurricane Katrina. *Public Administration Review*, 67, 160–170. doi:10.1111/j.1540–6210.2007.00825.x.

Eisinger, P. (2006). Imperfect federalism: The intergovernmental partnership for homeland security. *Public Administration Review*, 66(4), 537–545.

Eller, W.S., & Gerber, B.J. (2010). Voluntary nonprofit organizations and disaster recovery: Assessing the value of the nonprofit contribution to the 2009 Alaskan rivers flood recovery effort. Unpublished report prepared for National Voluntary Organizations Active in Disaster.

Elliott, J., & Pais, J. (2006). Race, class, and Hurricane Katrina: Social differences in human responses to disaster. *Social Science Research*, 35, 295–321.

Elliott, J., & Pais, J. (2010). When nature pushes back: Environmental impact and the spatial redistribution of socially vulnerable populations. *Social Science Quarterly*, 91, 1187–1202.

Elsner, J.B., Kossin, J.P., & Jagger, T.H. (2008). The increasing intensity of the strongest tropical cyclones. *Nature*, 455, 92–95.

Emanuel, K.A. (2005). Increasing destructiveness of tropical cyclones over the past 30 years. *Nature*, 436, 686–688.

Emanuel, K.A. (2007). Environmental factors affecting tropical cyclone power dissipation. *Journal of Climate*, 20(22), 5497–5509.

EMAP (Emergency Management Accreditation Program). (2012). Who is accredited? Retrieved March 1, 2012 from http://www.emaponline.org/index.php?option=com_content&view=article&id=175&Itemid=139.

Enarson, E. (1999). Women and housing issues in two U.S. disasters: Hurricane Andrew and the Red River Valley Flood. *International Journal of Mass Emergencies and Disasters*, 17(1), 39–63.

Enarson, E. (2001). What women do: Gendered labor in the red river valley flood. *Environmental Hazards*, 3, 1–18.

Enarson, E., & Fordham, M. (2001). From women's needs to women's rights in disasters. *Environmental Hazards*, 3, 133–136.

Enarson, E., Fothergill, A., & Peek, L. (2006). Gender and disaster: Foundations and directions. In H. Rodriguez, E.L. Quarantelli, & R. Dynes (Eds.), *Handbook of disaster research*. United States of America: Springer.

Enarson, E., & Scanlon, J. (1999). Gender patterns in flood evacuation: A case study in Canada's Red River Valley. *Applied Behavioral Science Review*, 7, 103–124.

Erdik, M. (2000). Report on 1999 Kocaeli and Duzce (Turkey) earthquakes. Retrieved from http://www.koeri.bound.edu.tr/depremmuh/Kocaelireport.pdf.

Ergunay, O. (1999) A Perspective of Disaster Management in Turkey: Issues and Prospects. Urban Settlements and Natural Disasters, Proceedings of UIA Region II Workshop, June 1999, pp.1–9 .

Erickson, B. (2003). Social networks: The value of variety. *Contexts*, 2(1), 25–31.

Ersing, R. L. & Kost, K. A. (2012). Approaching practice: Social networks in the context of disasters. In Ersing, R. L. & Kost, K. A. (Eds.) *Surviving disaster: The role of social networks*. (p. 1–14). Chicago, IL: Lyceum Books, Inc.

EU-ASSRBCVUL Group & Lemyre, L. (2007). Final report: Assessment of the vulnerabilities of modern societies to terrorist acts employing radiological, biological or chemical agents with the view to assist in developing preventive and suppressive crisis management strategies. Project 502 476. European Union Research Commission 6th Framework Report.

European-Mediterranean Seismological Center. (2011). Earthquake details. Retrieved from http://www.emsc-csem.org/Earthquake/202/Earthquake-M7-2-Eastern-Turkey.

Falk, W.W., Hunt, M.O., & Hunt, L.L. (2006). Hurricane Katrina and New Orleanians' sense of place: Return and reconstitution or "Gone with the Wind"? *Du Bois Review*, 3(1), 115–128.

Fankhauser, S. (2010). The costs of adaptation. *WIRES Climate Change*, 1(1), 23–30.

Fankhauser, S., Smith, J.B., & Tol, R.S.J. (1999). Weathering climate change: Some simple rules to guide adaptation decisions. *Ecological Economics*, 30, 67–78.

FBRM (State Of Florida Department of Community Affairs Bureau of Recovery & Mitigation). (n.d.). *Wildfires*. Retrieved October 30, 2009 from http://www.sjcemergencymanagement.org/wildfires.html#Wild.

FDEM (Florida Division of Emergency Management). (2002). *Florida hazard watch: Tornadoes*. Retrieved November 12, 2009 http://www.floridadisaster.org/bpr/EMTOOLS/severe/tornadoes.html.

FDEM (Florida Division of Emergency Management). (2009). Retrieved November 15, 2009 from http://www.floridadisaster.org/bpr/EMTOOLS/Severe/tornadoes.htm.

FDOF (Florida Division of Forestry). (2004a). *Wildfire mitigation in Florida.* Retrieved October 30, 2009 from http://www.fl-dof.com/wildfire/wf_pdfs/Wildfire_Mitigation_in_FL.pdf.

FDOF (Florida Division of Forestry). (2004b). *Wildland fire.* Retrieved October 30, 2009 from http://www.fl-dof.com/wildfire/stats_significant_fires.html.

FDOF (Florida Division of Forestry). (2004c). *Wildland fire and fire protection.* Retrieved October 30, 2009 from http://www.fl-dof.com/wildfire/index.html.

FEMA (Federal Emergency Management Agency). (2002). Disaster aid totals more than $12.8 million. Release no. 1439–17. November 22. Retrieved from http://www.fema.gov/news/newsrelease.fema?id=2685.

FEMA (Federal Emergency Management Agency). (2004). *Emergency Management Institute catalog of activities.* Retrieved March 10, 2012 from http://training.fema.gov/EMIWeb/downloads/EMICatalog0304PartI.pdf.

FEMA (Federal Emergency Management Agency). (2005a). *Adoption of flood insurance rate maps by participating communities.* Publication no. 495. September. Retrieved April 16, 2012 from www.fema.gov/library/file?type=publishedFile&file=fema.

FEMA (Federal Emergency Management Agency). (2005b). *Principles of Emergency Management.* 3.1–3.2. Retrieved March 15, 2005 from http://training.fema.gov/EMIWeb/Downloads/is230.doc.

FEMA (Federal Emergency Management Agency). (2010). *Developing and maintaining emergency operations plans: Community preparedness guide (CPG) 101.* November. Washington, DC: FEMA.

FEMA (Federal Emergency Management Agency). (2011a). Housing recovery support function. In *National Disaster Recovery Framework.* Retrieved from http://www.fema.gov/pdf/recoveryframework/housing_rsf.pdf.

FEMA (Federal Emergency Management Agency). (2011b). *National Disaster Recovery Framework.* September. Retrieved from http://www.fema.gov/recoveryframework/.

FEMA (Federal Emergency Management Agency). (2011c). *Whole community.* December. Retrieved from http://www.fema.gov/about/wholecommunity.shtm.

FEMA (Federal Emergency Management Agency). (2011d). *A whole community approach to emergency management: Principles, themes, and pathways for action.* FDOC 104–008–1. January. Washington, DC: FEMA. Retrieved January 12, 2012 from http://www.pascenter.org/publications/publication_home.php?id=1256&focus=PAS%20Library.

FEMA. (2012). National Mitigation Framework Review Package. Retrieved July 26, 2012 http://www.google.com/url?sa=t&rct=j&q=&esrc=s&source=web&cd=2&ved=0CGMQFjAB&url=http%3A%2F%2Fwww.fema.gov%2Flibrary%2Ffile%3Bjsessionid%3D9A7BEC5D10B1DE5E5A3B9337EA1EA309.WorkerLibrary%3Ftype%3DpublishedFile%26file%3Dpeo_nationalmitigationframework_20120302_v3.pdf%26fileid%3D58c40a70-6638-11e1-948f-001cc456982e&ei=no8RUKKpAYSY9QSY-YHYCA&usg=AFQjCNHEZFQarieWrFbWsqlVdQgV8fog7g&sig2=a8ItlwoC7jN4xaNcUz-6kgFEMA (Federal Emergency Management Agency). (n.d.). Basic disaster supplies kit. Retrieved April 3, 2012 from http://www.ready.gov/basic-disaster-supplies-kit.

Few, R. (2003). Flooding, vulnerability and coping strategies: Local responses to a global threat. *Program Developement Studies*, 3, 43–58.

Fitzpatrick, P. (1999). *Natural disasters: Hurricanes.* Santa Barbara, CA: ABC-CLIO.

Fleury, J., & Lee, S.M. (2006). The social ecological model and physical activity in African American women. *American Journal of Community Psychology*, 37, 129–140.

Folke, C. (2006). Resilience: The emergence of a perspective for social-ecological system analyses. *Global Environmental Change*, 16, 253–267.

Folke, C., Hahn, T., Olsson, P., & Norberg, J. (2005). Adaptive governance of social-ecological systems. *Annual Review of Environmental Resources*, 30, 441–473.

Forbes, G. (2007). *Deadly Florida tornado outbreak*. Retrieved November 16, 2009 from http://www.weather.com/blog/weather/8_11687.html.

Fordham, M. (1999). The intersection of gender and social class in disaster: Balancing resilience and vulnerability. *International Journal of Mass Emergencies and Disasters*, 17, 15–37.

Foster, J., Winkelman, S., & Lowe, A. (2011). *Lessons learned on local climate adaptation from the Urban Leaders Adaptation Initiative*. Washington, DC: Center for Clean Air Policy.

Fothergill, A., Maestas, E.G., & Darlington, J.D. (1999). Race, ethnicity and disasters in the United States: A review of the literature. *Disasters*, 23(2), 156–173.

Fothergill, A., & Peek, L.A. (2004). Poverty and disasters in the United States: A review of recent sociological findings. *Natural Hazards*, 32, 89–110.

Frank, K.A. (2009). Quasi-ties: Directing resources to members of a collective. *American Behavioral Scientist*, 52(12), 1613–1645.

Freitag, B., Bolton, S., Westerlund, F., & Clark, J. (2009). *Floodplain management: A new approach for a new era*. Washington, DC: Island Press.

Friedland, R., & Alford, R.R. (1991). Bringing society back in: Symbols, practices and institutional contradictions. In W.W. Powell & P.J. DiMaggio (Eds.), *The new institutionalism in organizational analysis* (pp. 232–263). Chicago: University of Chicago Press.

Fritz, C. (1961). Disaster. In R.K. Merton & R.A. Nisbet (Eds.), *Contemporary social problems*. New York: Harcourt.

Fugate, C. (2010). Opening plenary speech at the national conference of the International Association of Emergency Managers, San Antonio, TX, October 31.

Fujita, K., & Shaw, R. (2010). Environmental changes and Japan's forest management. *Asian Journal of Environment and Disaster Management*, 2(2), 153–178.

Fujita, K., Takeuchi, Y., & Shaw, R. (2011). Voluntary self-help organization and fire volunteer for mountain disaster risk reduction in Reihoku area, Japan. *Asian Journal of Environment and Disaster Management*, 3(3), 317–337.

Fussel, H.M. (2007). Adaptation planning for climate change: Concepts, assessment approaches, and key lessons. *Sustainability Science*, 2, 265–275.

Ganapati, N.E., & Ganapati, S. (2009). Enabling participatory planning in post-disaster contexts: A case study of World Bank's housing reconstruction in Turkey. *Journal of the American Planning Association*, 75(1), 41–59.

GAO (Government Accountability Office). (2008). *National disaster response: FEMA should take action to improve capacity and coordination between government and voluntary sectors*. Washington, DC: GAO.

GAO (Government Accountability Office). (2008). FEMA should take action to improve capacity and coordination between government and voluntary sectors. Report no. GAO-08–369. February 27. Retrieved from http://www.gao.gov/products/GAO-08–369.

Garmezy, N. (1974). The study of competence in children at risk for severe psychopathology. In A. Koupernik (Ed.), *The child and his family: Children at psychiatric risk*, vol. 3 (pp. 77–97). New York: Wiley.

Garmezy, N. (1991). Resiliency and vulnerability to adverse developmental outcomes associated with poverty. *American Behavioral Scientist*, 34, 416–430.

Garmezy, N. (1993). Vulnerability and resilience. In D.C. Funder, R.D. Parker, C. Tomlinson Keesey, & K. Widaman (Eds.), *Studying lives through time:*

Approaches to personality and development (pp. 377–398). Washington, DC: American Psychological Association.

Garmezy, N., & Masten, A. (1991). The protective role of competence indicators in children at risk. In E. Cummings et al. (Eds.), *Perspective on Stress and Coping* (pp. 151–174). New Jersey: Hilldale, Erlbaum.

Gazley, B. (2008). Inter-sectoral collaboration and the motivation to collaborate: Toward an integrated theory. In L.B. Bingham & R. O'Leary (Eds.), *Big ideas in collaborative public management* (pp. 36–54). Armonk, NY: M.E. Sharpe.

Gazley, B. (2010). Linking collaborative capacity to performance measurement in government-nonprofit partnerships. *Nonprofit and Voluntary Sector Quarterly*, 39(4), 653–673.

Gazley, B., & Brudney, J.L. (2007). The purpose (and perils) of government-nonprofit partnership. *Nonprofit and Voluntary Sector Quarterly*, 36(3), 389–415.

Gell-Mann, M. (1994). *The quark and the jaguar: Adventures in the simple and the complex.* New York: WH Freeman and Co.

Gerber, B.J. (2007). Disaster management in the United States: Examining key political and policy challenges. *Policy Studies Journal*, 35(2), 227–238.

Gibson, S., & Lemyre, L. (2012). A multi-level analysis of preparedness behaviours. Manuscript under review.

Girard, C., & Peacock, W.G. (1997). Ethnicity and segregation: Post-hurricane relocation. In W.G. Peacock, B.H. Morrow, & H. Gladwin (Eds.), *Hurricane Andrew: Ethnicity, gender and the sociology of disasters* (pp. 191–205). London and New York: Routledge Press.

Godschalk, D.R. (2003). Urban hazard mitigation: Creating resilient cities. *Natural Hazards Review*, 4(3), 136–143.

Godschalk, D.R., Beatley, T., Berke, P., Brower, D., & Kaiser, E. (1999). *Natural hazard mitigation: Recasting disaster policy and planning.* Washington, DC: Island Press.

Godschalk, D.R., Brower, D.J., & Beatley, T. (1989). *Catastrophic coastal storms.* Durham, NC: Duke University Press.

Godschalk, D.R., Kaiser, E.J., & Berke, P.R. (1998). Integrating hazard mitigation and local land use planning. In R.J. Burby (Ed.), *Cooperating with nature: Confronting natural hazards with land-use planning for sustainable communities* (pp. 85–118). Washington, DC: Joseph Henry Press.

Goldstein, B.E. (2011). *Collaborative resilience: Moving through crisis to opportunity.* Cambridge, MA: MIT Press.

Goodnough, A. (2006). Chertoff pushes for more hurricane readiness. *New York Times*, April 13. Retrieved from http://select.nytimes.com/gst/abstract.html?res=F10A13F934540C708DDDAD0894DE404482.

Government of Japan. (2012). *Summary of disaster countermeasures and disaster management plan of 2011.* In Japanese. Retrieved from www.bousai.go.jp/1info/pdf/saigaipanf.pdf.

Grannis, J. (2011). *Adaptation tool kit: Sea-level rise and coastal land use.* Washington, DC: Georgetown Climate Center. Retrieved from http://www.georgetownclimate.org/sites/default/files/Adaptation_Tool_Kit_SLR.pdf.

Granovetter, M. (1985). Economic action and social structure: The problem of embeddedness. *American Journal of Sociology*, 91(3), 481–510.

Green, J.J., Gill, D.A., & Kleiner, A.M. (2006). From vulnerability to resiliency: Assessing impacts and responses to disaster. *Southern Rural Sociology*, 21, 89–99.

Green, J.J., Kleiner, A.M., & Montgomery, J.P. (2007). The texture of local disaster response: Service providers' views following Hurricane Katrina. *Southern Rural Sociology*, 22, 28–44.

Gregg, R.M., Hansen, L.J., Feifel, K.M., Hitt, J.L., Kershner, J.M., Score, A., & Hoffman, J.R. (2011). *The state of marine and coastal adaptation in North America: A synthesis of emerging ideas*. Bainbridge Island, WA: EcoAdapt.

Grinsted, A., Moore, J.C., & Jevrejeva, S. (2010). Reconstructing sea level from paleo and projected temperatures 200 to 2100 AD. *Climate Dynamics*, 34, 461–472.

Gulati, R., & Gargiulo, M. (2001). Where do interorganizational networks come from? *American Journal of Sociology*, 104(5), 1439–1493.

Gulkan, P. (2000). Recent natural disasters in Turkey: An overview of the national technological capacity and its utilization. Unpublished essay. Middle East Technical University, Ankara.

Gurwitch, R.H., Pfefferbaum, B., Montgomery, J.M., Klomp, R.W., & Reissman, D.B. (2007). *Building community resilience for children and families*. Oklahoma City: Terrorism and Disaster Center at the University of Oklahoma Health Sciences Center.

Haas, P.M. (1992). Introduction: Epistemic communities and international policy coordination. *International Organization*, 46(1), 1–36.

Hakim, D. (2011). Cuomo takes tough stance on nuclear reactors. *New York Times*, June 28. Retrieved from http://www.nytimes.com/2011/06/29/nyregion/cuomo-emphasizes-aim-to-close-indian-point-plant.html.

Hallegatte, S. (2009). Strategies to adapt to an uncertain climate change. *Global Environmental Change*, 19(2), 240–247.

Hammer, R.B., Stewart, S.I., & Radeloff, V.C. (2009) Demographic trends, the wildland-urban interface, and wildfire management. *Society & Natural Resources*, 22(8), 777–782.

Haney, T.J., Elliot, J.R., & Fussell, E. (2007). Families and hurricane response: Evacuation, separation, and the emotional toll of Hurricane Katrina. In D.B. Brunsma, D. Overfelt, & J.S. Picou (Eds.), *The sociology of Katrina: Perspectives on a modern catastrophe* (pp. 71–90). New York: Rowman and Littlefield.

Hanneman, R.A. (2001). *Introduction to social network methods*. Riverside: University of California. Retrieved from http://faculty.ucr.edu/~hanneman/nettext/.

Hansen, J.E. (2007). Scientific reticence and sea level rise. *Environmental Research Letters*, 2(2), 024002 (6 pp.).

Hansen, J.E., & Sato, M. (2011). Paleoclimate implications for human-made climate change. Retrieved December 29, 2011 from http://arxiv.org/ftp/arxiv/papers/1105/1105.0968.pdf.

Harris, J., & Clements, B. (2007). Using social network analysis to understand Missouri's system of public health emergency planners. *Public Health Reports*, 122, 488–498.

Hartman, C.W., & Squires, G.D. (2006). *There is no such thing as a natural disaster: Race, class, and Hurricane Katrina*. New York: Routledge.

Hausman, J.A. (1978). Specification tests in econometrics. *Econometrica*, 46(6), 1251–1271.

Hawkins, C., & Wang, X. (2012). Sustainable development governance: Citizen participation and support networks in local sustainability initiatives. *Public Works Management & Policy*, 17(1), 7–29.

Hays, J. (2012). *The community resilience framework: A pathway toward resilience*. Oakland, CA: FEMA Region IX.

Hazards and Vulnerability Research Institute. (2010). *The spatial hazard events and losses database for the United States, Version 8.0* [online database]. Columbia, SC: University of South Carolina. Retrieved December 17, 2010 from http://www.sheldus.org.

Henry, A.D. (2011). Ideology, power, and the structure of policy networks. *Policy Studies Journal*, 39(3), 361–383.

Hey, D.L. (2002). Modern drainage design: The pros, the cons, and the future. *Hydrological Science and Technology*, 18(14), 89–99.

HFA. (2005). *Hyogo framework for action 2005–2015: Building the resilience of nations and communities to disasters*. Geneva, Switzerland: UN ISDR.

Hilhorst, D. (2003). Responding to disasters: Diversity of bureaucrats, technocrats and local people. *International Journal of Mass Emergencies and Disasters*, 21(1), 37–55.

Hill, M. (1968). A goals-achievement matrix for evaluating alternative plans. *Journal of the American Planning Association*, 34(1), 19–29.

Hirsch, R.M., Walker, J.F., Day, J.C., & Kallio, R. (1990). The influence of main on hydrological systems. In M.G. Wolman & H.C. Riggs (Eds.), *Surface water hydrology* (pp. 329–359). Boulder, CO: Geological Society of America.

Hofferth, S.L., & Iceland, J. (1998). Social capital in rural and urban communities. *Rural Sociology*, 63(4), 574–598.

Hoffman, S.M., & Oliver-Smith, A. (Eds.) (2002). *Catastrophe and culture: The anthropology of disaster*. Santa Fe, NM: School of American Research Press.

Holland, J.H. (1995). *Hidden order*. Reading, MA: Addison-Wesley.

Holling, C.S. (1973). Resilience and stability of ecological systems. *Annual Review of Ecology and Systematics*, 4, 2–23.

Holling, C.S. (1978). *Adaptive environmental assessment and management*. New York: John Wiley and Sons.

Homeland Security Institute. (2006). *Heralding unheard voices: The role of faith-based organizations and nongovernmental organizations during disasters*. Arlington, VA: Department of Homeland Security Science and Technology Directorate.

Hopkins, A. (1999). Counteracting the cultural causes of disaster. *Journal of Contingencies and Crisis Management*, 37, 141–149.

Hsu, M.H., Chen, S.H., & Chang, T.J. (2000). Inundation simulation for urban drainage basin with storm sewer system. *Journal of Hydrology*, 234, 21–37.

Huber, P.W. (1986). The Bhopalization of American tort law. In National Academy of Engineering, *Hazards: Technology and fairness* (pp. 89–110). Washington, DC: The National Academies Press.

Hughey, J., Speer, P.W., & Peterson, A. (1999). Sense of community in community organizations: Structure and evidence of validity. *Journal of Community Psychology*, 27(1), 97–113.

Hurriyet daily newspaper. (2011). Istanbul, Turkey, October 23–November 8, 2011 issues.

Hutchins, R., & Speevak-Sladowski, P. (2009). Community resiliency handbook: Enhancing the role of the voluntary sector on health emergencies. Centre for Voluntary Sector Research and Development, Carleton University.

IBHS (Insurance Institute for Business and Home Safety). (2011). Rating the states: An assessment of residential building code and enforcement systems for life safety and property protection in hurricane-prone regions. Tampa, FL: IBHS, December 31.

Innes, J.E. (1996). Planning through consensus building: A new view of the comprehensive planning ideal. *Journal of the American Planning Association*, 62(4), 460–472.

Innes, J., & Booher, D.E. (2003). *Collaborative policymaking: Governance through dialogue in deliberative policy analyses*. Cambridge: Cambridge University Press.

Innes, J.E., & Booher, D.E. (2010). *Planning with complexity: An introduction to collaborative rationality for public policy*. New York: Routledge Press.

Isserman, A.M., & Merrifield, J.D. (1987). Quasi-experimental control group methods for regional analysis: An application to an energy boomtown and growth pole theory. *Economic Geography*, 63(1), 3–19.

Isserman, A.M., & Rephann, T. (1995). The economic effects of the Appalachian Regional Commission: An empirical assessment of 26 years of regional development planning. *Journal of the American Planning Association*, 61(3), 345–364.

Iwami, T. (1997). A study on community self-help organization for disaster prevention. *Journal of the Japan Association for Social and Economic Systems Studies*.

Janssen, M.A., & Ostrom, O. (2006). Resilience, vulnerability, and adaptation: A cross-cutting theme of the international human dimensions program on global enviromental change. *Global Environmental Change*, 16, 237–239.

Jensen, J.K. (2009a). Climate change and rural communities in the US. Briefing paper. Rural Policy Research Institute. Retrieved from http://www.rupri.org/Forms/Climate_Change_Brief.pdf.

Jensen, J. (2009b). NIMS in rural America. *Journal of Homeland Security and Emergency Management*, 27(3).

Jepson, E.J. (2004). The adoption of sustainable development policies and techniques in U.S. cities: How wide, how deep, and what role for planners? *Journal of Planning Education and Research*, 23, 229–241.

JMA (Japan Meteorological Agency). (2010). Abnormal weather risk map. 3–1 day precipitation with more than 100mm. In Japanese. Retrieved October 31, 2011 from http://www.data.kishou.go.jp/climate/riskmap/index.html.

JMA (Japan Meteorological Agency). (2012). *Observed earthquakes larger than seismic intensity of five: Within aftershocks, after the initial shake*. In Japanese. Retrieved from http://www.seisvol.kishou.go.jp/eq/2011_03_11_tohoku/i5.pdf.

Jones, C., Hesterly, W.S., & Borgatti, S.P. (1997). A general theory of network governance: Exchange conditions and social mechanisms. *Academy of Management Review*, 22(4), 911–945.

Kadushin, C. (2004). Too much investment in social capital? *Social Networks*, 26, 75–90.

Kahneman, D., & Klein, G. (2009). Conditions for intuitive expertise: A failure to disagree. *American Psychologist*, 64(6), 515–526.

Kaiser, E.J., Godschalk, D.R., & Chapin, F.S. (1995). *Urban land use planning*. 4th edition. Chicago, IL: University of Illinois Press.

Kamel, N.M.O., & Loukaitou-Sideris, A. (2004). Residential assistance and recovery following the Northridge earthquake. *Urban Studies*, 41(3), 533–562.

Kapucu, N. (2005). Interorganizational coordination in dynamic context: Networks in emergency response. *Connections*, 26(2), 33–48.

Kapucu, N. (2006a). Interagency communication networks during emergencies: Boundary spanners in multi-agency coordination. *American Review of Public Administration*, 36(2), 207–225.

Kapucu, N. (2006b). Public–nonprofit partnerships for collective action in dynamic contexts of emergencies. *Public Administration*, 84(1), 205–220.

Kapucu, N. (2007). Building community capacity to respond. *The Public Manager*, 36(3), 21–25.

Kapucu, N. (2008a). Collaborative emergency management: Better community organizing, better public preparedness and response. *Disasters*, 32(2), 239–262.

Kapucu, N. (2008b). Culture of preparedness: Household disaster preparedness. *Disaster Prevention and Management*, 17(4), 526–535.

Kapucu, N. (2009). Interorganizational coordination in complex environments of disasters: The evolution of intergovernmental disaster response systems. *Journal of Homeland Security and Emergency Management*, 6(1), 1–26.

Kapucu, N. (2012). *The network governance in response to acts of terrorism: Comparative analyses*. New York: Routledge.

Kapucu, Naim. (2012). Disaster Resilience and Adaptive Capacity in Central Florida, US, and in Eastern Marmara Region, Turkey. *Journal of Comparative Policy Analysis: Research & Practice*. 14(3): 202–216.

Kapucu, N., Arslan, T., & Collins, M. (2010). Examining intergovernmental and interorganizational response to catastrophic disasters. *Administration & Society*, 42(2), 222–247.

Kapucu, N., Arslan, T., & Demiroz, F. (2010). Collaborative emergency management and national emergency management network. *Disaster Prevention and Management*, 19(4), 452–468.

Kapucu, N., & Demiroz, F. (2011). Measuring performance for collaborative public management using network analysis methods and tools. *Public Performance Management Review*, 34(4), 551–581.

Kapucu, N., Garayev, V., & Augustin, M. (2009). Interstate partnerships in emergency management: Emergency management assistance compact (EMAC) in response to catastrophic disasters. *Public Administration Review*, 69(2), 297–313.

Kapucu, N., & Özerdem, K. (2012). *Managing emergencies and crises*. Burlington, MA: Jones & Bartlett Learning.

Kapucu, N., & Van Wart, M. (2006). The evolving role of the public sector in managing catastrophic disasters: Lessons learned. *Administration and Society*, 38(3), 279–308.

Kara Kuvvetleri Komutanligi Egitim ve Doktrin Komutanligi. (2000). *Depremler ve alinmasi gereken dersler* [Earthquakes and lessons need to be taken]. Ankara.

Karl, T.R., Meehl, G.A., Miller, C.D., Hassol, S.J., Waple, A.M., & Murray, W.L. (2008). *Weather and climate extremes in a changing climate; Regions of focus: North America, Hawaii, Caribbean, and U.S. Pacific Islands*. Synthesis and Assessment Product 3.3. Washington, DC: U.S. Climate Change Science Program and the Subcommittee on Global Change Research.

Karl, H., Susskind, L., & Wallace, K. (2007). A dialogue, not a diatribe: Effective integration of science and policy through joint fact finding. *Environment*, 49(1), 20–34.

Kauffman, S.A. (1993). *The origins of order: Self-organization and selection in evolution*. New York: Oxford University Press.

Khademian, A.M. (2005). The politics of homeland security. In D.G. Kamian (Ed.), *The McGraw-Hill homeland security handbook*. Columbus, OH: McGraw-Hill.

Kiel, L.D. (1994). *Managing chaos and complexity in government*. San Francisco: Jossey-Bass Publishers.

Kiker, G.A., Bridges, T., Varghese, A., Seager, T., & Linkov, I. (2005). Application of multicriteria decision analysis in environmental decision making. *Integrated Environmental Assessment and Management*, 1(2), 95–108.

Killian, L.M. (1955). The significance of multiple-group membership in disaster. *The American Journal of Sociology*, 57(4), 309–314.

Kilmer, R.P., Gil-Rivas, V., Tedeschi, R.G., & Calhoun, L.G. (2010). *Helping families and communities recover from disaster: Lessons learned from Hurricane Katrina and its aftermath*. Washington, DC: American Psychological Association.

Kim, Y.C., & Kang, J. (2010). Communication, neighborhood belonging and household hurricane preparedness. *Disasters*, 34, 470–488.

King, D.M., & Herbert, L.W. (1997). The fungibility of wetlands. *National Wetlands Newsletter*, 19(5), 10–13.

Kneese, A.V. (1984). *Measuring the benefits of clean air and water*. Washington, DC: Resources for the Future.

Kobasa, S.C.O. (1987). Stress responses and personality. In R.C. Barnett, L. Biener, & K. Baruch (Eds.), *Gender and stress*. New York: Free Press.

Koliba, C.J., Zia, A., & Mills, R.M. (2011). Accountability in governance networks: An assessment of public, private, and nonprofit emergency management practices following Hurricane Katrina. *Public Administration Review*, 71(2), 210–220.

Koppenjan, J., & Klijn, E.J. (2004). *Managing uncertainties in networks*. London, UK: Routledge.

Koski, C. (2011). Committed to protection? Partnerships in critical infrastructure protection. *Journal of Homeland Security and Emergency Management*, 8(1), 1860–1879.

Kretzman, J.P., & McKnight, J.L. (1993). *Building communities from the inside out: A path toward finding and mobilizing a community's assets*. Chicago: ACTA Publications.

Kroll, C.A., Lands, J.D., Shen, Q., & Stryker, S. (1991). *Economic impacts of the Loma Prieta earthquake: A focus on small business*. Working paper no. 91–187. Berkeley: University of California at Berkeley, University of California Transportation Center and the Center for Real Estate and Economics.

Kunreuther, H. (2006). Disaster mitigation and insurance: Learning from Katrina. *The Annals of the American Academy of Political and Social Science*, 604, 208–227.

Kusenbach, M. (2003). Street phenomenology: The go-along as ethnographic research tool. *Ethnography*, 4, 449–479.

Kusenbach, M. (2008). A hierarchy of urban communities: Observations on the nested character of place. *City and Community*, 7, 225–249.

Kusenbach, M., Simms, J.L., & Tobin, G.A. (2010). Disaster vulnerability and evacuation readiness: Coastal mobile home residents in Florida. *Natural Hazards*, 52, 79–95.

Kusenbach, M., & Taylor, C.K. (2012). Hurricane evacuation among mobile home residents in Florida: The complex role of social networks. In R. Ersing & K.A. Kost (Eds.), *Surviving disaster: The role of social networks* (pp. 63–83). Chicago, IL: Lyceum.

Lai, A. (2012). Organizational collaborative capacity in fighting pandemic crises: A literature review from the public management perspective. *Asia-Pacific Journal of Public Health*, 24(1), 7–20.

Lannoo, S., Verhaeghe, P.-P., Vandeputte, B., & Devos, C. (2012). Differences in social capital between urban and rural environments. *Journal of Urban Affairs*. (Early view http://onlinelibrary.wiley.com/doi/10.1111/j.1467–9906.2011.00592.x/).

Larson, L., & Plasencia, D. (2001). No adverse impact: A new direction in floodplain management policy. *Natural Hazard Review*, 2(A), 167–181.

Lazarus, R., & Folkman, S. (1984). *Stress, appraisal and coping*. NewYork: Springer.

Leary, N., Adjuwon, J., V. Barros, P. Batima, B. Biagini, I. Burton, . . . & Webbe, M. (2008). A stitch in time: General lessons from specific cases. In N. Leary, J. Adejuwwon, V. Barroos, I. Burton, J. Kulkarni & R. Lasco (Eds.), *Climate Change and Adaptation*, 1–27. United Kingdom: Earthscan.

Lebel, L., Anderies, J.M., Campbell, B., Folke, C., Hatfield-Dodds, S., Hughes, T.P., & Wilson, J. (2006). Governance and the capacity to manage resilience in regional social-ecological systems. *Ecology and Society*, 11(1), 19, 1–15.

Lee, J.E.C., Gibson, S., Markon, M.P.L., & Lemyre, L. (2009). A preventive coping perspective of individual response to terrorism in Canada. *Current Psychology*, 28, 69–84.

Lee, J.E.C., Lemyre, L., & Krewski D. (2010). A multi-method, multi-hazard approach to explore the uniqueness of terrorism risk perceptions and worry. *Journal of Applied Social Psychology*, 40(1), 241–272.

Lee, K.N. (1993). *Compass and gyroscope: Integrating politics and science for the environment*. Washington, DC: Island Press.

Lemyre, L., Clément, M., Corneil, W., Craig, L., Boutette, P., Tyshenko, M.G., Karyakina, N., Clarke, R., & Krewski, D. (2005). A psychosocial risk assessment and management framework to enhance response to CBRN terrorism threats and attacks. *Biosecurity and Bioterrorism*, 3(4), 316–330.

Lemyre, L., Gibson, S., Zlepnig, J., Meyer-Macleod, R., & Boutette, P. (2009). Emergency preparedness for higher risk populations: Psychosocial considerations. *Radiation Protection Dosimetry*, 134, 207–214.

Lemyre, L., Johnson, C., & Corneil, W. (2010). Psychosocial considerations for mass decontamination. *Radiation Protection Dosimetry*, 142, 17–23.

Lemyre, L., & Orpana, H. (2002). Integrating population health into social ecology: Role of family medicine researchers. *Canadian Family Physician*, 48, 1349–1350.

Lemyre, L., Pinsent, C., Boutette, P., Corneil, W., Riding, J., Riding, D., Johnson, C., Lemus, C., Blust, S., Kitchener, H.M., Markon, M.-P.L., & Gibson, S. (2011). *Research using in vivo simulation of meta-organizational shared decision making (SDM)*. Published report for Defence Research and Development Canada, no. CR-2011–31.

Lemyre, L., Pinsent, C., Johnson, C., Boutette, P., Corneil, W., Riding, J., Lemus, C., Blust, S., & Riding, D. (2012, in press). Developing collaboration in complex events: A model for civil-military inter-organizational problem-solving and decision-making. Proceedings of the 16th ICCRTS, Collective C2 in multinational civil-military operations.

Lemyre, L., & Tessier, R. (2003). Measuring psychological stress: Concept, model, and measurement instrument in primary care research. *Canadian Family Physician*, 49, 1159–1160.

Levine, J., Esnard, A-M., & Sapat, A. (2007). Population displacement and housing dilemmas due to catastrophic disasters. *Journal of Planning Literature*, 22(1), 3–15.

Levy, D.A. (2011). It takes a village: Community approach to managing large scale incidents. Presentation to the U.S. HHS Integrated Training Summit, May 3. PowerPoint slides. Retrieved from http://www.integratedtrainingsummit.org/presentations/2011/main_training_summit/17_-_it_takes_a_village_community_approach_to_managing_large_scale_incidents.pdf.

Lewis, W.M. (2001). *Wetlands explained: Wetland science, policy, and politics in America*. New York: Oxford University Press.

Lindell, M.K., & Perry, R.W. (2007). Planning and preparedness. In W. Waugh and K. Tierney (Eds.), *Emergency management: Principles and practice for local government*. Washington, DC: ICMA.

Linder, S.H. (2000). Coming to terms with the public-private partnership. In P.V. Rosenau (Ed.), *Public-private policy partnerships*. Cambridge, MA: MIT Press.

Lindseth, G. (2004). The cities for climate protection campaign (CCPC) and the framing of local climate policy. *Local Environment*, 9, 325–336.

Lobao, L., & Kraybill, D.S. (2005a). The emerging roles of county governments in metropolitan and nonmetropolitan areas: Findings from a national survey. *Economic Development Quarterly*, 19(3), 245–259.

Lobao, L., & Kraybill, D. (2005b). Rural county governments, economic development, and public services: The impact of decentralization over time. USDA Research, Education, and Economics Information System.

Lobao, L., & Kraybill, D.S. (2009). Poverty and local governments: Economic development and community service provision in an era of decentralization. *Growth and Change*, 40, 418–451.

Longhurst, R. (1995). The assessment of community vulnerability in hazard prone areas. *Disasters*, 19, 269–270.

Longstaff, P.H., Armstrong, N.J., Perrin, K., Parker, W.M., & Hidek, M.A. (2010). Building resilient communities: A preliminary framework for assessment. *Homeland Security Affairs*, 6(3), 1–23.

Lowe, A., Foster, J., & Winkelman, S. (2009). *Ask the climate question: Adapting to climate change impacts in urban regions*. Washington, DC: Center for Clean Air Policy.

Lu, J-C., Peacock, W.G., Yang, Z., & Dash, N. (2007). A comparative study of single family and multifamily housing recovery following 1992 Hurricane Andrew in Miami-Dade County, Florida. In *Proceedings of the Hazards and Disasters Researchers Meeting* (pp. 22–26). Boulder, CO.

Luthar, S.S. (1993). Annotation: Methodological and conceptual issues in the study of resilience. *Journal of Child Psychology and Psychiatry*, 34, 441–453.

Luthar, S.S. (1999). *Poverty and children's adjustment*. Thousand Oaks, CA: Sage.

Luthar, S.S. (Ed.) (2003). *Resilience and vulnerability: Adaptation in the context of childhood adversities*. New York: Cambridge University Press.

Magsino, S.L. (2009). *Applications of social network analysis for building community disaster resilience*. Washington, DC: The National Academies Press.

Majchrzak, A., Jarvenpaa, S., & Hollingshead, A. (2007). Coordinating expertise among emergent groups responding to disasters. *Organization Science*, 18(1), 147–161.

Manyena, B., O'Brien, G., O'Keefe, P., & Rose, J. (2011). Disaster resilience: A bounce back or bounce forward ability? *Local Environment: The International Journal of Justice and Sustainability*, 16(5), 417–424.

Marcy, D., Allen, A., Sweet, W., Gill, S., Luscher-Aissaoui, A., Myers, E., & Zervas, C. (2012). Incorporating sea level change scenarios at the local level. National Oceanic and Atmospheric Administration.

Marion, R. (1999). *The edge of organization*. Thousand Oaks, CA: Sage Publications.

Marsalek, J., & Chocat, B. (2002). International report: Stormwater management. *Water Science and Technology*, 46(6/7), 1–17.

Masanori, M., Sassa, K., Yokoyama, S., Hiura, H., & Nakaya, S. (2006). Estimation of the flood peak by using the trace of debris flow: Cases of debris flow disasters in Reihoku District, Kochi Prefecture, induced by the heavy rainfall of Typhoon Megi in 2004. *Journal of Japan Landslide Society*, 42(6), 478–492.

Masek, J., & Ridge, R. (2009). *Identification of methods to achieve successful implementation of nonstructural and equipment seismic restraints*. Oakland, CA: Earthquake Engineering Research Institute.

Mason, R.O., & Mitroff, I.I. (1973). A program for research on management information systems. *Management Science*, 19(5), 475–487.

Masten, A.S. (1999). Resilience comes of age: Reflections on the past and outlook for the next generation of research. In M.D. Glantz, J. Johnson, & L. Huffman (Eds.), *Resilience and development: Positive life adaptations* (pp. 289–296). New York: Plenum.

Masten, A.S. (2001). Ordinary magic: Resilience processus in development. *American Psychologist*, 56, 227–238.

Masten, A.S. (2006). Promoting resilience in development: A general framework for children care. In R.J. Flynn, P.M. Dudding, & J.G. Barber (Eds.), *Promoting resilience in child welfare* (pp. 3–17). Ottawa, ON: University of Ottawa Press.

Masten, A.S., Best, K., & Garmezy, N. (1990). Resilience and development: Contributions from the study of children who overcome adversity. *Developmental Psychopathology*, 2, 425–444.

Mathez, E.A. (2009). *Climate change: The science of global warning and our energy future.* New York: Columbia University Press.

Mattson, G. (1997). Small community governance: Some impediments to policy making. In J.J. Gargan, *Handbook of local government administration.* New York: Marcel Dekker Publishers.

May, P.J. (1991a). Addressing public risks: Federal earthquake policy design. *Journal of Policy Analysis and Management*, 10(2), 263–285.

May, P.J. (1991b). Reconsidering policy design: Policies and publics. *Journal of Public Policy*, 11(2), 187–206.

May, P.J. (1994). Analyzing mandate design: State mandates governing hazard-prone areas. *Publius*, 24(2), 1–16.

May, P.J. (1995). Can cooperation be mandated? Implementing intergovernmental environmental management in New South Wales and New Zealand. *Publius: The Journal of Federalism*, 25(1), 89–113.

May, P.J. (1997). State regulatory roles: Choices in the regulation of building safety. *State and Local Government Review*, 29(2), 70–80.

May, P.J., & Birkland, T.A. (1994). Earthquake risk reduction: An examination of local regulatory efforts. *Environmental Management*, 18(6), 923–937.

May, P.J., & Burby, R.J. (1998). Making sense out of regulatory enforcement. *Law & Policy*, 20(2), 157–182.

May, P.J., Burby, R.J., Ericksen, N.J., Handmer, J.W., Dixon, J.E., Michaels, S., & Smith, D.I. (1996). *Environmental management and governance: Intergovernmental approaches to hazards and sustainability.* New York: Routledge.

May, P.J., & Deyle, R. (1998). Governing land use in hazardous areas with a patchwork system. In R.J. Burby (Ed.), *Confronting natural hazards: Land-use planning for sustainable communities* (pp. 57–82). Washington, DC: The National Academies Press.

May, P.J., & Feeley, T.J. (2000). Regulatory backwaters: Earthquake risk reduction in the western United States. *State and Local Governmental Review*, 32(1), 20–33.

May, P.J., & Koski, C. (2012). *Addressing public risks: Extreme events and critical infrastructures.* Seattle: Center for American Politics and Public Policy, University of Washington.

May, P.J., & Williams, W. (1986). *Disaster policy implementation: Management strategies under shared governance.* New York: Plenum Press.

Mayhew, B.H. (1980). Structuralism versus individualism; Part I: Shadowboxing in the dark. *Social Forces*, 9, 305–339.

McCaffrey, S., Velez, A.L., & Briefel, J. (under review). Difference in information needs for wildfire evacuees and non-evacuees. Submitted to *International Journal of Mass Emergencies and Disasters* on Dec 31, 2011.

McDonald, N. (2006). Organizational resilience and industrial risk. In E. Hollnagel, D.D. Woods, & N. Leverson (Eds.), *Resilience engineering, concepts and precepts.* Brookfield, VT: Ashgate Publishing Company.

McEntire, D.A. (2004). The status of emergency management theory: Issues, barriers and recommendations for improved scholarship. Proceedings from the FEMA Higher Education Conference, Emmitsburg, MD, June.

McEntire, D.A., Fuller, C., Johnston, C.W., & Weber, R. (2002). A comparison of disaster paradigms: The search for a holistic policy guide. *Public Administration Review*, 62(3), 267–281.

McEntire, D.A., Myers, A., & Foss, T. (2003). Disaster preparedness. *IQ Report*, 35(11).

McFarlane, A.C., & Norris, F.H. (2006). Definitions and concepts in disaster research. In F.H.

Norris, S. Galea, M.J. Friedman, & P.J. Watson (Eds.), *Methods for disaster mental health research* (pp. 3–19). New York: Guilford Press.

McGuire, M. (2006). Collaborative public management: Assessing what we know and how we
know it. Special issue, *Public Administration Review*, 66, 33–43.

McGuire, M.C., Brudney, J.L., & Gazley, B. (2010). The "New Emergency Management": Applying the lessons of collaborative governance to 21st century emergency planning. In R. O'Leary, D. Van Slyke, & S.H. Kim (Eds.), *The future of public administration, public management and public service around the world: The Minnowbrook perspective* (pp. 117–128). Washington, DC: Georgetown University Press.

McLoughlin, D. (1985). A framework for integrated emergency management. *Public Administration Review*, special issue, 165–172.

McManus, S.S., Seville, E., Vargo, J., & Brunsdon, D. (2008). Facilitated process for improving organizational resilience. *Natural Hazards Review*, 9, 81–90.

McMillan, B., & Chavis, D. (1986). Sense of community: A definition and theory. *Journal of Community Psychology*, 14(1), 6–23.

Meehl, G.A., Stocker, T.F., Collins, W.D., Friedlingstein, P., Gaye, A.T., Gregory, J.M., . . . & Zhao, Z.-C. (2007). Global climate projections. In S. Solomon, D. Qin, M. Manning, Z. Chen, M. Marquis, K.B. Averyt, M. Tignor, & H.L. Miller (Eds.), *Climate change 2007: The physical science basis; Contribution of Working Group I to the fourth assessment report of the Intergovernmental Panel on Climate Change* (pp. 747–845). New York: Cambridge University Press.

METU (Middle East Technical University Earthquake Engineering Research Center). (2011). Field observations about Mw 7.2-October 23, 2011 Van earthquake seismic and structural damage. Retrieved from http://www.imo.org.tr.

MEU (The Ministry of Environment and Urbanism). (2011). http://www.cevresehircilik.gov.tr/turkce/sayfa.php?Sayfa=bakanlikbasin&ID=716.

Meyer, A.D. (1982). Adapting to environmental jolts. *Administrative Science Quarterly*, 27, 515–537.

Meyer, A.D., Brooks G.R., & Goes, J.B. (1990). Environmental jolts and industry revolutions: Organizational responses to discontinuous change. *Strategic Management Journal*, 11, 93–110.

Miami-Dade County Climate Change Advisory Task Force. (2008). *Second report and initial recommendations.* Miami: Miami-Dade County Climate Change Advisory Task Force.

Mileti, D.S. (1999). *Disasters by design: A reassessment of natural hazards in the United States.* Washington, DC: Joseph Henry Press.

Mileti, D.S., Cress, D.M., & Darlington, J.D. (2002). Earthquake culture and corporate action. *Sociological Forum*, 17(1), 161–180.

Milward, H.B., & Provan, K. (2001). Do networks really work? A framework for evaluating public sector organizational networks. *Public Administration Review*, 61(4), 414–423.

Mimaki, J. (2009). Role of community based organization and social capital in disaster preparedness: Perspective from rural areas in Japan. Doctoral thesis. Kyoto University.

Mimaki, J., Takeuchi, Y., & Shaw, R. (2009). The role of community-based organization in the promotion of disaster preparedness at the community level: A case study of a coastal town in the Kochi Prefecture of the Shikoku Region, Japan. *Journal of Coastal Conservation*, 13(4), 207–215.

Minkel, J.R. (2008). The 2003 Northeast blackout—five years later. *Scientific American*, August 13. Retrieved January 31, 2012 from http://www.scientificamerican.com/article.cfm?id=2003-blackout-five-years-later.

Mintzberg, H. (2009). Rebuilding companies as communities. *Harvard Business Review*, July–August.

Missouri Disaster Recovery Partnership. (2002). *Missouri COAD guidance manual.* Revised January 1. Retrieved from http://sema.dps.mo.gov/docs/programs/Executive/Disaster%20Recovery%20Partnership/COAD.pdf.

Mitch, W.J., & Gosselink, J.G. (2000). *Wetlands.* 3rd edition. New York: John Wiley and Sons.

Mitchell, J.K. (2006). The primacy of partnership: Scoping a new national disaster recovery policy. *Annals of the American Academy of Political and Social Science*, 604(1), 228–255.

Mitchum, G.T. (2011). Sea level changes in the southeastern United States: Past, present, and future. Southeast Climate Consortium and the Florida Climate Institute. Retrieved from http://www.seclimate.org/pdfpubs/201108mitchum_sealevel.pdf.

Miyaguchi, T., & Shaw, R. (2007). Corporate community interface in disaster management: A preliminary study from Mumbai, India. *Risk Management*, 9, 209–222.

MOA. (2006). White paper on forest and forestry in 2006. In Japanese. Ministry of Agriculture, Forest and Fisheries (p. 67).

Moglen, G.E., & Kim, S. (2007). Limiting imperviousness: Are threshold-based policies a good idea? *Journal of the American Planning Association*, 73(2), 161–171.

Montz, B.E., & Tobin, G.A. (2005). Socio-economic patterns of vulnerability: Senior living developments and Hurricane Charley. *Papers of the Applied Geography Conferences*, 28, 216–225. Retrieved April 26, 2012 from http://applied.geog.kent.edu/AGCPapers/2005/AGC2005.htm.

Moote, A. (2011). *Multi-party monitoring and stewardship contracting: A tool for adaptive management.* Portland, OR: Sustainable Northwest. Retrieved February 9, 2012 from http://www.sustainablenorthwest.org/resources/publications/Multiparty%20Monitoring%20Guidebook%202012_finalV3_links.pdf.

Morel, B., & Ramanujam, R. (1999). Through the looking glass of complexity: The dynamics of organizations as adaptive and evolving systems. *Organization Science*, 10(3), 278–293.

Morgan, G. (1997). *Images of organization.* Thousand Oaks, CA: Sage Publications.

Morgan, G. (2006). *Images of organization.* 3rd edition. Thousand Oaks, CA: Sage Publications.

Morrow, B.H. (1999). Identifying and mapping community vulnerability. *Disasters*, 23(1), 1–18.

Morrow, B.H. (2008). Community resilience: A social justice perspective. Community and Regional Resilience Institute (CARRI) research report no. 4. Retrieved April 26, 2012 from http://www.resilientus.org/publications/reports.html.

Morrow, B.H., & Enarson, E. (1996). Hurricane Andrew through women's eyes: Issues and recommendations. *International Journal of Mass Emergencies and Disasters*, 14(1), 5–22.

Morrow, B.H., & Phillips, B. (1999). What's gender got to do with it? *International Journal of Mass Emergencies and Disasters*, 17, 5–11.

Morse, R. (2008). Federal recovery is off the rails. *Social Policy*, Spring, 34–37.

Moseley, C., & Toth, N. (2004). Fire hazard reduction and economic opportunity: How are the benefits of the national fire plan distributed? *Society and Natural Resources*, 17(8), 701–716.

Moynihan, D. (2008). Learning under uncertainty: Networks in crisis management. *Public Administration Review*, 68(2), 350–365.

Moynihan, D.P. (2009). The network governance of crisis response: Case studies of incident command systems. *Journal of Public Administration Research and Theory*, 19(4), 895–915.

MTA (General Directorate of Mineral Research and Exploration). (2011). 2011 Van earthquake information note. Retrieved from http://www.mta.gov.tr/v2.0/deprem/BilgiNotu_2011-10_Tabanli-Van.pdf.

Mueller, E.J., Bell, H., Chang, B.B., & Henneberger, J. (2011). Looking for home after Katrina: Postdisaster housing policy and low-income survivors. *Journal of Planning Education and Research*, 31(3), 291–307.

Nahapiet, J., & Ghoshal, S. (1998). Social capital, intellectual capital, and the organizational advantage. *The Academy of Management Review*, 23(2), 242–266.

Nakagawa, Y., & Shaw, R. (2004). Social capital: A missing link to disaster recovery. *International Journal of Mass Emergencies and Disasters*, 22(1), 5–34.

Nakamura, A., & Kikuchi, M. (2011). What we know, and what we have not yet learned: Triple disasters and the Fukushima Nuclear fiasco in Japan. *Public Administration Review*, 71(6), 893–899.

Namba, Y., & Furukawa, Y. (2006). *Present condition grasp of the disaster weak and evacuation at the time of disaster: Case study of Higashihiroshima City.* Research Reports of the School of Engineering, Kinki University, no. 40 (pp. 41–46).

National Atlas of the United States. (2012). *Major dams of the United States: National atlas of the United States.* Reston, VA. Retrieved from http://nationalatlas.gov/atlasftp.html?openChapters=chpwater#chpwater.

National Climatic Data Center. (n.d.) *Storm events for Florida.* Retrieved November 12, 2009 from http://www4.ncdc.noaa.gov/cgi-win/wwcgi.dll?wwevent~storms.

National Commission on Children and Disasters. (2010). Progress report on children and disasters. May 11. Retrieved from http://www.acf.hhs.gov/nccd/20100511_NCCD_Progress_Report_FINAL.pdf.

National Council of La Raza. (2011). Emergency manager's toolkit: Meeting the needs of Latino communities. Retrieved from http://www.nclr.org/index.php/publications/emergency_managers_tool_kit_meeting_the_needs_of_latino_communities/.

National Institute of Population and Social Security Research. (2003). Demography of cities, towns and villages in Japan. In Japanese.

National Weather Service Program Office. (2011). *NTHMP semi-annual report.* Retrieved from nthmp.tsunami.gov/ . . . /Semi-Annual_Report_8_30_11_FINAL-CA.doc.

Neal, D.M. (1997). Reconsidering the phases of disaster. *International Journal of Mass Emergencies and Disasters*, 15(2), 239–264.

Neal, D., & Phillips, B. (1995). Effective emergency management: Reconsidering the bureaucratic approach. *Disasters*, 19(4), 327–337.

Nelson, A., & French, S. (2002). Plan quality and mitigating damage from natural disasters: A case study of the Northridge Earthquake with planning policy considerations. *Journal of the American Planning Association*, 68(2), 194–207.

Neuendorf, K.A. (2002). *The content analysis guidebook.* Thousand Oaks, CA: Sage Publications.

Newman, J. (2004). Numbers tell story of season's destruction. *Orlando Sentinel*, November 29. A10.

Nicholls, R.J. (2011). Planning for the impacts of sea level rise. *Oceanography*, 24(2), 144–157. doi:10.5670/oceanog.2011.34.

Nicholls, R.J., Wong, P.P., Burkett, V.R., Codignotto, J.O., Hay, J.E., McLean, R.F., Ragoonaden, S., & Woodroffe, C.D. (2007). Coastal systems and low-lying areas. In M. Parry, O.F. Canziani, J.P. Palutikof, P. van der Linden, & C.E.

Hanson (Eds.), *Climate change 2007: Impacts, adaptation and vulnerability.* Cambridge: Cambridge University Press.

Nicolis, G., & Prigogine, I. (1989). *Exploring complexity: An introduction.* New York: W.H. Freeman.

NIFC (National Interagency Fire Center). (2009). *Quadrennial Fire Review*, January. Retrieved from http://www.nifc.gov/QFR/QFR2009 Final.pdf.

Ning L., Emanuel, K., Oppenheimer, M., & Vanmarcke, E. (2012). Physically based assessment of hurricane surge threat under climate change. *Nature Climate Change*, 2, 426–467.

NOAA (National Oceanic and Atmospheric Administration). (2007). *Known Florida tornado deaths 1882–2000.* Retrieved from http://www.srh.noaa.gov/mlb/PDFs/florida_tornado_death_stats.pdf.

NOAA (National Oceanic and Atmospheric Administration). (2004). *Population trends along the coastal United States: 1980–2008.* Retrieved April 27, 2012 from http://oceanservice.noaa.gov/programs/mb/pdfs/coastal_pop_trends_complete.pdf.

NOAA (National Oceanic and Atmospheric Administration). (n.d.). *Event summary.* Retrieved November 16, 2009 from http://www.nws.noaa.gov/om/data/pdfs/ctrlfl.pdf.

NOAA CSC (National Oceanic and Atmospheric Administration Coastal Services Center). (2012). Sea level rise and coastal flooding impacts viewer. Retrieved from http://www.csc.noaa.gov/tools/.

Nolan, B. (2005). In storm, N.O. wants no one left behind; Number of people without cars makes evacuation difficult. *New Orleans Times-Picayne*, July 24.

Nolte, I.M., & Boenigk, S. (2011). Public-nonprofit partnership performance in a disaster context: The case of Haiti. *Public Administration*, 89(4), 1385–1402.

Nordstrom, K.F. (2000). *Beaches and dunes of developed coasts.* Cambridge: Cambridge University Press.

Norman, S. (2004). Focus on recovery: A holistic framework for recovery. In S. Norman (Ed.), *New Zealand Recovery Symposium 2004.* Wellington, New Zealand: Ministry of Civil Defense and Emergency Management. Retrieved March 1, 2012 from http://www.civildefence.govt.nz/memwebsite.NSF/Files/Recovery/$file/Recovery_symposium2004.pdf#page=91.

Norris, F. (2009). Looking for resilience: Understanding the longitudinal trajectories of responses to stress. *Social Science & Medicine*, 68, 2190–2198.

Norris, F.H., Stevens, S.P., Pfefferbaum, B., Wyche, K.F., & Pfefferbaum, R.F. (2008). Community resilience as a metaphor, theory, set of capacities and strategy for disaster readiness. *American Journal of Community Psychology*, 41(1–2), 127–150.

Norris, F.H., Tracy, M., & Galea, S. (2009). Looking for resilience: Understanding the longitudinal trajectories of responses to stress. *Social Science & Medicine*, 68, 2190–2198.

Nowell, B., & Boyd, N. (2010). Viewing community as responsibility as well as resource: Deconstructing the theoretical roots of psychological sense of community. *Journal of Community Psychology*, 38(7), 828–841. doi:10.1002/jcop.20398.

Nowell, B., & Foster-Fishman, P. (2011). Examining multi-sector community collaboratives as vehicles for building organizational capacity. *American Journal of Community Psychology*, 48(3–4), 193–207.

Nowell, B., & Steelman, T. (under review). Familiarity, similarity, and performance

effectiveness in disaster response. 2012 Academy of Management annual meeting, Boston, MA.

NRC (National Research Council). (1995). *Beach nourishment and protection.* Washington, DC: The National Academies Press.

NRC (National Research Council). (2004). *Adaptive management for water resources project planning.* Washington, DC: The National Academies Press. Retrieved January 23, 2012 from http://www.nap.edu/catalog/10972.html.

NRC (National Research Council). (2006). *Facing hazards and disasters: Understanding human dimensions.* Washington, DC: The National Academies Press.

NRC (National Research Council). (2009). Applications of social network analysis for building community disaster resilience. Washington, DC: The National Academies Press.

NRC (National Research Council). (2010a). Building community disaster resilience through private-public collaboration. Washington, DC: The National Academies Press.

NRC (National Research Council). (2010b). *Private-public sector collaboration to enhance community disaster resilience: A workshop report.* Washington, DC: The National Academies Press.

NRC (National Research Council). (2011). *Building community disaster resilience through public-private collaboration.* Committee on Private-Public Sector Collaboration to Enhance Community Disaster Resilience, Geographical Science Committee. Washington, DC: The National Academies Press.

NRC (National Research Council), Committee on Private-Public Sector Collaboration to Enhance Community Disaster Resilience, & Geographical Science Committee. (2011). *Building community disaster resilience through private-public collaboration.* Washington, DC: The National Academies Press. Retrieved March 1, 2012 from http://www.nap.edu/catalog/13028.html.

O'Brien, G., O'Keefe, P., Rose, J., & Wisner, B. (2006). Climate change and disaster management. *Disasters*, 30, 64–80.

O'Brien, K. (2008). *Disaster risk reduction, climate change adaptation and human security: A commissioned report for the Norwegian Ministry of Foreign Affairs.* Global Environmental Change and Human Security report no. 2008:3.

OCOEM (Orange County Office of Emergency Management). (2008). *Orange County comprehensive emergency management plan.*

OCOEM (Orange County Office of Emergency Management). (2009). *Orange County local mitigation strategy.*

Office of Economic and Demographic Research. (2011). Sumter County. Retrieved July 22, 2011 from http://edr.state.fl.us/Content/area-profiles/county/sumter.pdf.

Ogawa, H. (2008). *Study on substance and subject of the emergency countermeasure of the old people against the natural disaster.* In Japanese. Tokyo: Architectural Institute of Japan.

Ohta, T. (2006). Sediment-related disasters and new forest management. In Japanese. *Forest Science*, 4.

Ojerio, R., Moseley, C., Lynn, K., & Bania, N. (2011). Limited involvement of social vulnerable populations to mitigate wildfire risk in Arizona. *Natural Hazards Review*, 12(1), 28–36.

Okhuysen, G.A., & Bechky, B.A. (2009). Coordination in organizations: An integrative
perspective. *The Academy of Management Annals*, 3(1), 463–502.

Olshansky, R., & Johnson, L. (2010). *Clear as mud: Planning for the rebuilding of New Orleans.* Chicago: American Planning Association.

Olshansky, R.B., & Kartez, J.D. (1998). Managing land use to build resilience. In J.R. Burby (Ed.), *Cooperating with nature: Confronting natural hazards with land-use planning for sustainable communities.* Washington, DC: Joseph Henry Press.

Ono, A. (1998). Sansonkankyoshakaigakujosetsu [Environmental sociology in mountainous villages]. In Japanese. Tokyo: Nousangyosonbunkakyokai.

Ono, A. (2005). The present state of less-favored areas and the direct payment system. In S. Yoshizawa (Ed.), *Japanese less favored areas and regional revitalization*. Tokyo: Tsukuba-Shobo.

Orange County Planning Division. (2003). *Rural resettlement study*. Retrieved 3 May, 2009 from http://www.orangecountyfl.net/Portals/0/Resources/Internet/DEPARTMENTS/Growth_Management/Planning/compplan/ruralsettlement.pdf.

Orlando Sentinel. (2007). February 2–9 issues.

Osborne, D., & Gabler, T. (1992). *Reinventing government*. Reading, MA: Addison Wesley.

Oscasio, W. (1995). The enactment of economic diversity: A reconciliation of theories of failure-induced change and threat rigidity. In L.L. Cummings & B.M. Staw (Eds.), *Research in organizational behavior*, vol. 17. Greenwich, CT: JAI Press.

O'Sullivan, T.L., Kuziemsky, C.E., Toal-Sullivan, D., & Corneil, W. (2012). Unraveling the complexities of disaster management: A framework for critical social infrastructure to promote population health and resilience. Manuscript submitted for evaluation.

O'Toole, L.J., Jr. (2003). Interorganizational relations in implementation. In B.G. Peters & J. Pierre (Eds.), *Handbook of public administration* (pp. 234–244). Thousand Oaks, CA: Sage.

Oxfam. (2009). *Exposed: Social vulnerability and climate change in the US Southeast*. Washington, DC: Oxfam.

Pahl-Wostl, C., Sendzimir, J., Jeffrey, P., Aerts, J., Berkamp, G., & Cross, K. (2007). Managing change toward adaptive water management through social learning. *Ecology and Society*, 12(2), 30. Retrieved January 26, 2012 from http://www.ecologyandsociety.org/vol12/iss2/art30/.

Pais, J., & Elliott, J. (2008). Places as recovery machines: Vulnerability and neighborhood change after major hurricanes. *Social Forces*, 86, 1415–1453.

Paquet, G. (1999). *Governance through social learning*. Ottawa: University of Ottawa Press.

Paquet, G. (2005). *Gouverance: Une invitation à la subversion*. Montreal: Liber.

Paquet, G. (2011). *Gouvernance collaborative*. Montreal: Liber.

Parry, M., Canziani, O., Palutikof, J., van der Linden, P., & Hanson, C. (2007). *Climate change 2007: Impacts, adaptation, and vulnerability. Contribution of Working Group II to the fourth assessment report of the Intergovernmental Panel on Climate Change* [electronic version]. Retrieved May 23, 2007 from http://www.ipcc.ch.

Passerini, E. (2010). The nature of human communities. In N.D. Phillips, D.S.K. Thomas, A. Fothergill, & L. Blinn-Pike (Eds.), *Social vulnerability to disasters*. Boca Raton, FL: CRC Press, Taylor and Francis Group.

Paton, D., & Johnston, D. (Eds.) (2006). *Disaster resilience: An integrated approach*. Springfield, IL: Charles Thomas.

Paul, M.J., & Meyer, J.L. (2001). Streams in the urban landscape. *Annual Review of Ecological Systems*, 32, 333–365.

Peacock, W.G., Brody, S.D., Seitz, W.A., Merrell, W.J., Vedlitz, A., Zahran, S., Harriss, R.C., & Stickney, R. (2010). Advancing the resilience of coastal localities: Developing, implementing and sustaining the use of coastal resilience indicators. Final report for NOAA CSC grant no. NA07NOS4730147.

Peacock, W.G., Dash, N., & Zhang, Y. (2007). Sheltering and housing recovery following disaster. In H. Rodríguez, E.L. Quarantelli, & R.R. Dynes (Eds.), *Handbook of disaster research* (pp. 258–274). New York: Springer.

Peacock, W.G., & Girard, C. (1997). Ethnic and racial inequalities in hurricane damage and insurance settlements. In W.G. Peacock, B.H. Morrow, & H. Gladwin (Eds.), *Hurricane Andrew: Ethnicity, gender and the sociology of disasters* (pp. 171–190). London and New York: Routledge Press.

Peacock, W.G., Zhang, Y., & Dash, N. (2005). *Single family housing recovery after Hurricane Andrew in Miami-Dade County*. College Station, TX: Hazard Reduction and Recovery Center, Texas A&M University.

Pear, R. (2012). In congress, storm relief is grist for dispute. *New York Times*, September 7. Retrieved from http://www.nytimes.com/2011/09/08/us/politics/08fema.html.

Pelling, M. (2003). *The vulnerability of cities: Natural disasters and social resilience*. London: Earthscan.

Pelling, M., & High, C. (2005). Understanding adaptation: What can social capital offer assessment of adaptive capacity? *Global Environmental Change*, 15(4), 308–319.

Perrow, C. (2007). *The next catastrophe: Reducing our vulnerabilities to natural, industrial, and terrorist disasters*. Princeton, NJ: Princeton University Press.

Peterson, G., Cumming, G., & Carpenter, S. (2003). Scenario planning: A tool for conservation in an uncertain world. *Conservation Biology*, 17(2), 358–366.

Pfeffer, W.T., Harper, J.T., & O'Neel, S. (2008). Kinematic constraints on glacier contributions to 21st-century sea-level rise. *Science*, 321, 1340–1343.

Phillips, B. (1993). Cultural diversity in disaster: Shelter, housing, and long-term recovery. *International Journal of Mass Emergencies and Disasters*, 11(1), 99–110.

Phillips, B.D. (2009). *Disaster recovery*. Boca Raton, FL: CRC Press.

Piaget, J. (1975). *L'équilibration des structures cognitives: Problème central du développement*. Paris: Presses Universitaires de France.

Pielke, R.A. (1999). Nine fallacies of floods. *Climatic Change*, 42(2), 413–438.

Pielke, R.A., & Downton, M.W. (2000). Precipitation and damaging floods: Trends in the United States, 1932–1997. *Journal of Climate*, 13(20), 3625–3637.

Pipa, T. (2006). *Weathering the storm: The role of local nonprofits in the Hurricane Katrina relief effort*. Washington, DC: Aspen Institute.

Pittman, E. (2012). How recovery is ingrained in Waffle House's culture. *Emergency Management*, January 19. Retrieved March 1, 2012 from http://www.emergencymgmt.com/disaster/How-Recovery-Is-Ingrained-in-Waffle-Houses-Culture.html.

Platt, R.H. (1999). *Disaster and democracy: The politics of extreme natural events*. Washington, DC: Island Press.

Podolny, J., & Page, K. (1998). Network forms of organization. *Annual Review of Sociology*, 24, 57–76.

Porter, D.R. (1997). *Managing growth in America's communities*. Washington, DC: Island Press.

Portes, A., & Sensenbrenner, J. (1993). Embeddedness and immigration: Notes on the social
determinants of economic action. *American Journal of Sociology*, 98(6), 1320–1350.

Possekel, A.K. (1999). *Living with the unexpected: Linking disaster recovery to sustainable development*. New York: Springer; Berlin: Heidelberg.

Prigogine, I., & Stengers, I. (1984). *Order out of chaos*. New York: Bantam Press.

Provan, K., & Kenis, P. (2008). Modes of network governance: Structure, management, and effectiveness. *Journal of Public Administration Research and Theory*, 18(2), 229–252.

Quarantelli, E.L. (1982). *Sheltering and housing after major community disasters: Case studies and general observations*. Columbus: Ohio State University.

Quarantelli, E.L. (1995). Patterns of sheltering and housing in U.S. disasters. *Disaster Prevention and Management*, 4(3), 43–53.

Quay, R. (2010). Anticipatory governance: A tool for climate change adaptation. *Journal of the American Planning Association*, 76(4), 496–511.

Quigley, W.P. (2007). Obstacle to opportunity: Housing that working and poor people can afford in New Orleans since Katrina. *Wake Forest Law Review*, 42, 393–419.

Raphael, B. (1986). *When disater strikes: How individuals and communities cope with catastrophe*. New York: Basic Books.

Rasmussen, P.P., & Perry, C.A. (2000). Estimation of peak streamflows for unregulated rural streams in Kansas. U.S. Geological Survey, Water-Resources Investigations, report no. 00–4079.

Raths, D. (2010). Working together. *Emergency Management*, May–June. http://api.ning.com/files/OxOZ*R4YvzBlxE3bhI-FYyoERYfCQoLoJ9yAedwP-0PVPpA8ea0vobBQpPhgNVhd06460T9RSmxE8WF89x4*IkxoV1n9HFLyg/EmergMgtMagWorkingTogetherMay2010.pdf.

Royal Commission Report. (2010). *The Fires and the Fire Related Deaths*. Parliament of Victoria: Government Printer of the State of Victoria.

Reay, T., & Hinings, C.R. (2009). Managing the rivalry of competing institutional logics. *Organization Studies*, 30(6), 629–652.

Régnier, P., Neri, B., Scuteri, S., & Miniati, S. (2008). From emergency relief to livelihood recover: Lessons learned from post-tsunami experiences in Indonesia and India. *Disaster Prevention and Management*, 17(3), 410–429.

Resnick, B. (2011). FEMA's Craig Fugate reflects on a year of disasters. *The Atlantic*, September. Retrieved September 15, 2011 from http://www.theatlantic.com/national/print/2011/09/femas-craig-fugate-reflects-on-a-year-of-disasters.

Rhodes, R.A.W. 2008. Policy network analysis. In M. Moran, M. Rein, & R.E. Godin (Eds.), *The Oxford handbook of public policy* (pp. 425–447). Oxford: Oxford University Press.

Richardson, H.W., Gordon, P., & Moore, J. (2008). *Natural disaster analysis after Hurricane Katrina: Risk assessment, economic impacts and social implications*. Northampton: Edward Elgar.

Ritchie, L.A., & Gill, D.A. (2007). Social capital theory as an integrating theoretical framework in technological disaster research. *Sociological Spectrum*, 27, 103–129.

Robinson, M. (2010). *Making adaptive resilience real*. London: Arts Council of England. Retrieved February 10, 2012 from http://www.britishcouncil.org/making_adaptive_resilience_real-2.pdf.

Robinson, S.E., & Gerber, B. (2007). A seat at the table for nondisaster organizations. *The Public Manager*, 36(3), 4–7. Retrieved from http://www.feiaa.org/files/File/TPM_v36_n3.pdf#page=6.

Rolfe, R.E. (2006). *Social cohesion and community resilience: A multi-disciplinary review of literature for rural health research*. Halifax, NS: Rural Center.

Ronan, R.K., & Johnson, D.M. (2005). *Promoting community resiliency in disasters: The role for schools, youth, and families*. New York: Springer.

Rossi, P., Wright, J.D., & Weber-Burdin, E. (1982). *Natural hazards and public choice: The state and local politics of hazard mitigation*. New York: Academic Press.

Rothman, D.S. (2000). Measuring environmental values and environmental impacts: Going from the local to the global. *Climatic Change*, 44, 351–376.

Rubin, C. (2007). *Emergency management: The American experience, 1900–2005*. Fairfax, VA: PERI.

Rubin, C.B., & Popkin, R. (1990). *Disaster recovery after Hurricane Hugo in South Carolina*. Working paper no. 69. January. Boulder, CO: Natural Hazards Center.

Rubin, C.B., Saperstein, M.D., & Barbee, D.G. (1985). *Community recovery from a major natural disaster.* Monograph no. 41. Boulder, CO: University of Colorado, Program on Environment and Behavior, Institute of Behavioral Science. Retrieved November 27, 2009 from http://desastres.unanleon.edu.ni/pdf/2002/diciembre/pdf/eng/doc2510/doc2510.htm.

Ruhl, J.B., & Salzman, J. (2006). The effects of wetland mitigation banking on people. *National Wetlands Newsletter*, 28(2), 1, 9–14.

Rutter, M. (1987). Psychosocial resilience and protective mechanisms. *American Journal of Orthopsychiatry*, 57(3), 16–331.

Rutter, M. (1993). Resilience: Some conceptual considerations. *Journal of Adolescent Health*, 14, 626–631.

Rutter, M. (2000). Resilience reconsidered: Conceptual considerations, empirical findings, and policy implications. In J.P. Shonkoff & S.J. Meisels (Eds.), *Handbook of early childhood intervention*, 2nd edition (pp. 651–682). New York: Cambridge University Press.

Saenz, R., & Peacock, W.G. (2006). Rural people, rural places: The hidden costs of Hurricane Katrina. *Rural Realities*, 1(2), 1–11.

Sarason, S. (1974). *Psychological sense of community: Prospects for a community psychology.* San Francisco: Jossey Bass.

Saulny, S. (2006). Clamoring to come home to New Orleans projects. *New York Times*, June 6. Retrieved January 17, 2011 from http://www.nytimes.com/2006/06/06/us/nationalspecial/06housing.html.

Saxton, K.E., & Shiau, S.Y. (1990). Surface waters of North America: Influence of land and vegetation on streamflow. In M.G. Wolman & H.C. Riggs (Eds.), *The geology of North America*, vol. 0–1, *Surface water hydrology* (pp. 55–80). Washington, DC: The Geologic Society of America.

SCEM (Sumter County Emergency Management). (2011) *Sumter County comprehensive emergency management plan.*

Schneider, M., Scholz, J.T., Lubell, M., Mindruta, D., & Edwardsen, M. (2003). Building consensual institutions: Networks and the national estuary program. *American Journal of Political Science*, 47(1), 143–158.

Schneider, S.K. (1995). *Flirting with disaster: Public management in crisis situations.* Armonk, NY: M.E. Sharpe.

Schusler, T.M., Decker, D., & Pfeffer, M. (2003). Social learning for collaborative natural resource management. *Society and Natural Resources*, 16(4), 309–326.

Schwab, J.C. (2010). *Hazard mitigation: Integrating best practices into planning.* Planning Advisory Service report no. 560. Chicago: American Planning Association.

Schwab, J.C., Topping, K.C., Eadie, C.C., Deyle, R.E., & Smith, R.A. (1998). *Planning for post disaster recovery and reconstruction.* Planning Advisory Service report no. 483/484. Chicago: American Planning Association Publication.

Scott, W.R. (1992). *Organizations: Rational, natural, and open systems.* Englewood Cliffs, NJ: Prentice Hall.

Sears, D.W., Redman, J.M., Gardner, R.L., & Adams, S.J. (1993). Designing successful state rural development strategies. In R.P. McGowan & E.J. Ottensmeyer (Eds.), *Economic development strategies for state and local governments.* Chicago: Nelson Hall Publishers.

Selye, H. (1956). *The stress of life.* New York: McGraw-Hill.

Sharp, E.B., Daley, D.M., & Lynch, M.S. (2010). Understanding local adoption and implementation of climate change mitigation policy. *Urban Affairs Review*, 47(3), 433–457.

Shaw, R. (2009). Role of local actors in community based disaster risk reduction. In *Perspectives in disaster management* (pp. 123–145). METU, ISDR, WBI.

Shaw, R., & Goda, K. (2004). From disaster to sustainable community planning and development: The Kobe experiences. *Disaster*, 28(1), 16–40.

Shaw, R., Pulhin J., & Pereira, J. (2010). Climate change adaptation and disaster risk reduction: Overview of issues and challenges. In R. Shaw, J. Pulhin, & J. Pereira (Eds.), *Climate change adaptation and disaster risk reduction: Issues and challenges*. Bingley, UK: Emerald Publishers.

Shaw R., & Takeuchi, Y. (2011). Role of local institutions in climate related disasters in small and medium sized cities of Japan. *Asian Journal of Environment and Disaster Management*, 3(2), 137–150.

Shaw, R., & Takeuchi, Y. (2012). *East Japan earthquake and tsunami: Evacuation, communication, education and volunteerism*. Singapore: RPS Publishers.

Shaw, R., Takeuchi, Y., & Badaoui, R. (2008): Education, capacity building, public awareness for disaster reduction. In K. Sassa & P. Canuti (Eds.), *Landslides disaster risk reduction*. Berlin: Springer-Verlag.

Shinohara, S. (1991). Kasochiiki no Henbo to Sanson no Doukou [Changes in under populated region and movement in mountainous areas]. In Japanese. Tokyo: Taimeido.

Siegel, J.M., Shoaf, K.I., & Afifi, A.A. (2003). Surviving two disasters: Does reaction to the first predict response to the second? *Environment and Behavior*, 35, 637–654.

Sigo, S. (2005). Mississippi: Casino insurance. *The Bond Buyer*, October 6.

Simmons, K.M., & Sutter, D. (2008). Manufactured home building regulations and the February 2, 2007 Florida tornadoes. *Natural Hazards*, 46(3), 415–425.

Smith, G. (2011). *Planning for post-disaster recovery: A review of the United States disaster assistance framework*. Washington, DC: Island Press.

Smith, G., Ward, W., & Berke, P.R. (n.d.) The role of hazard mitigation planning in building local capacity and commitment: A tale of six states. Working paper. Chapel Hill: University of North Carolina. Retrieved April 2012 from http://www.ie.unc.edu/cscd/projects/pdf/Smith_Lyles_Berke_State_Case_Working_Paper.pdf.

Smith, J.B. (1997). Setting priorities for adapting to climate change. *Global Environmental Change*, 7, 251–264.

Smith, K., & Petley, D.N. (2001). *Environmental hazards: Assessing risk and reducing disaster*. New York: Routledge.

Sokolow, A.D. (1997). Town and township government: Serving rural and suburban communities. In J.J. Gargan, *Handbook of local government administration*. New York: Marcel Dekker Publishers.

Stacey, R.D. (2000). *Complexity and management*. London: Routledge.

Stacey, R.D. (2001). *Complex responsive processes in organizations*. London: Routledge.

Stallings, R.A. (1995). *Promoting risk: Constructing the earthquake threat*. New York: Aldine DeGruyter.

Stallings, R.A. (2002). Weberian political sociology and sociological disaster studies. *Sociological Forum*, 17(2), 281–305.

Stallings, R., & Quarantelli, E. (1985). Emergent citizen groups and emergency management. *Public Administration Review*, 45, 93–100.

Stark, N. (n.d.). Effective rural governance: What is it? Why does it *matter*? Washington, DC: Rural Governance Initiative, Rural Policy Research Institute.

Statistics Bureau. (2008). Chapter 2: Population. In *Statistical Handbook of Japan*. Tokyo: Statistics Bureau. Retrieved May 28, 2012 from http://www.stat.go.jp/english/data/handbook/c02cont.htm#cha2_2.

Staw, B., Sandelands, L., & Dutton, J. (1981). Threat rigidity effects in organizational behavior: A multilevel analysis. *Administrative Science Quarterly*, 26, 501–524.

Steelman, T., & Burke, C. (2007). Is wildfire policy in the united states sustainable? *Journal of Forestry*, 105(2), 67–72.

Steelman, T., & McCaffrey, S. (2011). What is limiting more flexible fire management—public
or agency pressure? *Journal of Forestry*, 109(8), 454–461.

Steelman, T., Nowell, B., Bayoumi, D., & McCaffrey, S. (under review). Understanding information exchange during disasters: Methodological insights from info-centric social network analysis.

Steffen, K., Clark, P.U., Cogley, J.G., Holland, D., Marshall, S., Rignot, E., & Thomas, R. (2008). Rapid changes in glaciers and ice sheets and their impacts on sea level. In J.P. McGeehin (Ed.), *Abrupt climate change*, Synthesis and Assessment Product 3.4 (pp. 29–66). Washington, DC: U.S. Climate Change Science Program and the Subcommittee on Global Change Research.

Stein, J., Moreno, P., Conrad, D., & Ellis, S. (2000). *Troubled waters: Congress, the Corps of Engineers, and wasteful water projects*. Washington, DC: Taxpayers for Common Sense and National Wildlife Federation.

Stephens, C. (2007). Social capital in its place: Using social theory to understand social capital and inequalities in health. *Social Science and Medicine*, 66(5), 1174–1184.

Stern, N. (2006). *Stern review on the economics of climate change: Report to HM Treasury*. London: HM Treasury.

Stone, B. (2004). Paving over paradise: How land use regulations promote residential imperviousness. *Landscape and Urban Planning*, 69(1), 101–113.

St. Petersburg Times. (2007). Storm Q&A. February 2. Retrieved November 16, 2009 from http://www.sptimes.com/2007/02/02/State/Storm_QA.shtml.

Sumter County. (n.d.). *Sumter County, Florida population table*. Retrieved July 22, 2011 from http://sumtercountyfl.gov/DocumentView.aspx?DID=2271.

Sumter County Chamber of Commerce. (2011). *Sumter for business*. Retrieved July 22, 2011 from http://www.sumterchamber.org/sumter-business.asp.

Sumter County Sheriff's Office. (2005). *Sumter County emergency management*. Retrieved July 22, 2011 from http://www.sumtercountysheriff.org/emergency-management.asp.

Sylves, R. (2008). *Disaster policy and politics: Emergency management and homeland security*. Washington, DC: CQ Press.

Takeuchi, Y., & Shaw, R. (2010). Climate change adaptation and disaster risk reduction experiences in Japan. In R. Shaw, J. Pulhin, & J. Pereira (Eds.), *Climate change adaptation and disaster risk reduction: Issues and challenges*. Bingley, UK: Emerald Publishers.

Taleb, N.N. (2010). *The black swan: The impact of the highly improbable*. 2nd edition. New York: Random House.

Tang, Z. (2009). How are California local jurisdictions incorporating a strategic environmental assessment in local comprehensive land use plans? *Local Environment*, 14(4), 313–328.

Taylor-Powell, E., B. Rossing, et al. (1998). *Evaluating Collaboratives: Reaching the potential. Program Development and Evaluation*. Madison, Wisconsin: University of Wisconsin-Extension: 190

TBMM (Turkiye Buyuk Millet Meclisi). (1997). *Dogal afetlerde can ve mal kaybinin en aza indirilmesi icin alinmasi gereken tedbirlerin belirlenmesine iliskin meclis arastirmasi* [Turkey Grand National Assembly special committee on natural disasters]. Ankara.

TBMM (Turkiye Buyuk Millet Meclisi). (1999). *Deprem arastirma raporu* [Earthquake investigation report]. Report nos. 10/66, 67, 68, 69, 70. Ankara.

The Infrastructure Security Partnership (TISP). (2011). *Regional disaster resilience guide: 2011 edition*.

Thomson, A.M., & Perry, J.L. (2006). Collaboration processes: Inside the black box. *Public Administration Review*, 66(6), 20–32.

Thomson, A.M., Perry, J.L., & Miller, T.K. (2009). Conceptualizing and measuring collaboration. *Journal of Public Administration Research and Theory*, 19(1), 23–56.

Thurston, H.W. (2006). Opportunity costs of residential best management practices for stormwater runoff control. *Journal of Water Resources Planning and Management*, 132(2), 89–96.

Thurston, H.W., Goddard, H.C., Szlag, D., & Lemberg, B. (2003). Controlling storm-water runoff with tradable allowances for impervious surfaces. *Journal of Water Resources Planning and Management*, 129(5), 409–418.

Tierney, K. (2006). *Facing hazards and disasters: Understanding human dimensions*. Washington, DC: The National Academies Press.

Tierney, K.J., Lindell, M.K., & Perry, R.W. (2001). *Facing the unexpected: Disaster preparedness and response in the United States*. Washington, DC: Joseph Henry Press.

Tierney, K.J., Nigg, J.M., & Dahlhamer, J.M. (1996). The impact of the 1993 Midwest floods: Business vulnerability and disruption in Des Moines. In R.T. Sylves & W.L. Waugh (Eds.), *Disaster management in the US and Canada*, 2nd edition (pp. 214–233). Springfield, IL: Charles C. Thomas.

Titus, J.G., & Craghan, M. (2009). Shore protection and retreat. In J.G. Titus (Ed.), *Coastal sensitivity to sea-level rise: A focus on the mid-Atlantic region*, Synthesis and Assessment Product 4.1 (pp. 87–103). Washington, DC: U.S. Climate Change Science Program and the Subcommittee on Global Change Research.

Titus, J.G., Hudgens, D.E., Trescott, D.L., Craghan, M . . . Tanski, J., & Wang, J. (2009). State and local governments plan for development of most land vulnerable to rising sea level along the US Atlantic coast. *Environmental Research Letters*, 4, 1–7.

Titus, J.G., & Neumann, J.E. (2009). Implications for decisions. In J.G. Titus (Ed.), *Coastal sensitivity to sea-level rise: A focus on the mid-Atlantic region*, Synthesis and Assessment Product 4.1 (pp. 141–156). Washington, DC: U.S. Climate Change Science Program and the Subcommittee on Global Change Research.

Tobin, G.A. (1995). The levee love-affair: A stormy relationship. *Water Resources Bulletin*, 31(3), 359–367.

Tobin, G.A., & Montz, B.E. (1997). *Natural hazards: Explanation and integration*. New York: The Guilford Press.

Tollan, A. (2002). Land-use change and floods: What do we need most, research or management? *Water Science and Technology*, 45(8), 183–190.

Toman, M. (2006). Values in the economics of climate change. *Environmental Values*, 15, 365–379.

Tornado Project. (2000). *Most important US tornadoes by states*. Retrieved November 16, 2009 from http://www.tornadoproject.com/alltorns/worstts.htm#FL.

Tourbier, J.T., & Westmacott, R. (1981). *Water resources protection technology: A handbook of measures to protect water resources in land development*. Washington, DC: The Urban Land Institute.

Turner, B. (1978). *Man-made disasters*. London: Wykeham.

Turner, B.A., & Pidgeon, N.F. (1997). *Man-made disasters*. 2nd edition. Butterworth-Heinemann.

Turner, R.H. (1978). Earthquake prediction volunteers: What can the United States learn from China? *Mass Emergencies*, 3(2), 143–160.

UCF OEM&RS (University of Central Florida Office of Emergency Management and Natural Resources). (2011). *University of Central Florida wildfire response plan*. Retrieved May 14, 2012 from http://emergency.ucf.edu/Plans/UCF%20Wildfire%20Response%20Plan%202011.pdf.

UN (United Nations). (1992). Proceedings of the United Nations framework convention on climate change. Retrieved from http://unfccc.int/resource/docs/convkp/conveng.pdf.

UN (United Nations). (2012). *2011 revision of the world urbanization prospects*. New York: United Nations, Department of Economic and Social Affairs, Population Division. Retrieved April 23, 2012 from http://esa.un.org/unpd/wup/index.htm.

UN-ISDR (United Nations International Strategy for Disaster Reduction) (2005). Hyogo framework for action, 2005–2015: Building the resilience of nations and communities to disasters, international, world conference on disaster reduction, January 2005, Kobe, Hyogo, Japan. www.unisdr.org.

United States Global Change Research Program. (2009). *Global climate change impacts in the United States: A state of knowledge report from the U.S. global change research program*. Cambridge and New York: Cambridge University Press.

United States Indian Ocean Tsunami Warning System Program. (2007). *How resilient is your coastal community? A guide for evaluating coastal community resilience to tsunamis and other coastal hazards*. Bangkok, Thailand: U.S. Indian Ocean Tsunami Warning System Program.

United States Landfalling Hurricane Probability Project. (2012). Retrieved April 26, 2012 from http://www.e-transit.org/hurricane/welcome.html.

United States Office of the President. (2003). *National strategy for the protection of infrastructures and key assets*. Washington, DC: White House.

Unlu, A., Kapucu, N., & Sahin, B. (2010). Disaster and crisis management in Turkey: A need for a unified crisis management system. *Disaster Prevention and Management*, 19(2), 155–174.

Ursano, R., Fullerton, C.S., & Terhakopian, A. (2008). Disasters and health: Distress, disorders, and disaster behaviors in communities, neighborhoods, and nations. *Social Research*, 75, 1015–1028.

USACE (United States Army Corps of Engineers). (1983). Economic and environmental principles for water and related land resources implementation studies. Retrieved February 8, 2012 from http://planning.usace.army.mil/toolbox/library%5CGuidance%5CPrinciples_Guidelines.pdf.

USACE (United States Army Corps of Engineers). (2011). Coastal storm risk management national economic development manual. Retrieved February 9, 2012 from http://www.corpsnedmanuals.us/Includes/PDFs/2011-R-09.pdf.

USACE (United States Army Corps of Engineers). (2012). National economic development (NED) procedures manual for flood risk management. Retrieved February 13, 2012 from http://www.corpsnedmanuals.us/FloodDamageReduction/FDRID001Welcome.asp.

USEPA (United States Environmental Protection Agency). (2008). *2008 nationwide survey of local emergency planning committees (LEPCs)*. Retrieved from http://www.epa.gov/osweroe1/docs/chem/2008_lepcsurv.pdf.

USEPA (United States Environmental Protection Agency). (2010). Climate change indicators in the United States. EPA no. 430-R-10–007. Retrieved July 2010 from http://epa.gov/climatechange/indicators.html.

U.S. Census Bureau. (2010). Physical Housing Characteristics For Occupied Housing Units: 2010 American Community Survey 1-Year Estimates. Retrieved July 25, 2012 http://factfinder2.census.gov/faces/tableservices/jsf/pages/productview.xhtml?pid=ACS_10_1YR_S2504&prodType=table

USGS (United States Geological Survey) & Department of the Interior. (2006). *Earthquake hazards: A national threat*. Fact sheet 2006–3016. Retrieved April 26, 2012 from http://pubs.usgs.gov/fs/2006/3016/2006–3016.pdf.

USGS (United States Geological Survey). (2009). Top earthquake states. October 21. Retrieved April 23, 2012 from http://earthquake.usgs.gov/earthquakes/states/top_states.php.

Vale, L.J., & Campanella, T.J. (2005). *The resilient city: How modern cities recover from disaster.* New York: Oxford University Press.

Van Bueren, E.M., Klijn, E., & Koppenjan, J. (2003). Dealing with wicked wicked problems, knowledge challenges, and collaborative capacity problems in networks: Analyzing an environmental debate from a network perspective. *Journal of Public Administration Research and Theory*, 13(2), 193–212.

Van Raalten, D., van der Laan, T., Wijsman, P., Boeije, L., Schellekens, E., Dircke, P., . . . & Schoots, M. (2009). San Francisco Bay: Preparing for the next level. San Francisco Bay Conservation and Development Commission. Retrieved from www.bcdc.ca.gov/planning/climate_change/SFBay_preparing_%20for_the_next_Level.pdf.

VanSusteren, D. (2011). Vermont officials assess the risks in towns lacking water and power. *New York Times*, August 31. Retrieved September 1, 2011 from http://www.nytimes.com/2011/09/01/us/01flood.html?nl=todaysheadlines&emc=tha23.

Varghese, J.,Krogman, N.T., Beckley, T.M., & Nadeau, S. (2006). Critical analysis of the relationship between local ownership and community resiliency. *Rural Sociology*, 71, 505–527.

Vermeer, M., & Rahmstorf, S. (2009). Global sea level linked to global temperature. *Proceedings of the National Academy of Sciences of the United States of America*, 106(51), 21527–21532. doi:10.1073/pnas.0907765106.

Vigdor, J. (2008). The economic aftermath of Hurricane Katrina. *The Journal of Economic Perspectives*, 22(4), 135–154.

Visit Orlando. (2012). *Visitor volumes.* Retrieved May 14, 2012 from http://corporate.visitorlando.com/research-and-statistics/orlando-visitor-statistics/visitor-volumes/.

Walker, B., Holling, C.S., Carpenter, S.R., & Kinzig, A. (2004). Resilience, adaptability and transformability in socialecological systems. *Ecology and Society*, 9(2), 5.

Walker, B., & Salt, D. (2006). *Resilience thinking: Sustaining ecosystems and people in a changing world.* Washington, DC: Island Press.

Wasserman, S., & Faust, K. (1994). *Social network analysis: Methods and applications.* Cambridge: Cambridge University Press.

Watanabe, A., & Fukuoka, S. (2003). Characteristics of the heavy rain in Hiroshima disaster 1999 and the significance of information of radar rain gage. *Japan Society for Natural Disaster Science, Journal of Natural Disaster Science*, 22(2), 166–185.

Waugh, W.L. (1988). Current Policy and Implementation Issues in Disaster Preparedness. in L.K. Comfort (ed) *Managing Disaster: Strategies and Policy Perspectives*, pp. 111–125. Durham: Duke University Press.

Waugh, W.L. (1990). Emergency management and the capacities of state and local governments. In R.T. Sylves & W.L. Waugh (Eds.), *Cities and disaster.* Springfield, IL: Charles C. Thomas.

Waugh , W.L. (1993) . Coordination or Control: Organizational Design and the Emergency Management Function . *International Journal of Disaster Prevention and Management*, 2(4), 17—31.

Waugh, W.L. (1997). County government as the "administrative arm" of state government. In J.J. Gargan, *Handbook of local government administration.* New York: Marcel Dekker Publishers.

Waugh, W.L. (2000). *Living with hazards, dealing with disasters: An introduction to emergency management.* Armonk, NY: M.E. Sharpe.

Waugh, W.L. (2006). The political costs of failure in the Katrina and Rita disasters. *The Annals of the American Academy of Political and Social Science*, 604, 10–25.

Waugh, W.L. (2009). Mechanisms for collaboration in emergency management: ICS, NIMS, and the problem with command and control. In R. O'Leary & L.B. Bingham (Eds.), *The collaborative public manager: New ideas for the 21st century* (pp. 157–176). Washington, DC: Georgetown University Press.

Waugh, W.L. (2011) Risk-based planning at the community level. Paper presented at the Southeastern Conference on Public Administration, New Orleans, LA, September 22–24.

Waugh, W.L., & Hy, R.J. (1988). The policymaking, administrative, and fiscal capacities of county government. *State and Local Government Review*, 20, 28–31.

Waugh, W.L., Rotanz, R., & Sylves, R.T. (forthcoming). Resilience and emergency management in New York City. In P. Barnes & A. Nakamura (Eds.), *Resilience and crisis planning in mega-cities: Issues, opportunities, and uncertainties.* Edward Elgar Publishing Ltd.

Waugh, W.L., & Streib, G. (1993). County capacity and intergovernmental relations. In D.R. Berman (Ed.), *County governments in an era of change.* Westport, CT: Greenwood Press.

Waugh, W.L., & Streib, G. (2006). Collaboration and leadership for effective emergency management. *Public Administration Review*, 66(6), 131–140.

Waugh, W.L., & Tierney, K. (Eds.) (2007). *Emergency management: Principles and practice for local government.* 2nd edition. Washington, DC: ICMA.

Waugh, W.L., & Waugh, D.M. (1990). The political economy of seduction: Promoting business relocation & economic development in nonindustrial states. In A. Gitelson & M. Dubnick (Eds.), *Public policy and economic institutions.* Greenwich, CT: JAI Press.

WCED. (1987). *Our common future.* London: Oxford University Press.

Weber, E.P. (1998). *Pluralism by the rules: Conflict and cooperation in environmental regulation.* Washington, DC: Georgetown University Press.

Weber, E.P. (2003). *Bringing society back in: Grassroots ecosystem management, accountability, and sustainable communities.* Cambridge, MA: MIT Press.

Webster, P.J., Holland, G.J., Curry, J.A., & Chang, H.-R. (2005). Changes in tropical cyclone number, duration, and intensity in a warming environment. *Science*, 309, 1844–1846.

Weichselgartner, J. (2001). Disaster mitigation: The concept of vulnerability revisited. *Disaster Prevention and Management*, 10(2), 85–95.

Weick, K., Sutcliffe, K., & Obstfeld, D. (2005). Organizing and the process of sensemaking. *Organization Science*, 16(4), 409–421.

Weil, J.H. (2009). Finding housing: Discrimination and exploitation of Latinos in the post-Katrina rental market. *Organization and Environment*, 22(4), 491–502.

Werner, E.E., & Smith, R.S. (1982). *Vulnerable but invincible.* New York: McGraw Hill.

Werner, E.E., & Smith, R.S. (1992). *Overcoming the odds: High risk children from birth to adolescence.* Ithaca, NY: Cornell University Press.

Wheeler, S.M. (2000). Planning for metropolitan sustainability. *Journal of Planning Education and Research*, 20, 133–145.

Western Governors' Association (WGA). (2008). Western Wildlife Habitat Council Established. Jackson, Wyoming, June 29. Retrieved July 26, 2012 http://www.westgov.org/wga/publicat/wildlife08.pdf

White, G.W., Fox, M.H., & Rooney, C. (2007). Nobody left behind: Report on exemplary and best practices in disaster preparedness and emergency response for people with disabilities. March 15. Retrieved from http://www.nobodyleftbehind2.org/powerpoint/index.shtml.

Wigley, T.M.L. (2005). The climate change commitment. *Science*, 307(5716), 1766–1769.

Wilkinson, K.P. (1991). *The community in rural America*. New Work: Greenwood Press.

Wisner, B. (2003). Disaster risk reduction in megacities: Making the most of human and social capital. In A. Kreimer, M. Arnold, & A. Carlin (Eds.), *Building safer cities: The future of disaster risk* (pp. 181–196). Washington, DC: World Bank.

Wisner, B., Blaikie, P., & Cannon, T. (2004). Vulnerability of coastal communities to sea level rise: A case study of Cape May County, New Jersey. *Climate Research*, 22, 255–270.

Woolcock, M. (1998). Social capital and economic development: Toward a theoretical synthesis and policy framework. *Theory and Society*, 27(2), 151–208.

Wright, J.D., & Jasinski, J.L. (2005). The ties that bind: The Central Florida social capital survey. Institute for Social and Behavioral Science, Department of Sociology, University of Central Florida.

Wu, J-Y., Chen, L-C., Liu, Y.C., & Tszeng, J-S. (2007). Relationship between the speed of post-disaster condominium housing reconstruction and household characteristics. In *Proceedings of the Hazards and Disasters Researchers Meeting* (pp. 35–37). Boulder, CO.

Wu, J-Y., & Lindell, M.K. (2004). Housing reconstruction after two major earthquakes: The 1994 Northridge Earthquake in the United States and the 1999 Chi-Chi Earthquake in Taiwan. *Disasters*, 28(1), 63–81.

Xiao, Y. (2011). Local economic impacts of natural disasters. *Journal of Regional Science*, 51(4), 804–820.

Yin, R.K. (1993). *Applications of case study research*. Thousand Oaks, CA: Sage Publications.

Yin, R.K. (1994). *Case study research*. Thousand Oaks, CA: Sage Publications.

Yoe, C. (2004). Scenario planning literature review. Retrieved February 13, 2012 from http://www.corpsnedmanuals.us/FloodDamageReduction/FDRIncludes/FDRYoeLitReviewSept2204PG102504.pdf.

Yohe, G., & Tol, R. (2002). Indicators for social and economic coping capacity: Moving toward a working definition of adaptive capacity. *Global Environmental Change*, 12(1), 25–40.

Yomiuri Newspaper. Tsunami Warning, In Need of System Improvement (in Japanese). (2011, June 19).

Yoshida Y., Takeuchi Y., & Shaw, R. (2009). Town watching as the useful tool of urban risk reduction: Case of Saijo, Japan. In R. Shaw, H. Srinivas, & A. Sharma (Eds.), *Urban risk: An Asian perspective*. Bingley, UK: Emerald Publishers.

Zahran, S., Brody, S.D., Vedlitz, A., Grover, H., & Miller, C. (2008). Explaining local commitment to climate change policy in the United States. *Environment and Planning C*, 26(3), 544–562.

Zellner, M., Hock, C.J., & Welch, E.W. (2011). Leaping forward: Building resilience by communicating vulnerability. In Bruce Goldstein (Ed.), *Collaborative resilience: Moving through crisis to opportunity*. Cambridge, MA: MIT Press.

Zhang, Y., & Peacock, W.G. (2010). Planning for housing recovery? *Journal of the American Planning Association*, 76(1), 5–24.

Contributors

EDITORS

Naim Kapucu, Ph.D., is a professor of public policy and administration at the School of Public Administration at the University of Central Florida (UCF). He is also the founding director of the Center for Public and Nonprofit Management (CPNM) at UCF (2008–2011). He has developed Emergency Management and Homeland Security graduate certificate and undergraduate minor programs at UCF. He is chair of the Section on Emergency and Crisis Management of the American Society for Public Administration. He has published widely in areas of public policy and administration, crisis leadership, nonprofit management, and disaster management. His work has been published in *Public Administration Review, Administration & Society*; *Journal of Public Administration Research and Theory*; the *American Review of Public Administration*; and *Disasters*. His main research interests are emergency and crisis management, network leadership and governance, decision making in complex environments, and collaborative governance. He is the author of *Network Response to Acts of Terrorism: Comparative Analyses*; and *Managing Emergencies and Crises* (with Alp Özerdem).

Christopher V. Hawkins, Ph.D., is an assistant professor in the School of Public Administration at the University of Central Florida. His research on metropolitan governance, local smart growth, and planning has been published in *Journal of Urban Affairs*; *Urban Affairs Review*; *Public Administration Review*; and *Policy Studies Journal*. His current work focuses on sustainable development and application of network analysis to understanding planning and policy decisions. He received his Ph.D. in urban and regional planning from The Florida State University.

Fernando I. Rivera, Ph.D., is an associate professor of Sociology at the University of Central Florida. His research interests and activities fall under staple areas of sociology, primarily the sociology of mental health, race and ethnicity, medical sociology, and the family. Within these areas he has a particular interest in Latino populations. This particular interest has allowed him to expand his research to other areas of social inquiries such as: obesity/diabetes, disability, mental health, substance use, energy consumption, help-seeking in disaster situations, and rural emergency management, among others. His research has emphasized on the desegregation of Latino subgroup ethnicities within the aforementioned areas of sociological research. His published work has investigated how different mechanisms are related to certain health and mental health outcomes, particularly the influence of perceived discrimination and perceived social support.

CONTRIBUTORS

Christine Bevc, Ph.D., is a research associate at the University of North Carolina–Chapel Hill working in conjunction with the UNC Center for Public Health Preparedness at the North Carolina Institute for Public Health in the UNC Gillings School for Global Public Health and MEASURE Evaluation within the Carolina Population Center. Her research specializes in the application of social science research methods and social networks to improve interorganizational relationships and collaboration associated with emergency preparedness, response, and mitigation.

Mary M. Brown, Ph.D., is a professor of Public Administration at University of North Carolina, Charlotte, a senior fellow with the Center for Excellence in Municipal Management at George Washington University, and a visiting scientist at the Software Engineering Institute at Carnegie Mellon University. Her research interests center on participatory design, knowledge management, and joint problem solving and program planning.

Samuel D. Brody, Ph.D., is an associate professor of Landscape Architecture and Urban Planning at Texas A&M University. He is conducting research as part of the Sustainable Coastal Margins Program and teaching in the areas of environmental planning, coastal management, and dispute resolution. He holds the George P. Mitchell '40 Chair in Sustainable Coasts and directs the Center for Texas Beaches and Shores at the Galveston campus. He also heads the Environmental Planning and Sustainability Research Unit.

William Butler, Ph.D., is an assistant professor in the Department of Urban and Regional Planning at Florida State University (FSU). He teaches courses in collaborative governance, community involvement, environmental planning, and food systems planning. His research explores how to enhance socio-ecological resilience of human-natural systems. In particular, he is interested in how to develop adaptive capacity and transformative potential through collaborative governance in urban and regional systems.

Suleyman Celik, Ph.D., is an assistant professor of Political Science and Public Administration at Gazi University, Ankara, Turkey. Dr. Celik received his master's in Public Policy and Management from Carnegie Mellon University. He received his Ph.D. in Public and International Affairs from the Graduate School of Public and International Affairs of the University of Pittsburgh in 2005. His research interests are emergency and crisis management, network analysis, and comparative public administration.

Shaoming Cheng, Ph.D., is an assistant professor at the Department of Public Administration, Florida International University. He received his Ph.D. in Public Policy from George Mason University. His research focuses on firms' location strategy and the location of economic activities, the reasons why some places prosper and thrive while others decline, the ways in which local areas may capture the benefits of desired development strategies, and how entrepreneurship replaces traditional production assets in economic development.

Gabriela B. Christmann, Ph.D., is head of research department Dynamics of Communication, Knowledge and Spatial Development at Leibniz Institute for Regional Development and Structural Planning, Erkner (Berlin), and she is a private lecturer at Berlin University of Technology, Institute of Sociology. She conducts research on sustainable urban and regional development, climate change, vulnerability and resilience, civil society engagement, communication processes, and governance.

Louise L. Comfort, Ph.D., is a professor of Public and International Affairs and the director of Center for Disaster Management at the University of Pittsburgh. She is a fellow at National Academy of Public Administration, and author or co-author of five books, including *Designing Resilience: Preparing for Extreme Events*, University of Pittsburgh Press, 2010. She has engaged in field studies following nineteen earthquake disasters in fourteen countries, including the EERI reconnaissance study of the September 30, 2009 Padang earthquake, Indonesia, an NSF RAPID response study of the January 12, 2010 Haiti earthquake, and a recovery study of the March 11, 2011 Tohoku earthquake, Japan.

Sitki Corbacioglu, Ph.D., is an assistant professor in the Department of Business Administration at the Osmangazi University. He received his master's in Public Policy at Carnegie Mellon University in 1999, and finished his Ph.D. at the University of Pittsburgh in 2004. Dr. Corbacioglu's main research interests are emergency management, complex adaptive systems, and organizational learning. His research has been published in *International Journal of Mass Emergencies and Disasters*; *Disasters*; and *Journal of Business and Public Affairs*.

Fatih Demiroz, Ph.D., is a visiting assistant professor at Florida International University. His research interests are interorganizational networks, emergency management and homeland security, and social networks.

Robert Deyle, Ph.D., is a professor of Urban and Regional Planning at Florida State University where he teaches coastal planning and planning for climate change. His research focuses on coastal hazard mitigation and adaptive response planning for sea level rise. His most recent work concerns integrating hazard mitigation, including the effects of sea level rise, into long-range transportation planning.

Rebecca Dodson, MPA, is a research analyst for the School of Public Administration at the University of Central Florida. She is currently working on her graduate certificate in Emergency Management and Homeland Security.

Ann-Margaret Esnard, Ph.D., is a professor of Urban and Regional Planning at Florida Atlantic University, and director of the Visual Planning Technology Lab (VPT Lab). Her expertise encompasses hazard assessment; vulnerability reduction; GIS/spatial analysis; coastal mitigation planning; post-disaster recovery planning; and land use and environmental planning. She has been involved in a number of research initiatives, including two NSF-funded projects on topics of disaster-induced population displacement and long-term recovery, and has written several articles and book chapters on these topics.

Emel N. Ganapati, Ph.D., is an assistant professor in Public Administration at Florida International University (FIU). Her areas of interest include disaster management, citizen participation/community empowerment, and international development administration. She teaches emergency management, policy analysis, and public participation.

Sukumar Ganapati, Ph.D., is an associate professor in Public Administration at Florida International University (FIU). His research mainly deals with the role of institutions in the urban context, particularly with respect to housing, community development, and information technology. Before joining FIU, he taught at California State University, Northridge; California Polytechnic State University, Pomona; University of Southern California; and Manipal Institute of Technology (India).

Beth Gazley, Ph.D., is an associate professor in the School of Public and Environmental Affairs at Indiana University–Bloomington. She conducts research on public-private collaboration, public and nonprofit management, and volunteerism. She sits on the Monroe County (Indiana) COAD. In 2006, she collaborated with the National Association of Counties and three universities on the first comprehensive, national study of county-level emergency planning activity.

Joshua Gunn is a Ph.D. student in the Urban and Regional Sciences Program at Texas A&M University in the Department of Landscape Architecture and Urban Planning. He works in the Center for Texas Beaches and Shores at the Texas A&M University–Galveston Campus.

Maggie Kusenbach, Ph.D., is an associate professor of Sociology and incoming graduate director at the University of South Florida. Her research areas include qualitative methods and, most lately, rural poverty. Dr. Kusenbach has published numerous papers and chapters on her research in various journals and collections.

Louise Lemyre, Ph.D., University of Ottowa, Gap-Sante Research Unit, is a fellow of the Academy of Social Sciences of the Royal Society of Canada. Her work involves developing unique qualitative and quantitative methodologies for the triangulation between subjective appraisal and contextual evaluation of social environment, especially with respect to stress, risk, adaptation, and organizational learning, in either health or work settings. The psychosocial aspects of terrorism, emergency preparedness, and psychosocial management of risks are at the core of her applied research program.

Peter J. May, Ph.D., is the Donald R. Matthews Distinguished Professor of American Politics at the University of Washington and an adjunct professor at the Evans School of Public Affairs. Dr. May's research addresses policy processes, environmental regulation, and policy making for natural hazards and disasters. Dr. May has taught undergraduate and graduate courses addressing American politics, bureaucratic politics, environmental regulation, public policy processes, policy analysis, risk politics, and quantitative methods.

Branda Nowell, Ph.D., is an associate professor at North Carolina State University. She teaches courses in organizational behavior, change management, organizational theory, and program evaluation. Her research interests center on design and capacity building of multi-organizational systems for addressing complex community issues, community empowerment, organizational collaboration, public network management, and the process and evaluation of systems change. Focal content areas include domestic violence, disaster management, community development, and public health. As an interdisciplinary scholar, she integrates community and organizational psychology with public management to better understand community-based networks of public and nonprofit agencies working in a common problem domain.

Tracey O'Sullivan, Ph.D., is an associate professor in the Honours Bachelor of Health Sciences Program, in the Faculty of Health Sciences, and is cross-appointed with the School of Nursing. Her research program focuses on the social determinants of population health, with particular emphasis on support mechanisms to enhance coping and resilience among high-risk populations, using a function-based approach. Specific projects focus on professional and family caregivers, community supports to assist people with physical or cognitive limitations, and promoting health and resilience through the workplace.

Marc Settembrino is a doctoral student in the Department of Sociology at the University of Central Florida. He received his master's in Sociology at the University of South Florida. His research interests include urban and community sociology, and social stratification.

Rajib Shaw, Ph.D., is an associate professor in the Graduate School of Global Environmental Studies of Kyoto University, Japan. He worked closely with the local communities, NGOs, governments, and international organizations, including United Nations, especially in the Asian countries. He is currently the chair of the United Nations Asia Regional Task Force for Urban Risk Reduction. His research interests are: community-based disaster risk management, climate change adaptation, urban risk management, and disaster and environmental education. He is also the chief editor of *Asian Journal of Environment and Disaster Management*.

Toddi A. Steelman, Ph.D., is a professor of Environmental and Natural Resource Policy within the Department of Forestry and Environmental Resources at North Carolina State University. Her research focuses on improving the governance of environmental and natural resources, with an emphasis on science, policy, and decision-making interactions. She places special emphasis on the role of the public and community in decision making.

William Waugh Jr., Ph.D., is a professor of Public Management and Policy in the Andrew Young School at Georgia State University and an adjunct professor in the Executive Masters Program in Emergency and Crisis Management at the University of Nevada Las Vegas. He is the editor-in-chief of the Journal of Emergency Management. His recent work focuses on local capacity building, collaboration, cultural interoperability, and emergency management policy.

Index

U

V

W

An environmentally friendly book printed and bound in England by www.printondemand-worldwide.com

PEFC Certified

This product is from sustainably managed forests and controlled sources

www.pefc.org

PEFC/16-33-415

MIX

Paper from responsible sources

FSC® C004959

www.fsc.org

This book is made entirely of sustainable materials; FSC paper for the cover and PEFC paper for the text pages.

#0018 - 170914 - C0 - 229/152/23 [25] - CB